Birkhäuser

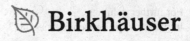

Progress in Mathematical Physics

Volume 65

For further volumes:
http://www.springer.com/series/4813

Victor Chulaevsky • Yuri Suhov

Multi-scale Analysis for Random Quantum Systems with Interaction

Birkhäuser

Victor Chulaevsky
Département de Mathématiques
Université de Reims Champagne-Ardenne
Reims, France

Yuri Suhov
Statistical Laboratory
University of Cambridge
Cambridge, UK

ISSN 1544-9998 ISSN 2197-1846 (electronic)
ISBN 978-1-4939-3952-7 ISBN 978-1-4614-8226-0 (eBook)
DOI 10.1007/978-1-4614-8226-0
Springer New York Heidelberg Dordrecht London

Mathematics Subject Classification (2010): 47B80, 47A75, 81Q10, 60H25, 35P10

Printed on acid-free paper

Springer is part of Springer Science+Business Media (www.birkhauser-science.com)

Preface

This book is about rigorous results on localization in multi-particle interactive Anderson models, mostly on a cubic lattice $\mathbb{Z}^d$ of dimension $d \geq 1$. The book emerged in the wake of recent publications on the mathematical theory of localization in such models (also called multi-particle Anderson tight-binding models). The aim of the book, as the authors see it, is to introduce the reader to a recent progress in this field and to attract attention to possible directions for future research.

The term *multi-particle* addresses here a model with N particles where $N > 1$ is an arbitrary given number. This is a natural generalization of the model with one particle, and it is still a far cry from the ultimate goal to construct a theory where particles are distributed with a positive spatial density. However, even this modest step required (and no doubt will further require) some new ideas and technical tools. It is the presence of interaction and a specific form of the external random potential field which makes the principal difference between the model with one particle and that with many particles. More precisely, the structure of the external random term in the multi-particle Hamiltonian prevents one from applying single-particle techniques in a direct fashion. Compared with its single-particle counterpart, the multi-particle localization theory is still at a very early stage of its development; a number of questions here remain open, whereas the single-particle versions of these questions have received answers. We hope that our book will help the multi-particle localization theory to take off strongly; as we said above, an ultimate goal would be to attack the case of infinitely many particles with a positive spatial density moving in a random environment.

On the other hand, this book was written with a degree of uncertitude, and we do not make an attempt to conceal it. The fact is that at present, the single-particle Anderson theory does not offer much of a clue as to what one should expect when dealing with multi-particle systems with interaction. More precisely, working on the rigorous multi-particle Anderson theory requires a careful review of all technical aspects of the single-particle theory in combination with a well-developed intuition. All this demands the use of Probability, Functional Analysis, Dynamical Systems, and Quantum Theory, including Quantum Statistical Mechanics, quite a formidable band of reputable mathematical disciplines. It is conceivable that the crucial progress in the development of the rigorous theory of multi-particle Anderson systems can only be achieved when a serious breakthrough (or rather a series of breakthroughs) is made, perhaps involving the above mathematical disciplines in the first place.

Methodologically, our book focuses mainly on the so-called *Multi-Scale Analysis* (MSA). In our context, the MSA represents a method that emerged in the single-particle Anderson theory after the 1983 paper by Fröhlich and Spencer [95] (for a brief and by no means complete historical outline, see Sect. 1.1). We believe that possibilities offered by this method are far from being exhausted, and the current and future generations of researchers would greatly benefit from a unified presentation where strong and weak points of the MSA method are discussed in a context of possible applications to complex systems. We decided to restrict the book to tight-binding Anderson models on a cubic lattice with independent, identically distributed random amplitudes of the external potential, as it allows us to free the presentation from a number of technical complications. (Models in a continuous space are mentioned in passing.) Multi-particle localization results and findings based on the MSA and related or applicable to the tight-binding Anderson model have been published in [58, 69–71], and, in part, in [49, 50, 72, 148]. However, it would be fair to say that the entire progress in the mathematical theory of disorder in random quantum systems contributed to the development of the multi-particle MSA.

We also mention in various places of the book the alternative *Fractional Moment Method* (FMM) proposed by Aizenman and Molchanov, in the set-up of single-particle systems, in 1993 (cf. [7]) and modified for multi-particle models in 2009 (the first preprint appeared in 2008); see [9, 10].

So far, progress in multi-particle theory has been achieved by using modifications of one or another of these two methods, allowing the passage from a single particle to N particles; as was said earlier, it requires a thorough scrutiny of all technical steps from the original method(s).

In the context of single-particle Anderson localization, various aspects of the MSA and FMM (as well as other technical means) have been discussed at the mathematically rigorous level in a number of previously written books and detailed reviews; cf. [39, 52, 53, 77, 113, 118, 144, 155, 158]. Section 1.2 of our book also gives a short version of such a review, with a view of possible extensions from one-particle systems to systems with several interacting particles, although such extensions in some instances can be quite challenging. It is worth noting that even in the case of a single-particle tight-binding model, the MSA requires some rather involved technical tools (albeit based, as a rule, on quite simple observations).

Before we proceed further, we would like to point out that localization can manifest itself in various ways, although in the physics community it is customary to consider these manifestations as different "signatures" of a single phenomenon. Until the late 1990s, rigorous results on single-particle localization have been about *exponential spectral localization* where the goal was to establish that, on the whole lattice, the spectrum (more precisely, the spectral measure) of a given (random)

Hamiltonian (where the kinetic energy part is represented by a discrete Laplacian) is pure point and its eigenfunctions decay exponentially fast at infinity. In this regard, we stress that when we speak in this book of localized eigenfunctions (or talk about localization loosely), we mean exponential localization.

However, in line with technical progress, there appeared possibilities to analyze *dynamical localization*, i.e., to assess the spread of an initial wave function (say, with a finite-point support on the lattice) in the course of the time evolution generated by the Hamiltonian. Dynamical localization, besides its purely theoretical value, is important in experiments as well. The link between the two forms of localization is provided by the so-called RAGE theorems (see [77]; the original publications that gave rise to the acronym are [19, 88, 147]) allowing to deduce spectral localization (more precisely, the pure-point nature of the spectral measure) from dynamical localization. However, the method developed in this book derives both forms of localization from the same (probabilistic) estimates, along the same path as in the single-particle theory. In our view, this fact puts these two manifestations of the localization, essentially, on equal footage.

By the time of the book's completion, some new results appeared or were expected to appear. We listed them (with various degrees of detail) in the concluding Sect. 4.7 of Chap. 4.

A couple of words about the style of the book: Due to a special character of the material (and restrictions upon the length of the book), we were not in position to derive all necessary concepts and facts from the first principles. However, most of the arguments used in the book are straightforward and, taken in isolation, should not be overtly difficult to follow. It is the combination of these arguments that makes the presentation technically involved. In this situation, we decided to follow the principle reflected in a popular Russian saying "*Repetition is the mother of learning,*" which is actively used in various educational systems. (In our own schooling experience, teachers sometimes abused this principle, which caused lack of enthusiasm among pupils, but the maxim itself should not have been blamed.) As far as this book is concerned, we systematically repeat formulas, equations, assertions, and comments. However, the authors hope that they have not exceeded reasonable levels of repetitiveness, beyond which it becomes irritating and counterproductive.

The book contains four chapters. Chapters 1 and 2 form Part I, on single-particle systems, while Chaps. 3 and 4 constitute Part II, about multi-particle systems. Chapter 1 is introductory and contains comments on a number of crucial results about localization and delocalization for various single-particle models of quantum disorder. Chapter 2 develops the theory of single-particle Anderson localization via the MSA. We pay a particular attention to key parts of the single-particle MSA and present them from the point of view of future extension of the MSA to multi-particle Anderson models. Chapters 3 and 4 are devoted to the multi-particle MSA, culminating in the proof of spectral and dynamical localization for large amplitudes of the external random potential field.

Taking into account an abundance of various constants and other minor protagonists in the presentation (a feature that is difficult to avoid when you speak about an asymptotical theory), we chose to vary the numeration of constants from one section to another. Besides, we decided to follow an Orwellian principle that "*all constants are constant but some are more constant than others.*" The lower caste is formed by those faceless values that appear in proofs of various technical statements or are mentioned in passing: These are simply marked as Const. The upper caste is formed by constants deserving an individual notation or a specific comment. For these we employ a variety of notation as we see fit in a particular context; we are sure that our system (if indeed it can be called a system at all) may be significantly improved. But, rephrasing Hitchcock[1]: "*All of them are only constants...*" (and in many instances they have been identified numerically or in terms of basic quantities).

The authors would like to acknowledge the hospitality, including financial support, at a number of universities and institutes they visited, individually or jointly, in recent years, where countless discussions took place with colleagues working on related subjects and with members of wider mathematics and physics groups. These include the University of Cambridge (particularly, the Isaac Newton Institute), Université de Reims Champagne-Ardenne, IHES (Bures-sur-Yvette), IAS (Princeton) and Princeton University, University of Alabama (Birmingham), University of California (campuses at Berkeley, Davis, Irvine, and Los Angeles), University of Toronto, Université Paris 6–7, Technion (Haifa), ETH (Zürich), EPF (Lausanne), DIAS (Dublin), IMPA (Rio de Janeiro), and University of Sao Paulo, the latter under the financial support from FAPESP. It would take up a lot of space to name all the individuals who generously shared their thoughts with us, either privately or in public, or simply listened patiently to what we wanted to say (sometimes in a rather incoherent fashion). However, we should particularly thank Michael Aizenman, Anne Boutet de Monvel (who made a considerable number of comments and suggestions which contributed to improving the quality of the manuscript), David Damanik, Efim I. Dinaburg, Shmuel Fishman, Jürg Fröhlich, Yan Fyodorov, François Germinet, Ilya Goldsheid, Igor Gornyy, Misha Goldstein, Gian Michele Graf, Lana Jitomirskaya, Yulia Karpeshina, Dmitrii Khmelniskii, Werner Kirsch, Abel Klein, Frédéric Klopp, Shinichi Kotani, Alexander Mirlin, Stanislav A. Molchanov, Peter Müller, Fumihiko Nakano, Leonid A. Pastur, Jeff Schenker, Boris Shapiro, Dima Shepelyansky, Hermann Schulz-Baldes, Sasha Sodin, Tom Spencer, Roman Shterenberg, Peter Stollmann, Günter Stolz, Ivan Veselić, Simone Warzel, and Martin Zirnbauer.

Our special thanks go to our teacher Yakov Grigorievich Sinai: one of the traits of his inimitable tutoring style, which helped with this project, was that he instilled in us a complete disregard for formal boundaries between different areas of mathematics.

Cambridge, UK-Reims, France- Sao Paolo, Brazil — Victor Chulaevsky
Cambridge, UK — Yuri Suhov

---

[1]Alfred Joseph Hitchcock (1899–1980) was an English film director and producer who invented novel techniques of suspense in the genre of a psychological thriller. "*It's Only a Movie*" is the title of his biography (written by Charlotte Chandler).

Contents

Part I
Single-Particle Localization

Chapter 1
A Brief History of Anderson Localization

1.1 Anderson Localization in Theoretical Physics

In the modern literature, the phenomenon of exponential decay of eigenfunctions of a quantum system in a disordered environment is called Anderson localization, after Philip W. Anderson who, at the age of 34, published a seminal paper [1] aiming, as he pointed out, "*to lay down the foundation for a quantum-mechanical theory of transport*"[1] of a new (and surprising) kind. (According to Anderson, he actually obtained the main result in 1956.) The paper discussed a quantum model on a regular three-dimensional lattice, describing random "impurities" in the spatial environment, where potentials generated at the lattice sites are independent and identically distributed, with mean zero and a finite variance. The hopping amplitude was assumed to be small compared to the variance of the on-site distribution, and a particular perturbation series was considered (later called a "locator series" by the author). The outcome of Anderson's calculation was that the series converges when the hopping amplitude is small compared with the above variance. The conclusion (also formulated somewhat later) was that quantum wave functions in strongly disordered random media become "localized"; the original text referred to the "absence of diffusion."

Anderson liked his paper and affectionately called it the "Nobel Prize One". However, the new approach to the quantum mechanical transport phenomenon was not accepted by the physics community overnight. Years later, to the question "*What was the response to this paper?*" Anderson, in his characteristic style, replied: "*Zero*" (meaning an immediate reaction of his colleagues, since the number of citations of [1] has exceeded 4,000 by now and keeps growing). Among the few people who showed keen interest in the paper from the beginning were Eliahu Abrahams in the United States and Nevill Mott in Great Britain. The former was

[1] A number of quotations in this section have been borrowed from Ref. [20]. The rest were taken from various Internet sources.

V. Chulaevsky and Y. Suhov, *Multi-scale Analysis for Random Quantum Systems with Interaction*, Progress in Mathematical Physics 65, DOI 10.1007/978-1-4614-8226-0_1,

an old acquaintance (discussions with him helped to shape paper [1]), while the latter, from 1949 onwards, was developing a theory of a metal–insulator transition (for a concise review, see [137]) and was eager to establish connections with the notion of localization emerging from Anderson's calculations. Mott was so keen to see Anderson that he arranged for Anderson to obtain a position at the Cavendish Laboratory in Cambridge, which brought the latter in contact with young British researchers, in particular, Josephson and Thouless.

But getting to what Anderson had in mind was not straightforward. As Anderson put it "*Mott kept asking what my 1958 paper meant.*" Perhaps this should not be surprising since, according to Thouless, the paper [1] was "*masterly but opaque.... It was no surprise to me that I found that Anderson was correct, but it was a surprise how often I struggled through obscure and apparently tentative arguments to find that the detailed conclusions were right.*"

One of the subsequent results, by Mott and Twose, was the absence of the metal–insulator transition in one-dimensional systems [138], considered as a verification of the Anderson localization in dimension one for any non-zero level of impurity. Anderson's next paper on localization, [4], came out 15 years later: it was about so-called self-consistent solutions to free energy equations; in [4] these equations were solved exactly for systems on Cayley trees (Bethe lattices). See also the follow-up [5].

Then came the Nobel Prize awarded to Anderson and Mott in 1977, in a large part (but not exclusively) in recognition of their above-mentioned achievements. The third Nobel recipient in that year was Van Vleck who was, incidentally, Anderson's graduate school advisor at Harvard. Next, in 1979, the so-called "Gang of Four" paper appeared [3], which, in Anderson's words, "*revitalized ... [the localization] theory ... [and] made it into a quantitative experimental science with precise predictions as a function of magnetic field, interactions, dimensionality, etc.*" As far as this book is concerned, the prediction made in [3]—that in a two-dimensional disordered system the eigenfunctions are localized for any (non-zero) strength of disorder—remains a challenging open problem for rigorous research. (Experimental works conducted so far reveal a complex picture, and, according to the paper [122], "there is currently no microscopic theory that can explain the whole range of observed phenomena.")

Later on, the random matrix theory attracted a lot of attention in the physics community as a tool of studying quasi-one-dimensional systems. Another emerging aspect was the interaction between quantum particles moving in a random environment—the topic that is central to this book. Interestingly, Anderson thought about this issue from the beginning: in his own words, "*[I] had always been worried that the model I did was strictly non-interacting electrons, hence, strictly linear equations. There was this question of what would the interactions do? And of course, in the presence of interactions the quantum states aren't exact; they fluctuate ... One day I realized that what Mott was telling me was that interactions would make things more local, rather than less local. And so I realized that my theorem had a much better chance of really being meaningful and exact. And at that point, I woke up and rejoined the subject.*" (He was speaking of the period of the late 1960s, early 1970s.) In 1980, the paper [94] argued that, in the presence of interaction,

Mott's original theory could not be applied when the temperature is low, unless the interaction is long-range. On the other hand, paper [18] claimed that if the temperature is high enough, electron–electron interactions destroy localization of the single-particle wave functions.

A wave of celebration events in 2008–2009 marking the 50th anniversary of Anderson localization revealed a rich panorama of research that had developed far beyond the limits of the particular system considered in [1] and its follow-up papers—the tight-binding model of motion of an electron in an imperfect crystal. We particularly refer the reader to the collection of reviews [2] and a concise feature article [125]. The localization theory actually became a paradigm in theoretical and experimental physics as well as in mathematics. In the words from [125], "*After more than a half century of Anderson localization, the subject is more alive than ever.*" And although these words were originally intended to address the situation in physics, the same can be said, in our view, about mathematics. Localization theory has brought to life a large variety of techniques, including, as an important part, rigorous (and powerful) mathematical methods and results, applicable to a variety of models. Such models cover quantum systems on periodic lattices and Cayley trees, in Euclidean spaces, in multi-dimensional tubular domains extended in a particular direction, and on so-called "quantum graphs"—locally one-dimensional structures made out of a countable set of intervals with contact points. The phenomenon of Anderson localization has also been discovered in propagation of acoustic and electromagnetic waves and in nonlinear disordered multi-component systems. Venturing into all these areas would have required a huge amount of work and, in our view, several monographs.

However, quite conspicuously, the level of research output on localization in many-particle disordered systems with interaction remains comparatively low. Only recently the flow of physical publications on this topic reached a notable rate, in part influenced by papers [31, 32] and [105].

The authors of the present book participated in one (but very impressive) event dedicated to the 50th anniversary of the Anderson paper, the six-month programme "*Mathematics and Physics of Anderson Localization: 50 Years After*", organized in 2008 at the Isaac Newton Institute for Mathematical Sciences in Cambridge. Although predominantly mathematical, the programme included a number of informal discussions between mathematicians and physicists on various problems related to localization. This, together with our own work, gave us an impetus to write the present book. We felt that at this stage of theoretical development characterized by numerous crossroads and junctions, the reader would benefit from a unified and consistent presentation of methods that can be used for rigorous analysis of interacting many-particle systems in a random environment.

1.2 Localization in an IID External Potential

The first mathematically rigorous result concerning the Anderson localization was obtained by Goldsheid and Molchanov in [101] (a part of this work was done in the framework of Goldsheid's PhD supervised by Sinai). The result of [101] was the existence of an infinite number of exponentially decaying eigenfunctions $\psi_n(x;\omega)$, for the random Sturm–Liouvillle operator in $L^2(\mathbb{R})$ given as the sum

$$-\frac{d^2}{dx^2}\phi(x) + V(x;\omega)\phi(x), \quad x \in \mathbb{R}, \ \phi \in L^2(\mathbb{R}).$$

Here the potential $V(x;\omega)$ is generated by a sufficiently "regular" Markov process $X_t(\omega)$ on a Riemannian manifold $\mathcal{X}$:

$$V(x;\omega) = F\big(X_x(\omega)\big), \ x \in \mathbb{R},$$

with a Morse function $F : \mathcal{X} \to \mathbb{R}$ (which is smooth and "non-flat": at every point of the manifold $\mathcal{X}$, some of the differentials of function F, of the first or a higher order, are non-degenerate). A simple example is where the function $F : (r,\theta) \mapsto \cos\theta$ is defined on the unit circle

$$\mathcal{X} = \{(r,\theta) : \ r = 1, \ 0 \le \theta < 2\pi\}$$

serving as the phase space for the Brownian motion X_t. Without going into details, we would like to emphasize that the "non-flatness" condition for F is more relevant than its smoothness.

Paper [101] stated that, for almost all samples of $X_t(\omega)$, the set of eigenvalues $E_j(\omega)$ associated with localized eigenfunctions of the above operator is dense in its spectrum. Although it is not difficult to construct artificial examples of a self-adjoint operator in a Hilbert space featuring such a property, the fact that it is generic, that is, holds with probability one, for a random ensemble of Schrödinger operators, was quite unusual from the traditional point of view in mathematical physics and functional analysis.

Shortly after, the (widely cited) paper [102] by Goldsheid et al. appeared, proving that with probability one the whole spectrum of the aforementioned operator is *pure point* (p.p), i.e., its eigenfunctions form an orthonormal basis in $L^2(\mathbb{R})$. See also [134]. Furthermore, these eigenfunctions have been shown to be exponentially localized (i.e., exponentially fast decaying at infinity). A similar property was announced in [102] for a lattice analog of such an operator, acting in $\ell^2(\mathbb{Z})$:

$$H(\omega)\phi(x) = H^0\phi(x) + V(x;\omega)\phi(x), \quad x \in \mathbb{Z}, \ \phi \in \ell^2(\mathbb{Z}). \qquad (1.1)$$

Here $H^0\phi(x) = 2\phi(x) - \phi(x-1) - \phi(x+1)$; the operator H^0 is the negative of the discrete Laplacian on $\mathbb{Z}$ and represents the kinetic energy term (in many publications, the diagonal part is absorbed in the potential). The random potential $V(x;\omega)$, $x \in \mathbb{Z}$, was assumed in [102] to be a discrete-time Markov chain on some compact set, was vaguely described as "good enough." The operator $H(\omega)$ from Eq. (1.1) is considered as a Hamiltonian in the *tight-binding* model on the one-dimensional lattice $\mathbb{Z}$.

For a certain period of time, subsequent publications on localization for one-dimensional tight-binding Hamiltonians (1.1), beginning with Kunz and Souillard [123], universally adopted an assumption that the random variables $V(x;\omega)$ in (1.1) are independent and identically distributed (IID) and have a common probability density or at least a non-zero absolutely continuous component. Formal conditions upon the distribution of the individual variable $V(x;\omega)$ (referred to as the on-site potential distribution) have been gradually extended in several aspects; see, e.g., Carmona et al. [54]. In particular, paper [54] managed, for the first time, to get rid of the assumption that the on-site distribution has a continuous component. This covered the one-dimensional *Bernoulli–Anderson*[2] tight-binding model, where the variables $V(x;\omega)$, $x \in \mathbb{Z}$, are IID and take finitely many values. The assertions, however, remained the same: the operator $H(\omega)$, either in $L^2(\mathbb{R})$ or in $\ell^2(\mathbb{Z})$, has a p.p. spectrum, and its eigenfunctions are localized.

The techniques of Refs. [101, 102], and [134] were specifically one-dimensional and did not work in $L^2(\mathbb{R}^d)$ or $\ell^2(\mathbb{Z}^d)$ for $d \geq 2$. To make progress in higher dimensions, one had to wait until Fröhlich and Spencer published their paper [95]. Fröhlich and Spencer considered a tight-binding model on the cubic lattice $\mathbb{Z}^d$, $d \geq 1$, with a random Hamiltonian $H(\omega)$ of the form

$$H(\omega)\phi(x) = H^0\phi(x) + gV(x;\omega)\phi(x),\ x \in \mathbb{Z}^d,\ \phi \in \ell^2(\mathbb{Z}^d). \tag{1.2}$$

Here and below, when working on $\mathbb{Z}^d$,

$$H^0\phi(x) = 2d\,\phi(x) - \sum_{j=1}^{d} \left(\phi(x+e_j) + \phi(x-e_j)\right) \tag{1.3}$$

where e_j stands for the unit vector $(0,\dots,0,1,0,\dots,0)$ (entry 1 at position j). We will refer to H^0 as the discrete Laplacian; it is useful to note that the $\ell^2(\mathbb{Z}^d)$-norm $\|H^0\| = 4d$.

The potential $V(x;\omega)$ represents a sample of a random field $V : \mathbb{Z}^d \times \Omega \to \mathbb{R}$ with real values, defined on a probability space $(\Omega, \mathfrak{B}, \mathbb{P})$. Here and below, $\mathbb{P}$ stands for the underlying probability distribution; from the physical point of view, Ω can be naturally identified with the sample space $\mathbb{R}^{\mathbb{Z}^d}$, and $\mathbb{P}$ is a probability distribution on Ω (more precisely, on the cylinder sigma-algebra $\mathfrak{B}$ of subsets of Ω). It was assumed in [95] that the variables $V(x;\omega)$, $x \in \mathbb{Z}^d$, are IID and have a common probability density which is bounded on $\mathbb{R}$.

The operator $H(\omega)$ from Eq. (1.2) is often called a (random) *lattice Schrödinger* operator (LSO) and represents the Hamiltonian of the tight-binding model on $\mathbb{Z}^d$. The outcome of paper [95] was that with probability one, the Green's functions

---

[2]Localization in the *lattice* Bernoulli–Anderson model in higher dimension, i.e. on $\mathbb{Z}^d$ with $d \geq 2$, remains a challenging open mathematical problem, while it was solved by Bourgain and Kenig [47] in $\mathbb{R}^d$.

$G(x, y; E, \omega)$ for the LSO $H(\omega)$ from Eq. (1.2) decay exponentially at large distances $|x - y|$ in two cases:

(i) $\forall\ E \in \mathbb{R}$, provided that the *disorder* is large enough (i.e., the constant g in Eq. (1.2) has the absolute value $|g| \geq g_0$), or
(ii) When the distribution of $V(x, \omega)$, $x \in \mathbb{Z}^d$, is Gaussian, and the value $|E - 2d|$ is sufficiently large.

This was considered (and indeed turned out to be) the first step towards a rigorous proof that the spectrum of $H(\omega)$ in $\ell^2(\mathbb{Z}^d)$, under conditions (i) or (ii), is p.p.

The approach adopted in [95] used an inductive procedure, later called the multi-scale analysis (MSA). In Sect. 2.1 we give a brief outline of this method.

In paper [96], Fröhlich et al. refined the earlier analysis[3] and proved the following result, on the random Hamiltonian $H(\omega)$ from (1.2), with $d \geq 1$. Under the assumption that the potential $V(x; \omega)$, $x \in \mathbb{Z}^d$, is a sample from an IID random field, and the on-site distribution has a bounded probability density, let $E^0 \geq 0$ be a given value. If $E^0 + |g|$ is large enough (that is, at least one of the quantities $|g|$ and E_0 is large enough), then the spectrum of $H(\omega)$ outside the interval $[-E^0, E^0]$ is p.p., and all eigenfunctions with eigenvalues from $\mathbb{R} \setminus [-E^0, E^0]$ are localized. In other words, the spectral projection $P_{\mathbb{R}\setminus[-E^0,E^0]}\big(H(\omega)\big)$ of operator $H(\omega)$ on the complement of the interval $[-E_0, E_0]$, restricted to its range, admits a basis of exponentially decaying eigenfunctions.

This result can be phrased as localization at strong disorder or "extreme" (in this case, large) energies.

On the other hand, it is expected that a "phase transition" should exist for $d \geq 3$, manifested in a change in spectral and transport properties of the system with Hamiltonian $H(\omega)$ (a metal–insulator transition predicted by the Mott theory).

The analysis of the spectrum of the LSO $H(\omega)$ for large $|g|$ was capped by von Dreifus and Klein in [161] where the MSA took its modern shape (see also [160]). An alternative approach, called Fractional-Moment Method (FMM), was initially proposed by Aizenman and Molchanov [7] and further developed in [6, 13, 14]. We remind the reader that in all these papers, the assumption that the family of variables $V(x, \omega)$ is "regular enough" was essential.

Most of the aforementioned results have been obtained by studying a finite-volume approximation $H_\Lambda(\omega)$ for the operator $H(\omega)$, as $\Lambda \nearrow \mathbb{Z}^d$.

1.3 Localization Versus Delocalization in a Quasiperiodic External Potential

So far our exposition has focused on the phenomenon of localization (in fact, the whole presentation in this book revolves around this phenomenon). However, we feel that, to convey a sense of perspective, existing results on (single-particle)

---

[3]Cf. the works by Martinelli and Scoppola [131] and Martinelli [129]. Martinelli and Holden, proving in [130] the absence of diffusion in Anderson-type models, considered also a continuous model in $\mathbb{R}^d$. See also [132].

delocalization should also be commented on. As we work with systems on a lattice $\mathbb{Z}^d$, we will only briefly mention here Anderson tight-binding models on trees (Bethe lattices): beginning with [124], there appeared a number of rigorous results on delocalization [116, 117] and localization [13] (see also [8, 11, 12, 15, 16]). On the other hand, we would like to discuss here LSO's with quasiperiodic potentials where notable results have been achieved over recent years showing various features of localization–delocalization transition. A more detailed discussion of the localization and delocalization phenomena in quasi-periodic media, along with an extensive bibliography, can be found, e.g., in works [23–27, 34, 40–48, 59, 64–67, 80, 85–87, 107–111, 136, 143, 152].

The first rigorous result on delocalization in a disordered environment was obtained in the paper by Dinaburg and Sinai [80], under influence of a pioneering work by Novikov [141] (on the Korteweg–de Vries equation with periodic boundary conditions). Answering Novikov's challenging question, Dinaburg and Sinai proved the existence and provided a description of the absolutely continuous (a.c.) spectrum for the Schrödinger operator in $L^2(\mathbb{R})$

$$H(\underline{\omega})\phi(x) = -\frac{\mathrm{d}^2}{\mathrm{d}x^2}\phi(x) + V(x,\underline{\omega})\phi(x),\ x \in \mathbb{R},\ \phi \in L^2(\mathbb{R}). \tag{1.4}$$

Here $\underline{\omega} = (\omega_1, \ldots, \omega_n)$ is a point in the n-dimensional torus $\mathbb{T}^n = \mathbb{R}^n/\mathbb{Z}^n$; $\underline{\omega}$ is often called the phase vector or the phase point. Some of the assertions below are stated for almost all $\underline{\omega}$ relative to the Lebesgue measure on $\mathbb{T}^n$. The potential was taken in [80] of the form

$$V(x,\underline{\omega}) = v(\omega_1 + \alpha_1 x, \ldots, \omega_n + \alpha_n x) \tag{1.5}$$

where $\underline{\alpha} = (\alpha_1, \ldots, \alpha_n) \in \mathbb{R}^n$ is a fixed vector (a frequency vector), the sums $\omega_j + \alpha_j$ are understood mod 1, and v is a real-valued function on $\mathbb{T}^n$ which admits an analytic continuation in a complex neighborhood[4] of $\mathbb{T}^n$. (The precise conditions on the function v adopted in [80] were in terms of its Fourier series.) The spectrum of $H(\underline{\omega})$ lies in the half-axis $[\underline{v}, +\infty)$ where $\underline{v} = \inf\,[v(\theta) : \theta \in \mathbb{T}^n]$.

The result of [80] was further generalized in [136]. The approach of [80] was based on a (linear) adaptation of the (nonlinear) KAM (Kolmogorov–Arnold–Moser) theory. The result was that, under an assumption of incommensurability of the components $\alpha_1, \ldots, \alpha_n$ of vector $\underline{\alpha}$, for Lebesgue-almost every (a.e.) phase points $\underline{\omega} \in \mathbb{T}^n$ the spectrum of $H(\underline{\omega})$ from Eq. (1.4), (1.5) is a.c. (of multiplicity 2) on a set of the form

$$[E^*, +\infty) \setminus \bigcup_{k=1}^{\infty} (a_k, b_k).$$

[4] We identify, where appropriate, the torus $\mathbb{T}^n$ with $[0, 1)^n \subset \mathbb{R}^n \subset \mathbb{C}^n$.

The excluded "dangerous" intervals (a_k, b_k) (not necessarily supporting any singular spectrum) have lengths $b_k - a_k$ rapidly decaying as $k \to \infty$. The values a_1, b_1, a_2, b_2, ... were determined by the function v and the vector $\underline{\alpha}$. Technically, these intervals appear in the course of an inductive KAM-style procedure and have to be excluded from the analysis for the same reason as in the general KAM setup. It was not clear at all if some of the excluded intervals could contain a singular spectrum.

Paper [80] opened a list of rigorous works on Schrödinger operators with quasiperiodic potentials, both on a lattice and on a continuous line. (About the same time the paper [109] appeared, drawing considerable attention to this area of research.) In the case of a scalar frequency α, the most accomplished series of results is available for the almost Mathieu operator in $\ell^2(\mathbb{Z})$:

$$\begin{aligned} H(\omega)\phi(x) =& \phi(x-1) + \phi(x+1) \\ &+ g\cos\left(2\pi(\omega + x\alpha)\right)\phi(x), \ x \in \mathbb{Z}, \ \phi \in \ell^2(\mathbb{Z}). \end{aligned} \tag{1.6}$$

The value ω is again called the phase and lies in the circle (one-dimensional torus) $\mathbb{T} = \mathbb{R}/\mathbb{Z}$. For $g = 2$, operator $H(\omega)$ in (1.6) is often called the Harper operator; cf. [108]. (The terms *almost Mathieu equation* and *Harper equation* are also in use.) As above, when α is rational, $\alpha = p/q$, function $x \in \mathbb{Z} \mapsto V(x;\omega)$ is periodic. In this case, $\forall\ \omega \in \mathbb{T}$, the solutions of the equation $H(\omega)\psi = E\psi$ for all but finitely many values of E are of the form

$$\psi(x;\kappa) = \varphi_{\pm}(x;E)\mathrm{e}^{\pm i\kappa(E,\omega,\alpha,g)x}$$

(the so-called Bloch waves). The spectrum of $H(\omega)$ lies in the segment $[-2(1+|g|), 2(1+|g|)]$ and consists of q disjoint intervals (spectral bands) $[a_1, b_1]$, ..., $[a_q, b_q]$ separated by "spectral gaps" (b_1, a_2), ..., (b_{q-1}, a_q), some of which can be empty (with $b_j = a_{j+1}$). As before, the spectrum of $H(\omega)$ is a.c. (of multiplicity 2) in these bands.

Now consider the operator $H(\omega)$ from Eq. (1.6) for α irrational. Here, the absolutely continuous, singular continuous, and pure point parts of the spectrum of $H(\omega)$ are almost surely independent[5] of $\omega \in \mathbb{T}$. (The whole spectrum, as a set, lies again in $[-2(1+|g|), 2(1+|g|)]$ and is the same for given g and α for a.e. $\omega \in \mathbb{T}$.) Furthermore, considering, for definiteness, $g > 0$ only, the following is true.

For $0 < g < 2$, the spectrum of $H(\omega)$ is a.c. $\forall\ \omega \in \mathbb{T}$. See [23] and references therein.

For $g = 2$, the spectrum of $H(\omega)$ is singular continuous for a.e. $\omega \in \mathbb{T}$; the question whether the p.p. component exists for some $\omega \in \mathbb{T}$ remains open. See [104] and references therein.

---

[5]This follows from the fact that the family of LSO $H(\omega)$ labeled by the points $\omega \in \mathbb{T}^1$ can be viewed as an ergodic family of operators; see, e.g., [53, 144].

For $g > 2$, the spectrum of $H(\omega)$ is p.p., and the eigenfunctions are localized. See [111] and references therein. (Earlier achievements in this direction can be found in [28, 29].) Moreover, the eigenvalues form a Cantor set, cf. [24, 25] (some earlier results on this can be seen in [33, 145]). The question of proving the latter property acquired a special name that should not be related with unreasonable alcohol consumption.

The proof of the above results uses specific properties of the potential $V(x,\omega) = g\cos\big(2\pi(\omega + x\alpha)\big)$. The p.p. spectrum of an LSO in one dimension with a more general scalar-frequency potential was studied in papers by Sinai [154] and by Fröhlich et al. [97] followed by a series of later works, notably Eliasson [86] (see also [87]). The operator $H(\omega)$ was assumed in [86] to be of the form

$$\begin{aligned} H(\omega)\phi(x) &= \phi(x-1) + \phi(x+1) \\ &\quad + gv(\omega + x\alpha)\phi(x),\ x \in \mathbb{Z},\ \phi \in \ell^2(\mathbb{Z}) \end{aligned} \tag{1.7}$$

generalizing Eq. (1.6). Here, the frequency $\alpha \in \mathbb{T}$ satisfies the Diophantine condition: for some $r \in (0, +\infty)$

$$\inf\Big[|k\alpha - 2\pi n|\cdot|k|^r\ :\ n \in \mathbb{Z}, k \in \mathbb{Z}\setminus\{0\}\Big] > 0. \tag{1.8}$$

The function $v : \mathbb{R} \to \mathbb{R}$ is smooth, periodic with period 1 and obeys

$$\sup\left[\left(\frac{1}{(n!)^2}\left|\frac{\mathrm{d}^n}{\mathrm{d}\theta^n}v(\theta)\right|\right)^{1/n}\ :\ \theta \in \mathbb{R},\ n = 1, 2, \dots\right] < \infty$$

(i.e., belongs to the Gevrey class) and also satisfies a "transversality condition" (i.e., obeys uniform lower bounds on the absolute values of the derivatives $\frac{\mathrm{d}^n}{\mathrm{d}\theta^n}\big(v(\theta + \varpi) - E(\theta)\big)$ and $\frac{\mathrm{d}^n}{\mathrm{d}\varpi^n}\big(v(\theta+\varpi) - E(\theta)\big)$, where $\theta, \varpi \in \mathbb{T}$). Under these assumptions, the result of [86] is that $\exists\, \overline{g}_0 \in (0,\infty)$ such that when $|g| \geq \overline{g}_0$, the spectrum of $H(\omega)$ is p.p. for a.e. ω.

However, the methods employed in [86] did not provide exponential localization; it was established, for Diophantine α and a.c. ω, by employing methods from [45], for nonconstant analytic functions v. See [40] and [41] for details.

On the other hand, Chulaevsky and Delyon [64] studied the a.c. spectrum of the almost Mathieu operator (1.6) with potential of small amplitude. Using the so-called Aubry duality argument (cf. [21, 22]), relating spectral properties of almost Mathieu operators with large and small values of the amplitude $|g|$, they proved that for almost all α $\exists\, g_*(\omega) \in (0,\infty)$ such that when $|g| \leq g_*$, the spectrum of $H(\omega)$ is a.c. for almost all ω.

Later, Bourgain and Jitomirskaya [46] analyzed the a.c. spectrum of more general operators $H(\omega)$. Assuming that v is a non-constant analytic function and α is Diophantine (cf. (1.8)), it was proven that $\exists\, g_* > 0$ such that when $|g| \leq g_*$, the spectrum of $H(\omega)$ is a.c. for almost all ω.

In the case of a quasiperiodic potential with a vector frequency $\underline{\alpha}$ (still in dimension one), the operator has the form

$$(H(\underline{\omega})\phi)(x) = \phi(x-1) + \phi(x+1) \\ +gv(\omega_1 + x\alpha_1, \ldots, \omega_n + x\alpha_n),\ x \in \mathbb{Z}. \tag{1.9}$$

Here, as in (1.5), v is a function of n variables, periodic with period 1 in each variable, and $\underline{\omega} = (\omega_1, \ldots, \omega_n)$ and $\underline{\alpha} = (\alpha_1, \ldots, \alpha_n)$ are points in an n-dimensional torus $\mathbb{T}^n = \mathbb{R}^n/\mathbb{Z}^n$. A Diophantine condition is imposed on $\underline{\alpha}$, and v is again assumed to be analytic (or quasianalytic). The study of this class of models was started in [66]. Later on, results similar to the above (localization for large g and a.c. spectrum for small g) have been successfully extended to this case; see [41] and references therein.

In parallel with the theory of LSOs with quasi-periodic potentials on the lattice $\mathbb{Z}$ there have been a stream of results published on operators on the line $\mathbb{R}$ (although not as complete as for operators on the lattice). After [80], an important step was made in [97]. Here, a specific operator $H(\omega)$ was considered, acting in $L^2(\mathbb{R})$:

$$H(\omega)\phi(x) = -\frac{\mathrm{d}^2}{\mathrm{d}x^2}\phi(x) \\ - K^2\big[\cos(2\pi x) + \cos\big(2\pi(\alpha x + \omega)\big)\big]\phi(x),\ x \in \mathbb{R},\ \phi \in L^2(\mathbb{R}). \tag{1.10}$$

As above, the phase $\omega \in \mathbb{T} = \mathbb{R}/\mathbb{Z}$ and the scalar frequency $\alpha \in \mathbb{T}$ is assumed irrational, satisfying a Diophantine condition[6]

$$\inf\Big[\big(|k\alpha| \bmod 1\big)k^2 : k \in \mathbb{Z}\Big] > 0. \tag{1.11}$$

The spectrum of the operator $H(\omega)$ from Eq. (1.10) lies in the half-axis $[-2K^2, +\infty)$, and $-K^2$ is the "lower edge". The result of [97] is that when K is large enough, the spectrum of $H(\omega)$ in $\Big[-2K^2, -2K^2 + 10K\sqrt{1+\alpha^2}\Big]$ is p.p., and the corresponding eigenfunctions are localized.

For more general quasiperiodic potentials on the line $\mathbb{R}$, the existing localization results are somewhat less precise. See Refs. [37, 38] and references therein.

On the other hand, the a.c. spectrum of an operator $H(\omega)$ on the line $\mathbb{R}$ had been studied in considerable detail in [85] and [86] for quite a general class of quasiperiodic potentials. More precisely, paper [85] assumed the operator $H(\underline{\omega})$ of the form as in [80]:

$$H(\underline{\omega})\phi(x) = -\frac{\mathrm{d}^2}{\mathrm{d}x^2}\phi(x) + v(\underline{\omega} + x\underline{\alpha})\phi(x),\ x \in \mathbb{R},\ \phi \in L^2(\mathbb{R}), \tag{1.12}$$

[6] This condition is essential: it follows from an elegant and elementary argument by Gordon [103] that for frequencies approximated by rational numbers abnormally quickly, the corresponding Schrödinger equation $H\psi = E\psi$ has no decaying solution, let alone a square-summable one.

where the phase vector $\underline{\omega} = (\omega_1, \ldots, \omega_n)$ is again a point in the n-dimensional torus $\mathbb{T}^n$, $\underline{\alpha} = (\alpha_1, \ldots, \alpha_n) \in \mathbb{T}^n$ is a frequency vector, and v is a real-valued function on $\mathbb{T}^n$ which admits an analytic continuation in a complex neighborhood of $\mathbb{T}^n$. As usual, it was assumed that the vector $\underline{\alpha}$ satisfies the Diophantine condition: for some $\tau > n - 1$,

$$\left|\sum_{j=1}^{n} \alpha_j k_j\right| \geq \left(\sum_{j=1}^{n} k_j^2\right)^{-\tau}, \quad \forall\ (k_1, \ldots, k_n) \in \mathbb{Z}^n \setminus \{\mathbf{0}\} \tag{1.13}$$

and that the function v is analytic in a complex neighborhood $\tilde{\mathbb{T}}_r^n$ of the torus $\mathbb{T}^n$:

$$\tilde{\mathbb{T}}_r^n = \left\{\underline{\theta} \in \mathbb{C}^n : \operatorname{Re} \underline{\theta} \in \mathbb{T}^n, \ |\operatorname{Im} \underline{\theta}| < r\right\}.$$

A particular sup-norm $\|v\|_{(r)}$ was considered:

$$\|v\|_{(r)} = \sup\left\{|v(\underline{\theta})| : \underline{\theta} \in \tilde{\mathbb{T}}_r^n\right\},$$

and the result of [85] was that $\forall\ a \in (0, \infty)\ \exists\ E^0(a) \in (0, \infty)$ such that, for a "generic" function v, in the sup-norm topology, with $\|v\|_{(r)} = a$, and all $\underline{\omega} \in \mathbb{T}^n$, the spectrum of $H(\omega)$ in $\big(E_0(a), \infty\big)$ (a) forms a Cantor set, and (b) is a.c. (of multiplicity 2). This was established with the help of a particular adaptation of the KAM theory.

So far, our discussion had been limited to the one-dimensional situation $d = 1$.

Before going into a discussion of general d-dimensional Anderson-type models in a quasiperiodic potential, we would like to mention a specific, completely solvable example proposed by Grempel et al. in paper [107]; being published initially as a preprint of the University of Maryland, it is called the "Maryland model." In this model the Hamiltonian $H(\omega)$ has a special potential term:

$$\big(H(\omega)\phi\big)(x) = (H^0\phi)(x) + g \tan \pi\big(\omega + \langle x, \underline{\alpha}\rangle\big)\phi(x), \qquad x \in \mathbb{Z}^d,\ \underline{\alpha} \in \mathbb{T}^d,\ \omega \in \mathbb{T}^1. \tag{1.14}$$

Here H^0 is as in Eq. (1.3) and

$$\langle x, \underline{\alpha}\rangle := x_1\alpha_1 + \cdots + x_d\,\alpha_d, \quad x = (x_1, \ldots, x_d),\ \underline{\alpha} = (\alpha_1, \ldots, \alpha_d),$$

with an incommensurate frequency vector $\underline{\alpha}$. The RHS in Eq. (1.14) is well-defined only for a.e. $\omega \in \mathbb{T}^1$. Clearly, for irrational frequencies the potential term is unbounded. It has been discovered that all eigenfunctions of $H(\omega)$ for a.e. ω are exponentially localized — for *any* nonzero amplitude g of the potential. Later on, this analysis was completed in independent rigorous mathematical works by Pastur–Figotin [143] and Simon [152] (see also [93]). On the other hand, paper [34] by Bellissard et al. partially extended this result to a class of Hamiltonians

$$\big(H(\omega)\phi\big)(x) = (H^0\phi)(x) + gv\big(\omega + \langle x, \underline{\alpha}\rangle\big)\phi(x), \tag{1.15}$$

with a meromorphic function $v : \mathbb{C} \to \mathbb{C}$ of period one, taking real values on the real line, having a single pole of order 1 at 0 and strictly monotone on $(0, 1)$. Here again, the potential is well-defined only for a.e. phase point ω and is necessarily unbounded when the frequency vector is incommensurate. For the localization to hold, this frequency vector must also fulfill a Diophantine condition, i.e., it should not be approximated by rational vectors too rapidly.

Note that [34] actually considered a larger class of quasiperiodic Hamiltonians where the kinetic energy operator H^0 was not limited to the second-order lattice Laplacian. For example, the methods of [34] allow H^0 to be of the form

$$(H^0\phi)(x) = \sum_{y \in \mathbb{Z}^d} a(x - y)\phi(y) \tag{1.16}$$

where

$$|a(x)| \leq \text{Const } \mathrm{e}^{-\rho|x|}, \quad \rho > 0. \tag{1.17}$$

Later, Chulaevsky and Dinaburg [65] combined methods of [154] and [34] and proved exponential localization for a class of Hamiltonians $H^0 + gV$ with non-local H^0 of the form (1.16) satisfying (1.17) and smooth bounded quasiperiodic potentials V, under the assumption that $|g|$ is large enough.

Since the main method of [34] is a (linear) version of the KAM procedure, the result is perturbative and establishes Anderson localization only for large $|g|$. It is interesting, however, that, owing to strict monotonicity of the function $v : (0, 1) \to \mathbb{R}$, Anderson localization was eventually proven here for all energies. That is, the spectrum of $H(\omega)$ from Eq. (1.15) is p.p. Furthermore, it turned out that the so-called "small denominators," inevitably appearing in the nonlinear KAM theory, can be completely avoided in the linear adaptation to operator $H(\omega)$. Such a situation is rather exceptional; for example, it never occurs in Anderson-type models with an IID random potential.

Apart from the above particular class of potentials (which, as we pointed out, must be unbounded), the only rigorous results on localization in a quasiperiodic potential were obtained for single-particle LSO in a series of works by Bourgain and co-workers; cf. in particular [48] and [43] (by employing distinctive elements of the MSA). Paper [48] dealt with the case $d = 2$ and [43] with an arbitrary d.

In dimension two, Ref. [48] considers an LSO of the form

$$\begin{aligned} H(\underline{\omega})\phi(x) &= H^0\phi(x) + gv(\omega_1 + \alpha_1 x_1, \omega_2 + \alpha_2 x_2)\phi(x), \\ & \qquad x = (x_1, x_2) \in \mathbb{Z}^2, \ \phi \in \ell^2(\mathbb{Z}^2). \end{aligned} \tag{1.18}$$

Here, as in (1.3), $H^0\phi(x) = -\sum_e \big(\phi(x+e) + \phi(x-e) - 2\phi(x)\big)$ where e stands for the vector $(1, 0)$ or $(0, 1)$. Next, $\underline{\omega} = (\omega_1, \omega_2)$ is a point in the two-dimensional torus $\mathbb{T}^2 = \mathbb{R}^2/\mathbb{Z}^2$. Further, the parameter $\underline{\alpha} = (\alpha_1, \alpha_2) \in \mathbb{T}^2$ and v is a real-analytic function on $\mathbb{R}^2$, periodic in each of the two variables, with period 1, and non-constant on every vertical or horizontal line. The main result of [48] is that, given an $\epsilon \in (0, 1)$, there exists a value $g_0 = g_0(\epsilon, v) \in (0, +\infty)$ such that $\forall$ $\underline{\omega} \in \mathbb{T}^2$ there exists a set $\mathbb{F}(\epsilon, \underline{\omega}) \subset \mathbb{T}^2$ such that the Lebesgue measure of the complement $\mathbb{T}^2 \setminus \mathbb{F}(\epsilon, \underline{\omega})$ is $< \epsilon$ and $\forall\ g > g_0$ and $\underline{\alpha} \in \mathbb{F}(\epsilon, \underline{\omega})$ the operator $H(\underline{\omega})$ has a p.p. spectrum, and its eigenfunctions are localized.

To clarify the role of the assumption of analyticity of the function $v : \mathbb{T}^2 \to \mathbb{R}$, without going into technical details, we would like to make here the following comment. The MSA inductive procedure, like the KAM induction, has to treat (and avoid) an infinite number of abnormally "small denominators"; in the case of the MSA, these are given by the distances $|E - E_j(H_\Lambda)|$, where $E_j(H_\Lambda)$ are the eigenvalues of the finite-volume approximants, H_Λ, of the operator H, and E is the spectral parameter varying in an interval $I \subset \mathbb{R}$ (also called an energy band). Naturally, eigenvalues are complicated implicit functions of the values of the potential v which, in turn, depend on the parameter $\omega \in \mathbb{T}^2 \subset \mathbb{C}^2$. If the graph of such a function has "almost flat" pieces, the set of points $\omega \in \mathbb{T}^2$ where $|E - E_j(H_\Lambda)|^{-1} \gg 1$ may not be small, preventing one from a successful scale analysis of localization. However, such anomalies turn out to be impossible for a non-constant analytic function v.

In [43] this approach was generalized to the case of an arbitrary dimension d. More precisely, consider a d-dimensional analog of the operator from Eq. (1.18):

$$\begin{aligned} H(\underline{\omega})\phi(x) = H^0\phi(x) + g v(\omega_1 + \alpha_1 x_1, \dots, \omega_d + \alpha_d x_d), \\ x = (x_1, \dots, x_d) \in \mathbb{Z}^d, \phi \in \ell^2(\mathbb{Z}^d), \end{aligned} \tag{1.19}$$

where $\underline{\omega} = (\omega_1, \dots, \omega_d)$ is a point in the d-dimensional torus $\mathbb{T}^d = \mathbb{R}^d/\mathbb{Z}^d$. Next, $\underline{\alpha} = (\alpha_1, \dots, \alpha_d) \in \mathbb{T}^d$ and v is a real analytic function $\mathbb{R}^d \to \mathbb{R}$, periodic in all variables with period 1, and satisfying the following condition of non-degeneracy:

$$\begin{aligned} &\forall\ j = 1, \dots, d \text{ and } (\theta_1, \dots, \theta_{j-1}, \theta_{j+1}, \dots, \theta_d) \in \mathbb{T}^{d-1}, \\ &\quad \text{the map } \theta \in \mathbb{T} \mapsto v(\theta_1, \dots, \theta_{j-1}, \theta, \theta_{j+1}, \dots, \theta_d) \text{ is nonconstant in } \theta. \end{aligned}$$

The result of [43] generalizes that of [48]: $\forall\ \epsilon \in (0, 1)$ and $\underline{\omega} \in \mathbb{T}^d$, $\exists\ g_0\big(= g_0(\epsilon, \underline{\omega}, v)\big) \in (0, +\infty)$ such that $\forall\ g \in (g_0, +\infty)$, $\exists$ a measurable set $\mathbb{A}\big(= \mathbb{A}(g, \epsilon, \underline{\omega}, v)\big) \subset \mathbb{T}^d$ of Lebesgue measure $< \epsilon$ such that $\forall\ \underline{\alpha} \in \mathbb{A}$, the spectrum of $H(\underline{\omega})$ is p.p., and its eigenfunctions are localized. Moreover, $H(\underline{\omega})$ satisfies the following (rather weak) dynamical localization[7] condition: if $\phi \in \ell^2(\mathbb{Z}^d)$ has $|\phi(x)| \le C|x|^{-A}$, $x \in \mathbb{Z}^d$, with $A > 0$ large enough then

[7] This is indeed a rather "weak" form of dynamical localization which does not require localization to occur with probability one, i.e., for a.e. sample of an ergodic external potential. See the discussion in Sect. 1.4 below.

$$\sup \left[\sum_{x \in \mathbb{Z}^d} |x|^2 \left| \left(e^{itH(\underline{\omega})} \psi \right)(x) \right|^2 : t \in \mathbb{R} \right] < +\infty.$$

The results for $d \geq 2$ are referred to as "perturbative", since the threshold g_0 depends on ϵ. (In the case $d = 1$, the above results are considered as "non-perturbative".)

For both $d = 2$ and $d > 2$, the approach adopted in [43] and [48] is based on an asymptotic analysis of the Green's functions

$$G_\Lambda(x, y, E, \underline{\omega}) = \left\langle \delta_x, (H_\Lambda(\underline{\omega}) - E)^{-1} \delta_y \right\rangle, \quad x, y \in \Lambda,$$

of the approximation $H_\Lambda(\underline{\omega})$ of the operator $H(\underline{\omega})$ on a bounded set $\Lambda \subset \mathbb{Z}^d$. The proof comprises two steps:

(I) Bounds on $G^\Lambda(x, y, E, \omega)$ for fixed E,
(II) Elimination of "bad" values of E.

There is, however, an important technical difference between the cases $d = 2$ and $d > 2$: in the former case step (I) was done in [43] by using arithmetic properties of the vector $\underline{\alpha}$ only, irrespective of specific properties of the potential function v, whereas in the latter case step (I) has been performed in [43] for a given v. Step (II) in both cases required some specific information about v. (Accordingly, the result for $d > 2$ can be qualified as "more perturbative" than for $d = 2$.)

Concluding this subsection, we would like to comment on connections with Sects. 2.7.1 and 2.7.2. Such connections can be made through the following observation. A potential $V(x, \omega)$ of the form $v(\omega + x\alpha)$ or $v(\underline{\omega} + x\underline{\alpha})$, $x \in \mathbb{Z}$ or $\mathbb{R}$, can be considered as a function along a trajectory

$$T^x\omega = \omega + x\alpha \text{ or } T^x\underline{\omega} = \underline{\omega} + x\underline{\alpha}$$

generated by iterations of a map T^1 that preserves a specific probability measure on the space of phase points (the normalized Haar, or Lebesgue, measure on $\mathbb{T}$ or $\mathbb{T}^n$). Thus, any property established for the family of operators $H(\omega)$ or $H(\underline{\omega})$, with ω or $\underline{\omega}$ running through a set of full Lebesgue measure, gives a characterization of the random operator where the "randomness" is represented by the Lebesgue measure on $\mathbb{T}$ or $\mathbb{T}^n$. This is of course a far cry from the assumption that the values $V(x, \omega)$ are IID, but the fact that the map T^1 is ergodic — when α is irrational—provides "enough randomness" for having some basic and useful properties of $H(\cdot)$ almost surely independent of the choice of the phase point. On the other hand, an "explicit form" of sample trajectories $T^x\omega$ and $T^x\underline{\omega}$ gives a powerful tool for a detailed analysis of a variety of delicate properties, which is impossible (or very difficult) in the case of IID potentials due to a much richer variety of their samples. This may be an explanation of why the theory for quasiperiodic potentials went much further than the theory for IID (or other similar) potentials in analyzing some (but not all) aspects of spectral properties of the operators $H(\omega)$.

In this connection, we refer to an alternative approach to Anderson localization in almost periodic media proposed recently by Chulaevsky [59]. The main analytic tool used in the proof of Anderson localization is again a variant of the MSA. In contrast with the methods developed by Bourgain et al. (cf., e.g., [43] and [48]), where the ergodic potential $V(x;\omega)$ is generated by a fixed function ("hull") $v : \mathbb{T}^\nu \to \mathbb{R}$,

$$V(x;\omega) = v(T^x\omega), \quad \omega \in \mathbb{T}^\nu,$$

one considers in [59] a parametric family of hulls, $v : \Omega \times \Theta \to \mathbb{R}$, where the parameter space Θ is endowed with the structure of a probability space $(\Theta, \mathfrak{B}_\Theta, \mu)$. The techniques of [59] allow to prove that, for μ-a.e. $\theta \in \Theta$ (i.e., for μ-generic hulls $v(\cdot,\theta) : \omega \to v(\omega,\theta)$), and $|g| \gg 1$, the spectrum of the operator $H(\omega,\theta) = H^0 + gV(\omega,\theta)$ is pure point with $\mathbb{P}$-probability one, and that the corresponding eigenfunctions decay exponentially. It is to be emphasized that the parametric ensemble of the potentials $V(\omega,\theta;x) = v(T^x\omega,\theta)$ can be constructed in various ways, and that the samples $v_\theta = v(\cdot,\theta) : \omega \to v(\omega,\theta)$ on the torus $\mathbb{T}^\nu$ (of arbitrary dimension ν) may have finite smoothness of any order. Moreover, one can also construct parametric ensembles with a.e. non-differentiable, and even a.e. discontinuous samples v_θ for which the operators $H(\omega,\theta)$ feature the complete exponential Anderson localization for $\mathbb{P}$-a.e. $\omega \in \Omega$.

In particular, the above-mentioned parametric ensemble $H(\omega,\theta) = H^0 + gV(\omega,\theta)$ can be constructed with the help of parametric Fourier series on the torus. For example, let $\nu = 1$, $d = 1$ and consider the Fourier series

$$v(\omega,\theta) = \sum_{n=1}^{\infty} n^{-M-1}\Big(a_n(\theta)\cos(2\pi n\omega) + b_n(\theta)\sin(2\pi n\omega)\Big), \quad \omega \in \mathbb{T}^1, \quad (1.20)$$

where $\{a_n(\theta), b_n(\theta), n \geq 1\}$ are IID Gaussian random variables with zero mean and unit variance relative to a probability space $(\Theta, \mathfrak{B}_\Theta, \mu)$ (not related to $\Omega = \mathbb{T}^1$). Taking arbitrarily large $M \geq 1$, one can guarantee the uniform convergence of the series (1.20) in $C^M(\mathbb{T}^1)$ for μ-a.e. $\theta \in \Theta$. In this example, the techniques of [59] allow to prove that μ-almost every choice of the Fourier coefficients in (1.20) gives rise to an operator $H(\omega,\theta)$ with p.p. spectrum in $\ell^2(\mathbb{Z}^1)$. At the same time, it is to be stressed that despite the use of the probabilistic language, the potential of the operator $H(\omega,\theta)$ remains quasi-periodic. The advantage of this approach is that it allows to establish Anderson localization — with $\mathbb{P}$-probability one — by means of a relatively simple variant of the MSA developed initially for LSO with a "truly random" (viz., IID) potential. Note also that the results of [59] complement those obtained by Bourgain et al., since the complete localization is proved with $\mathbb{P}$-probability one on a lattice of arbitrary dimension $d \geq 1$ and for an arbitrary number $\nu \geq 1$ of basic frequencies.

1.4 Spectral and Dynamical Manifestations of Anderson Localization

1.4.1 Spectral Localization

Historically, the starting point of the rigorous Anderson localization theory for a single-particle model on a lattice $\mathbb{Z}^d$ was the analysis of the decay properties of eigenfunctions of the tight-binding Hamiltonians $H(\omega)$ from (1.1) (for $d = 1$) and (1.2) (for a general d). Of course, an exponential decay of a function in $\mathbb{Z}^d$ automatically implies square summability. The existence of an exponentially decaying eigenfunction of an operator in the Hilbert space $\ell^2(\mathbb{Z}^d)$ implies, therefore, non-emptiness of its p.p. spectrum.

When the spectrum of a random operator $H(\omega)$ in some energy interval $I \subset \mathbb{R}$ is p.p., with $\mathbb{P}$-probability one, and its eigenfunctions show an exponential decay, we say that $H(\omega)$ features the phenomenon of exponential spectral localization in I.

As was pointed out in the Preface, the decay of eigenfunctions is considered in this book as one of the manifestations of the physical phenomenon of Anderson localization. Consequently, a part of the mathematical theory is to analyze how different forms of localization are related to one another. There are currently two principal rigorous approaches available to the Anderson localization: the MSA considered in detail in Chap. 2, and the FMM, upon which we touch in this book only briefly and rather informally.

Since properties of one-dimensional systems, as well as rigorous techniques used for their analysis, are special and usually cannot be extended to higher dimensions, we will focus almost exclusively on methods applicable in the case of a general dimension $d \geq 1$.

The basis of the MSA is the analysis of decay properties of the resolvent $(H_{\Lambda_L(u)}(\omega) - E)^{-1}$ of a finite-cube approximant, $H_{\Lambda_L(u)}(\omega)$, to the Hamiltonian $H(\omega)$ in the finite-dimensional Hilbert space $\ell^2(\Lambda_L(u))$ with some specified boundary conditions (most of the time Dirichlet's). The same can be said about the multi-particle version of the MSA developed in this book.

1.4.2 Dynamical Localization

Speaking loosely, one says that the disordered quantum system described by a random Hamiltonian $H(\omega)$ features the phenomenon of dynamical localization if, under the action of the unitary group $\mathrm{e}^{-itH(\omega)}$, initial wave packets (i.e., compactly supported wave functions) do not spread up. The opposite property (quantum diffusion) can be characterized as follows: for any function $\phi : \mathbb{Z}^d \to \mathbb{C}$ with finite support,

$$\lim_{|t|\to\infty} |\phi(x,t;\omega)| = 0, \ \forall\, x \in \mathbb{Z}^d,$$

where $\phi(x,t;\omega)$ stands for the time-shift of $\phi(x)$:

$$\phi(x,t;\omega) = \left(e^{-itH(\omega)}\phi\right)(x). \tag{1.21}$$

Since the operator $H(\omega)$ is random, dynamical localization can be defined in various ways (as far as reference to the probability distribution $\mathbb{P}$ of the random potential $V(\,\cdot\,;\omega)$ is concerned). To start with, one can require that, given a finitely supported function ϕ, with positive $\mathbb{P}$-probability $\exists$ a positive function $\chi(\,\cdot\,;\omega) \in \ell^2(\mathbb{Z}^d)$ such that

$$\sup_{t\in\mathbb{R}} |\phi(x,t;\omega)| \le \chi(x;\omega) < \infty, \quad x \in \mathbb{Z}^d. \tag{1.22}$$

A stronger form of this property is obtained by replacing the words "with positive $\mathbb{P}$-probability" by "with $\mathbb{P}$-probability one".

Furthermore, one could attempt to measure the rate of decay of the "limiting envelope" of the function $\phi(x,t;\omega)$ and introduce a number of related conditions.

Let M denote the (unbounded) operator in $\ell^2(\mathbb{Z}^d)$ acting as the multiplication by the max-norm $x \mapsto |x| = \max_{1\le i\le d} |x_i|$, $x = (x_1,\dots,x_d) \in \mathbb{Z}^d$:

$$(M\phi)(x) := |x|\,\phi(x).$$

We will say that a.s. dynamical localization occurs if for some $s > d/2$, the norm $\|M^s e^{-itH(\omega)}\phi\|$ obeys

$$\mathbb{P}\text{-a.s.} \sup_{t\in\mathbb{R}} \left\| M^s\phi(\cdot,t;\omega)\right\| < \infty. \tag{1.23}$$

(Such a form of dynamical localization was proved for random LSO by Germinet and De Bièvre [98].)

Next, one may replace the pointwise upper bounds by those in expectation.

We will say that strong dynamical localization holds if for some $s > d/2$, the norm $\|M^s e^{-itH(\omega)}\phi\|$ obeys

$$\mathbb{E}\left[\sup_{t\in\mathbb{R}} \left\| M^s\phi(\cdot,t;\omega)\right\|\right] < \infty \tag{1.24}$$

where $\mathbb{E}[\,\cdot\,]$ stands for the expectation relative to $\mathbb{P}$.

(Strong dynamical localization for random LSO was proved by Damanik and Stollmann [78] and, for a larger class of random operators in $\mathbb{R}^d$, by Germinet and Klein [99].)

An even stronger form of dynamical localization (strong Hilbert–Schmidt dynamical localization) was established for a large class of random operators in $\mathbb{R}^d$ by Germinet and Klein [99]. We do not discuss this form of localization in the present book.[8]

[8] Note, however, that recently Sabri [148] has proved the strong Hilbert–Schmidt localization for multi-particle Anderson-type Hamiltonians on the so-called quantum graphs. See the discussion in Sect. 4.7.6.

One also may be interested in replacing the polynomial rate of decay with an exponential one: viz., for some $a > 0$,

$$\sup_{t\in\mathbb{R}} \mathbb{E}\Big[\left\| e^{aM} \phi(\cdot, t; \omega) \right\| \Big] < \infty. \tag{1.25}$$

Note also that the above definitions refer to the phenomenon of complete localization (or localization on the whole spectral axis $\mathbb{R}$). However, from the physical point of view, it may be reasonable to restrict the analysis to a finite (and some times, small) energy interval. Indeed, as is widely believed, in weakly disordered systems of dimension $d \geq 3$ a part of spectrum of the Hamiltonian is a.c. On the other hand, it has been rigorously proven that localization persists in any dimension in some interval I of low energies. To include such a phenomenon, the propagator $e^{-itH(\omega)}$ should be replaced in the above equations by $P_I\big(H(\omega)\big)e^{-itH(\omega)}$, where $P_I\big(H(\omega)\big)$ is the spectral projection of $H(\omega)$ on a given interval $I \subset \mathbb{R}$. In this case, the study of dynamical localization is restricted to the energy band I.

It is worth mentioning that an exponential rate of decay of individual eigenfunctions of $H(\omega)$ is not necessarily equivalent to an exponential rate of decay of $|\phi(x, t; \omega)|$ when $|x| \to \infty$. In particular, the derivation of the strong dynamical localization from exponential bounds on the eigenfunctions, provided by the MSA, requires certain probabilistic inequalities. As a result, the rate of decay of $\phi(x, t; \omega)$ reflects the nature of the probabilistic bounds that one manages to obtain. Repeating what was said earlier, the MSA bounds imply, in a number of situations, dynamical localization with a polynomial[9] or sub-exponential (cf. [61, 99]) rate of decay of $|\phi(x, t; \omega)|$. However, an exponential decay requires stronger probabilistic bounds; such bounds can be obtained by using the FMM.

In Chap. 4 of this book, we present a multi-particle adaptation of the MSA and prove multi-particle dynamical localization similar to Eq. (1.24); cf. Theorem 4.3.13. Exponential dynamical localization similar to Eq. (1.25) for lattice multi-particle systems has been established by Aizenman and Warzel [9, 10] with the help of a multi-particle version of the FMM.

1.5 The N-Particle Model in a Random Environment

As was said before, in this book the term *multi-particle* refers to a system with a given number N of particles in the entire lattice $\mathbb{Z}^d$. Consequently, the use of this term may be misleading: in the literature, multi-particle systems (or, more generally, systems with many degrees of freedom) are often understood as systems where the

---

[9] See also [60] where dynamical localization has been established at a rate slightly faster than polynomial: $\sup_{t\in\mathbb{R}} \mathbb{E}\Big[\left\| e^{\ln^{1+c} M} \phi(\cdot, t, \omega) \right\| \Big] < \infty$ where c is a positive constant. A similar result is proven in [58] for multi-particle models.

number of particles (respectively, the number of individual degrees of freedom) is proportional to the volume of the "physical space" to which the system is confined. We therefore feel that a discussion (however brief and in places speculative) of some aspects of multi-particle localization may be instructive.

1.5.1 The Hamiltonian of the N-Particle System in $\mathbb{Z}^d$

From the point of view of the main results of this book, it makes sense to consider a system composed of distinguishable particles. Viz., spectral localization in the whole N-particle Hilbert space $\ell^2\left(\mathbb{Z}^d\right)^{\otimes N}$ would imply a similar phenomenon in both the bosonic and fermionic sectors.

Accordingly, for $N \geq 2$, the Hamiltonian $\mathbf{H}^{(N)} = \mathbf{H}^{(N)}(\omega)$ of an N-particle tight-binding model in $\mathbb{Z}^d$ acts on functions $\boldsymbol{\phi} \in \ell^2\left((\mathbb{Z}^d)^N\right)$, with the vector argument $\mathbf{x} = (x_1, \ldots, x_N)$ running over the N-particle configuration space $(\mathbb{Z}^d)^N$:

$$\left(\mathbf{H}^{(N)}\boldsymbol{\phi}\right)(\mathbf{x}) = -\sum_{i=1}^{N}\sum_{j=1}^{d}\left[\boldsymbol{\phi}(\mathbf{x}+\mathbf{e}_i^j) + \boldsymbol{\phi}(\mathbf{x}-\mathbf{e}_i^j) - 2\boldsymbol{\phi}(\mathbf{x})\right]$$
$$+\left[\sum_{k=2}^{N}\sum_{1\leq i_1<\ldots<i_k\leq N}\Phi^{(k)}(x_{i_1},\ldots,x_{i_k}) + g\sum_{j=1}^{N}V(x_j;\omega)\right]\boldsymbol{\phi}(\mathbf{x}). \tag{1.26}$$

Here the vector $\mathbf{e}_i^j = (0, \ldots, e^j, \ldots, 0) \in (\mathbb{Z}^d)^N$ has zeros in all positions except for the ith one, and $e^j = (0, \ldots, 1, \ldots, 0) \in \mathbb{Z}^d$ has the jth digit 1 and the rest 0. The term $\sum_{1\leq i\leq N}\sum_{1\leq j\leq d}$ represents the kinetic energy part of $\mathbf{H}^{(N)} = \mathbf{H}^{(N)}(\omega)$. Next, the expression

$$\mathbf{U}^{(N)}(\mathbf{x}) = \sum_{k=2}^{N}\sum_{1\leq i_1<\ldots<i_k\leq N}\Phi^{(k)}(x_{i_1},\ldots,x_{i_k})$$

yields the energy of interaction between particles forming a configuration $\mathbf{x} = (x_1, \ldots, x_N)$.

The energy of interaction $\mathbf{U}^{(N)}(\mathbf{x})$ comprises contributions from k-body interaction potentials $\Phi^{(k)}$ that are symmetric functions of their arguments. If one wishes to restrict the presentation to the case of a two-body interaction then all functions $\Phi^{(k)}$ with $k > 2$ are set to be zero. In general, an important role is played by conditions of decay of functions $\Phi^{(k)}$ at infinity; in this book we will assume at a certain point that the interaction has a finite range.

The sum $\sum\limits_{1\leq j\leq N} V(x_j;\omega)$ yields the energy of a (random) external potential field $\{V(x;\omega),\ x \in \mathbb{Z}^d\}$. Note that the argument ω (responsible for randomness) is the same in all summands; this yields a serious complication in the rigorous analysis of the spectrum of $\mathbf{H}^{(N)}$, even for the case where values $V(x;\omega)$ are IID. As above, the constant g measures the amplitude of the external potential (and alternatively is called the coupling constant).

A convenient form of writing is

$$\mathbf{H} = \mathbf{H}^0 + \mathbf{U} + g\mathbf{V} \tag{1.27}$$

where $\mathbf{H}^0 = \mathbf{H}^{0(N)}$ gives the kinetic term:

$$\left(\mathbf{H}^0\boldsymbol{\phi}\right)(\mathbf{x}) = -\sum_{i=1}^{N}\sum_{j=1}^{d}\left[\boldsymbol{\phi}(\mathbf{x}+\mathbf{e}_i^j) + \boldsymbol{\phi}(\mathbf{x}-\mathbf{e}_i^j) - 2\boldsymbol{\phi}(\mathbf{x})\right], \tag{1.28}$$

$\mathbf{U}$ is the term representing the energy of interaction:

$$\left(\mathbf{U}\boldsymbol{\phi}\right)(\mathbf{x}) = \sum_{k=2}^{N}\ \sum_{1\le i_1<\ldots<i_k\le N} \Phi^{(k)}(x_{i_1},\ldots,x_{i_k})\boldsymbol{\phi}(\mathbf{x}) \tag{1.29}$$

and $\mathbf{V} = \mathbf{V}(\omega)$ yields the energy of the (random) external field:

$$\left(\mathbf{V}\boldsymbol{\phi}\right)(\mathbf{x}) = \sum_{j=1}^{N} V(x_j;\omega)\boldsymbol{\phi}(\mathbf{x}). \tag{1.30}$$

As in the single-particle situation, the spectral localization for the Hamiltonian $\mathbf{H}^{(N)}$ means that the spectrum of $\mathbf{H}^{(N)}$ is p.p. and its eigenfunctions $\boldsymbol{\psi}_n(\mathbf{x})$ decay exponentially fast at infinity (e.g., as the norm $|\mathbf{x}| := \max\left[|x_j|,\ 1\le j\le N\right]\to\infty$). Properties of dynamical localization can also be reformulated in a similar manner. The proof of spectral and dynamical localization for $\mathbf{H}^{(N)}$ will be the main task in Part II. We emphasize that, as in the single-particle theory, localization phenomena are based on the fact that the system under consideration lives in the whole lattice $\left(\mathbb{Z}^d\right)^N$. The above-mentioned assumption that the interaction between particles has a finite range will be important here, although it can be weakened to cover potentials of an infinite range, with a certain rapid decrease (see the discussion in Sect. 4.7.3).

Given an N-particle cube $\boldsymbol{\Lambda}_L(0) = \left([-L,L]^d\cap\mathbb{Z}^d\right)^N$, the Hamiltonian $\mathbf{H}_{\boldsymbol{\Lambda}_L} = \mathbf{H}^{(N)}_{\boldsymbol{\Lambda}_L(0)}(\omega)$ in $\boldsymbol{\Lambda}_L = \boldsymbol{\Lambda}_L(0)$ is introduced by imitating (1.26):

$$\mathbf{H}_{\boldsymbol{\Lambda}_L} = \mathbf{H}^0_{\boldsymbol{\Lambda}_L} + \mathbf{U} + g\mathbf{V}. \tag{1.31}$$

Here the operators $\mathbf{H}^0_{\boldsymbol{\Lambda}_L}$, $\mathbf{U}$ and $\mathbf{V}$ act on functions $\boldsymbol{\phi}(\mathbf{x})$ where $\mathbf{x} = (x_1,\ldots,x_N)\in\boldsymbol{\Lambda}_L$. More precisely, $(\mathbf{U}\boldsymbol{\phi})(\mathbf{x})$ and $(\mathbf{V}\boldsymbol{\phi})(\mathbf{x})$ are defined as in (1.29) and (1.30) while $\mathbf{H}^0_{\boldsymbol{\Lambda}_L}$ is the version of $\mathbf{H}$ with the Dirichlet boundary condition:

$$\left(\mathbf{H}^0_{\mathbf{\Lambda}_L}\boldsymbol{\phi}\right)(\mathbf{x}) = -\sum_{i=1}^{N}\sum_{j=1}^{d}\Big[\boldsymbol{\phi}(\mathbf{x}+\mathbf{e}_i^j)\mathbf{1}_{\mathbf{\Lambda}_L}(\mathbf{x}+\mathbf{e}_i^j) \\ + \boldsymbol{\phi}(\mathbf{x}-\mathbf{e}_i^j)\mathbf{1}_{\mathbf{\Lambda}_L}(\mathbf{x}-\mathbf{e}_i^j) - 2\boldsymbol{\phi}(\mathbf{x})\Big],\ \mathbf{x}\in\mathbf{\Lambda}_L. \tag{1.32}$$

1.5.2 The Two-Particle Case

Features of the N-particle localization and difficulties that are encountered in the course of its analysis can be demonstrated already in the case $N = 2$. For simplicity we will also assume in this subsection that the dimension $d = 1$. The configuration space of a pair of distinguishable particles on $\mathbb{Z}^1$ is $\mathbb{Z}^1 \times \mathbb{Z}^1 \simeq \mathbb{Z}^2$. A point $\mathbf{x} = (x_1, x_2) \in \mathbb{Z}^1\times\mathbb{Z}^1$ corresponds to a quantum state where the first particle is located at x_1 and the second at x_2. The Hamiltonian $\mathbf{H} = \mathbf{H}^{(2)}(\omega)$ acts on functions $\boldsymbol{\phi}(x_1, x_2)$:

$$\mathbf{H}\boldsymbol{\phi}(x_1,x_2) = 4\boldsymbol{\phi}(x_1,x_2) - \boldsymbol{\phi}(x_1-1,x_2) - \boldsymbol{\phi}(x_1+1,x_2) \\ -\boldsymbol{\phi}(x_1,x_2-1) - \boldsymbol{\phi}(x_1,x_2+1) + \Phi^{(2)}(x_1,x_2)\boldsymbol{\phi}(x_1,x_2) \\ +g\big(V(x_1;\omega)+V(x_2;\omega)\big)\boldsymbol{\phi}(x_1,x_2). \tag{1.33}$$

Here $\Phi^{(2)}(x_1, x_2)$ is a two-body interaction potential (often assumed to be of the form $\Phi^{(2)}(x_1 - x_2)$) and $V(x;\omega)$ represents the potential of an external field, at points $x, x_1, x_2 \in \mathbb{Z}$, respectively. The aforementioned finite-range condition means that $\Phi^{(2)}(x_1, x_2) = 0$ when $|x_1 - x_2| > r_0$ where $r_0 \in (0, \infty)$ is a given value (the radius of two-body interaction; cf. Sect. 3.1.1).

When $\Phi^{(2)} \equiv 0$, the Hamiltonian $\mathbf{H}$ admits an algebraic decomposition:

$$\mathbf{H} = H_1 \otimes \mathbf{I} + \mathbf{I} \otimes H_2 \tag{1.34}$$

where each operator $H_j = H_j(\omega)$, $j = 1, 2$, acts on the variable x_j:

$$H_j\phi(x_j) = 2\phi(x_j) - \phi(x_j-1) - \phi(x_j+1) + gV(x_j;\omega)\phi(x_j).$$

In this situation it suffices to study the eigenfunctions and eigenvalues of the single-particle Hamiltonian $H(\omega)$. (Which we do, to some extent, in Chap. 2.) However, adding a non-zero interaction changes the nature of the operator. We will see in Chap. 3 that even relatively simple properties from the one-particle theory, about eigenvalue concentration bounds (which, in physical terms, estimate resonances between quantum states localized at given configurations), become considerably more involved for two particles in the presence of interaction.

Having known from the single-particle localization theory (cf. Chap. 2) that, under certain conditions, the eigenfunctions of the operators H_j in Eq. (1.34) decay exponentially fast when $x_j \to \infty$, it is natural to ask if switching on a non-zero interaction can destroy this phenomenon. We again stress that the system with Hamiltonian $\mathbf{H}^{(2)}(\omega)$ evolves in an infinitely extended configuration space $\mathbb{Z}^1 \times \mathbb{Z}^1$, and the interaction potential $\Phi^{(2)}$ is assumed to be of a finite range. In other words, the particles do not feel each other's presence when they are far apart. Logically, this class of systems seems to be a natural step beyond the one-particle models. However, from the physical point of view, the formal background for such systems is open to a criticism (although not very much more than in the single-particle case), since one would like to see grounds for localization or delocalization in a permanent presence of interacting particles.

Of course, such a criticism can be also addressed to general N-particle models in $\left(\mathbb{Z}^d\right)^N$. Here, one would like to ascertain that the number of moving particles in $\mathbb{Z}^d$ is infinite, and that they are distributed with a positive density. Alternatively, as it is usually done in statistical mechanics, one takes a "large volume" $\Lambda_L = [-L, L]^d \cap \mathbb{Z}^d$ and works with $\sim \rho(2L+1)^d$ particles where $\rho > 0$ represents the spatial density; see a discussion in the next subsection. On the other hand, we would like to observe that N-particle Hamiltonians similar to (1.26) appear naturally in a perturbative analysis of systems of positive density; cf. [128].

Concluding our brief debate about two-particle models, we note that their specific role has been recognized in a number of physical papers in the context of a parametric analysis of the so-called localization length; cf., e.g., [121, 149].

1.5.3 Systems of Positive Spatial Densities

As was stressed earlier, analysis of a system with a positive density of interacting particles moving in a random environment can be considered as the ultimate goal of localization theories. At the present moment, such a goal remains out of reach of a mathematically rigorous analysis; physical theories are still quite far from it as well. Nevertheless, it seems to us that sharing our informal thoughts on possible forms of localization in such type of systems might give further insight into the actual topic of this book: localization for N particles on lattice $\mathbb{Z}^d$.

We have already said above that two possible approaches are emerging[10]:

- A many-particle system in a large lattice cube $\Lambda_L = \Lambda_L(0)$ defined as $([-L, L] \cap \mathbb{Z})^d$, where the number of particles is $\sim \rho(2L+1)^d$ with $\rho \in (0, \infty)$ representing the particle density;
- A system of infinitely many particles in $\mathbb{Z}^d$, again with a (properly defined) positive spatial density.

The second category of systems (with infinitely many particles in $\mathbb{Z}^d$) is usually analyzed by means of a thermodynamical limit performed for systems of the

[10]We also recognize the importance of various approximation schemes, viz., mean-field, but choose not to dwell here on such models.

first category, in a chosen form of a canonical ensemble (e.g., grand or small), with $L \to \infty$. In such a limit, the Hamiltonian of the system becomes a formal expression[11] and although there exists a way to attribute to it a status of an operator (the Gelfand–Naimark–Segal construction, based on the concept of the Gibbs state), the physical picture looses clarity. For this reason, we focus on systems of the first category, in a large but finite cube Λ_L, with N particles where $N \sim \rho(2L+1)^d$.

As above, we will—for definiteness—consider distinguishable particles, although from the point of view of quantum Statistical Mechanics, one should work with bosons or fermions, i.e., focus on the eigenfunctions $\boldsymbol{\psi}(\mathbf{x}_1, \dots, \mathbf{x}_N)$ of the Hamiltonian $\mathbf{H}^{(N)}(\omega)$ that are symmetric or antisymmetric. Accordingly, the configuration space of the system is the Cartesian product $(\Lambda_L)^N \subset (\mathbb{Z}^d)^N$. Positivity of density $\rho \sim N/(2L+1)^d$ means that, typically, the functions $\boldsymbol{\psi}$ are supported by configurations $(\mathbf{x}_1, \dots, \mathbf{x}_N)$ where the points $\mathbf{x}_j \in \Lambda_L$ are at distances $\sim (1/\rho)^{1/d}$ (the specific volume per particle being $\sim \rho^{-1}$). Pictorially, the particles are "omnipresent" in the lattice cube Λ_L.

To open a discussion, let us put a coefficient η in front of the interaction energy part in Eq. (1.27):

$$\mathbf{H}_{\boldsymbol{\Lambda}_L} = \mathbf{H}^0_{\boldsymbol{\Lambda}_L} + \eta \mathbf{U} + g\mathbf{V}.$$

When $\eta = 0$, we have a system of non-interacting particles. The eigenfunctions $\boldsymbol{\psi}_j = \boldsymbol{\psi}^{(N)}_{\boldsymbol{\Lambda}_L}(\omega)$ are tensor products:

$$\boldsymbol{\psi}_j = \psi_{j_1} \otimes \dots \otimes \psi_{j_N}. \tag{1.35}$$

Here $\psi_{j_k} = \psi_{\Lambda_L, j_k}(\omega)$ are eigenfunctions for $H_{\Lambda_L} = H^0_{\Lambda_L} + gV(\cdot\,;\omega)$, the single-particle version of operator $\mathbf{H}^0_{\boldsymbol{\Lambda}_L} + \eta\mathbf{U} + g\mathbf{V}$. An outcome of a single-particle Anderson theory with an IID potential $V(x;\omega)$ is that if, for instance, the value g is large enough, the eigenfunctions ψ_{j_k} are concentrated near their "centers of localization"; see Sect. 2.7.2 for details. (The exact analysis of such spatial concentration leads to the spectral localization for a single particle in $\mathbb{Z}^d$.) It is conceivable (although it needs a formal justification) that, typically, the eigenfunctions $\boldsymbol{\psi}_j$ for $\eta = 0$, being projected to the single-particle cube Λ_L, have multiple peaks smeared over Λ_L more or less uniformly. A similar picture might emerge when we switch on the interaction, by taking $\eta \neq 0$. A logically natural conclusion is that, in the case of positive spatial density, one should not expect the N-particle eigenfunctions to be spatially localized in Λ_L in the same sense as their single-particle counterparts. In fact, it is quite likely that there is no physical *raison d'être* for the spatial concentration in systems with a positive density.

---

[11] A fruitful point of view may be to look at the differentiation of a (suitably chosen) quasilocal C*-algebra associated with the Hamiltonian in the thermodynamical limit; see, e.g., [51]. However, this is beyond the scope of this book.

The above heuristic argument should, of course be made more precise in the context of quantum statistical mechanics. A physical analysis of the (analog of the) Anderson localization for systems with large values of L and N can be found in recent works [31] and [105] (the order of citation is merely alphabetical). A general conclusion of such an analysis is that localization phenomena in this type of systems should be studied not in the configuration space but in the Fock space, in the framework of the second quantization. To put it simply, the role of a natural scene replacing the finite cube Λ_L should be played by the collection of eigenfunctions of the form (1.35).

The interaction operator $\eta\mathbf{U}$, considered as a perturbation of $\mathbf{H}^0_{\boldsymbol{\Lambda}_L} + g\mathbf{V}$ defines a graph structure on the collection of eigenfunctions (1.35); this graph structure depends upon the specific features of the interaction. Physical manifestations of localization are interpreted in terms of properties of the Hamiltonian $\mathbf{H}^0_{\boldsymbol{\Lambda}_L} + \eta\mathbf{H} + g\mathbf{V}$ in the space of functions on the above graph. A detailed discussion of this approach is beyond the limits of this book (the reader can find it in [30]).

A possible insight into properties of a system under consideration may be provided by an analysis of the current–current correlation measure for the operator $\mathbf{H}_{\boldsymbol{\Lambda}_L}(\omega)$. (For the definition and properties of current–current correlation measures in single-particle models, see a recent paper [76] and the bibliography therein.) However, such an analysis again should take into account the aforementioned features.

As we said earlier, rigorous mathematical techniques available at the time of writing this book do not allow us to analyze a physically meaningful N-particle localization in the cube Λ_L where $L \gg 1$ and $N \sim \rho(2L+1)^d$. Such a task requires developing a more direct and powerful approach, possibly treating the interaction operator $\mathbf{U}$ as a perturbation. However, the existing methodology of working with multi-particle systems only covers models where N is fixed (or grows "very slowly" with L) and $L \to \infty$. In essence, this methodology treats the kinetic energy operator $\mathbf{H}^0_{\boldsymbol{\Lambda}_L}$ as a perturbation of the potential energy operator $\eta\mathbf{H} + g\mathbf{V}$, following the same path as in the single-particle localization theory.

It is our hope that the present book will help to develop rigorous multi-particle techniques needed for an analysis of localization in systems with positive density of particles, although such time may not come any time soon.

Chapter 2
Single-Particle MSA Techniques

In this chapter we analyze a single-particle tight-binding Anderson model on $\mathbb{Z}^d$, with the random Hamiltonian $H(\omega)$:

$$\Big(H(\omega)\phi\Big)(x) = H^0\phi(x) + gV(x;\omega)\phi(x),\ x \in \mathbb{Z}^d,\ \phi \in \ell^2(\mathbb{Z}^d). \tag{2.1}$$

Here H^0 stands for the minus discrete Laplacian Δ:

$$(H^0\phi)(x) = -\sum_{i=1}^{d} \left(\phi(x+e_i) + \phi(x-e_i) - 2\phi(x)\right) \tag{2.2}$$

with $e_i = (0,\dots,0,1,0,\dots,0) \in \mathbb{Z}^d$ (all entries 0 except for one equal to 1, in position i). Next, $\{V(x;\omega),\ x \in \mathbb{Z}^d\}$ is a collection of real-valued random variables defined on a probability space $(\Omega, \mathfrak{B}, \mathbb{P})$. Here the set of outcomes

$$\Omega = \mathbb{R}^{\mathbb{Z}^d} = \{\omega = (\omega_x)_{x\in\mathbb{Z}^d}\},$$

so $V(x;\omega) = \omega_x$, $x \in \mathbb{Z}^d$, $\mathfrak{B}$ is the sigma-algebra of subsets of Ω generated by all cylinders, and $\mathbb{P}$ is a probability measure on $(\Omega, \mathfrak{B})$. Most of the time we require the random values $V(x;\omega)$, $x \in \mathbb{Z}^d$ to be IID (independent, identically distributed), and in that case $\mathbb{P}$ is the product-measure $\otimes_{x\in\mathbb{Z}^d}\mathbb{P}_x$ where the measures $\mathbb{P}_x$ on $\mathbb{R}$ have identical probability distribution functions:

$$\forall\, x \in \mathbb{Z}^d \qquad \mathbb{P}\{\,V(x;\omega) \le t\,\} = F(t).$$

The expectation relative to $\mathbb{P}$ will be denoted by $\mathbb{E}\,[\,\cdot\,]$.

V. Chulaevsky and Y. Suhov, *Multi-scale Analysis for Random Quantum Systems with Interaction*, Progress in Mathematical Physics 65, DOI 10.1007/978-1-4614-8226-0_2,

2.1 An Initiation into the Single-Particle MSA

2.1.1 *Technical Requisites*

We begin with the main assumptions upon the external random potential $V(\cdot\,;\omega)$. As was said above, throughout most of the book we assume that the values $V(x;\omega)$, $x \in \mathbb{Z}^d$, are IID random variables. (A notable exception is Sect. 4.7.3.)

Probabilistically, an IID sequence of random variables is characterized via the marginal distribution of a single variable. In the main body of the book the condition under which localization at large disorder will be proven is

Assumption A. *The common marginal distribution function*

$$F(a) = \mathbb{P}\{\, V(x;\omega) \le a\,\}, \quad x \in \mathbb{Z}^d,$$

of the IID random variables $\{V(x;\cdot), x \in \mathbb{Z}^d\}$ *satisfies a uniform Hölder continuity condition: there exist constants* $s \in (0,1]$, $C_s \in (0,\infty)$ *such that* $\forall\, \epsilon \in [0,1]$,

$$\varsigma(F,\epsilon) := \sup_{a\in\mathbb{R}} \big[F(a+\epsilon) - F(a)\big] \le C_s \epsilon^s. \tag{2.3}$$

In results on localization at the lower edge of the spectrum it is convenient to quote

Assumption B. *The common marginal distribution function* F *of the IID random variables* $\{V(x;\cdot), x \in \mathbb{Z}^d\}$ *vanishes on* $(-\infty,0)$ *and is strictly positive on* $(0,\infty)$*:*

$$\sup\big[a:\ F(a) = 0\big] = 0. \tag{2.4}$$

In other words, the value 0 *gives the infimum of values taken by the random potential* $V(\cdot\,;\omega)$ *with nonzero probability.*

Remark 2.1.1. A more general case where $\sup\big[a:\ F(a)=0\big] = E_* > -\infty$ can be easily reduced to the above by taking $\tilde V(x;\omega) := V(x;\omega) - E_*$. This results in a spectral shift, $\tilde H(\omega) = H(\omega) - E_*\mathbf{I}$, so the operators $\tilde H(\omega)$ and $H(\omega)$ share the same eigenvectors.

Remark 2.1.2. The condition of unform Hölder-continuity in Assumption A is used in this book in order to simplify the entire presentation. (This condition is used also to obtain slightly stronger probabilistic bounds than in the more conventional multi-scale analysis.) A uniform Hölder-continuity can be replaced, at a cost of complication of some technicalities, by a (uniform) log-Hölder continuity:

$$\sup_{a\in\mathbb{R}} \big[F(a+\epsilon) - F(a)\big] \le \frac{\mathrm{Const}}{|\ln \epsilon|^A} \tag{2.5}$$

with sufficiently large $A > 0$. This is where, in the context of this book, the MSA shows, at least for now, more flexibility than the FMM.

The multi-scale analysis method derives properties of the operator $H(\omega)$ from an asymptotical analysis of $H_\Lambda(\omega)$, a finite-volume counterpart of $H(\omega)$:

$$\big(H_\Lambda(\omega)\phi\big)(x) = \big(H^0_\Lambda\phi\big)(x) + gV(x;\omega)\phi(x), \ \ x \in \Lambda. \tag{2.6}$$

Here and below Λ is a finite subset in $\mathbb{Z}^d$, typically a (lattice) cube $\Lambda_L(u)$ centered at a given point $u \in \mathbb{Z}^d$ and with "radius" $L = 1, 2, \dots$ relative to the max-norm

$$|x| = \max\left[|x^{(j)}| : 1 \le j \le d\right], \text{ for } x = \left(x^{(1)}, \dots, x^{(d)}\right).$$

Explicitly,

$$\Lambda_L(u) = \{u + x \in \mathbb{Z}^d : |x| \le L\}; \tag{2.7}$$

the cardinality of $\Lambda_L(u)$ (or the lattice volume of $\Lambda_L(u)$) is denoted by $|\Lambda_L(u)|$ and equals $(2L+1)^d$. In the general case, $|\Lambda|$ will stand for the cardinality of a finite set $\Lambda \subset \mathbb{Z}^d$, and even more generally, $|\mathcal{J}|$ or card $\mathcal{J}$ will denote the cardinality of a given (finite) set $\mathcal{J}$.

We will use the distance $\text{dist}(\cdot,\cdot)$ induced by the max-norm $|\cdot|$, on lattices $\mathbb{Z}^d$ and Euclidean spaces $\mathbb{R}^d$, as well as on various subsets thereof (viz., on the line $\mathbb{R}$ where it coincides with the standard Euclidean metric).

It is also convenient to consider two other frequently used vector norms in $\mathbb{R}^d$ (and $\mathbb{Z}^d \hookrightarrow \mathbb{R}^d$):

$$|x|_1 := |x^{(1)}| + \dots + |x^{(d)}|, \ \text{ and } |x|_2 := \left((x^{(1)})^2 + \dots + (x^{(d)})^2\right)^{1/2}.$$

The distance induced by the norm $|\cdot|_1$ is denoted by $\text{dist}_1(\cdot,\cdot)$ and that induced by $|\cdot|_2$ by $\text{dist}_2(\cdot,\cdot)$.

Remark 2.1.3. It is useful to note that the lattice $\mathbb{Z}^d$ has a natural graph structure, with bonds (or edges) connecting the nearest neighbors in either of the distances $\text{dist}_i(\cdot,\cdot)$, $i = 1, 2$. Accordingly, two lattice points x and x' are "graph-neighbors" on $\mathbb{Z}^d$ if the difference $x - x'$ is a vector with $d - 1$ components 0 and 1 equal to ± 1. Keeping in mind this structure, we will call a lattice subset Λ *connected*, if the subgraph of $\mathbb{Z}^d$ with the vertex set Λ (and the edge set inherited from the lattice) is connected. In other words, Λ is connected if for any two points $x, y \in \Lambda$ there exists a path formed by lattice bonds (u_k, u_{k+1}), $1 \le k \le n$, $n \ge 1$, satisfying $u_k, u_{k+1} \in \Lambda$, with $u_1 = x$, $u_{n+1} = y$.

The term H^0_Λ in (2.6) represents the kinetic energy operator in Λ and stands for the negative of the discrete Laplacian $\Delta^{\mathrm{D},0}_\Lambda$ with Dirichlet's boundary conditions in Λ:

$$(H^0_\Lambda \phi)(x) = 2d\phi(x) - \sum_{y\in\Lambda:|y-x|_1=1} \phi(y), \quad x \in \Lambda. \tag{2.8}$$

In short, we will refer to H^0_Λ as the negative of the Dirichlet Laplacian in Λ.[1]

The (random) operator of multiplication by $V(x;\omega)$, $x \in \Lambda$, representing the second summand in the RHS of (2.6) will be often denoted by $V_\Lambda(\omega)$:

$$(V_\Lambda(\omega)\phi)(x) = V(x;\omega)\phi(x), \quad x \in \Lambda. \tag{2.9}$$

The operators $H_\Lambda(\omega)$, H^0_Λ and V_Λ act in the Hilbert space $\ell^2(\Lambda) \simeq \mathbb{C}^{|\Lambda|}$. In the standard delta-basis $\{\delta_x,\ x \in \Lambda\}$ the operator $V_\Lambda(\omega)$ is represented by a diagonal matrix of size $|\Lambda|$. Consequently, $V_\Lambda(\omega)$ has (non-random) eigenfunctions δ_x, with the random eigenvalues $V(x;\omega)$:

$$V_\Lambda(\omega)\delta_x = V(x;\omega)\delta_x, \quad x \in \Lambda.$$

The operators H^0_Λ and $H_\Lambda(\omega)$ are represented in the delta-basis by Hermitian square matrices of size $|\Lambda|$ (as was noted, when $\Lambda = \Lambda_L(u)$, the cardinality $|\Lambda| = (2L+1)^d$). Consequently, H^0_Λ and $H_\Lambda(\omega)$ have $|\Lambda|$ real eigenvalues (counted with their multiplicities).

The operator $H_\Lambda(\omega)$ will be interpreted as a (random) Hamiltonian in Λ (and alternatively called LSO in Λ). We will denote by $\sigma\left(H^0_\Lambda\right)$ and $\sigma\left(H_\Lambda(\omega)\right)$ the spectra of the operators H^0_Λ and $H_\Lambda(\omega)$ (i.e., the collections of the eigenvalues counted with their multiplicities); the same notation will be used for other operators under consideration. The eigenvalues of H^0_Λ and $H_\Lambda(\omega)$ are supposed to be listed in an increasing order and denoted by $E_j\left(H^0_\Lambda\right)$ and $E_j\left(H_\Lambda(\omega)\right)$, viz.:

$$E_1\left(H_\Lambda(\omega)\right) \le E_2\left(H_\Lambda(\omega)\right) \le \ldots \le E_{|\Lambda|}\left(H_\Lambda(\omega)\right). \tag{2.10}$$

The corresponding normalized eigenfunctions of H_Λ are denoted by $\psi_j(H_\Lambda)$, $1 \le j \le |\Lambda|$.

A different version of the kinetic energy operator is given by the Neumann (lattice) Laplacian; in the terminology of graph theory, it is the graph Laplacian in Λ. We define it in the following manner adapted to the view of $\Lambda \subset \mathbb{Z}^d$ as a graph, with the links structure inherited from that of $\mathbb{Z}^d$ (the links join the nearest neighboring sites x, y, with $|x-y|_1 = 1$, and go in both directions). Given a site $x \in \Lambda$, define the quantity $n_\Lambda(x)$ (called the coordination number of the point x in Λ) by

$$n_\Lambda(x) := \operatorname{card}\{y \in \Lambda : |y-x|_1 = 1\}. \tag{2.11}$$

[1]This terminology is not quite standard; in particular, it differs from that used in the review [113]. However, this allows to have an analog of Dirichlet–Neumann bracketing useful in the analysis of the Lifshitz tails phenomenon; cf. Sect. 2.3.5.

In other words, $n_\Lambda(x)$ is the number of nearest neighbors of x in Λ. Clearly, $n_\Lambda(x) \le 2d$.

The kinetic energy operator with Neumann boundary condition in Λ is introduced by

$$(H_\Lambda^{\mathrm{N},0}\phi)(x) = (H_\Lambda^0\phi)(x) - (2d - n_\Lambda(x)). \tag{2.12}$$

It is straightforward that $H_\Lambda^{\mathrm{N},0} \le H_\Lambda^0$ in the sense of quadratic forms, since

$$H_\Lambda^0 - H_\Lambda^{\mathrm{N},0} = 2d - n_\Lambda \tag{2.13}$$

is an operator of multiplication by a non-negative function $x \mapsto 2d - n_\Lambda(x)$. Further, it is straightforward that the quadratic form associated with $H_\Lambda^{\mathrm{N},0}$,

$$\langle \phi_1, H_\Lambda^{\mathrm{N},0}\phi_2\rangle = \frac{1}{2} \sum_{x,y\in\Lambda:|x-y|_1=1} \overline{\Big(\phi_1(x) - \phi_1(y)\Big)}\Big(\phi_2(x) - \phi_2(y)\Big),$$

is non-negative. Therefore, the operators $H_\Lambda^{\mathrm{N},0}$ and H_Λ^0 $\left(\ge H_\Lambda^{\mathrm{N},0}\right)$ are non-negative.

As a matter of fact, the kinetic energy operator with Neumann's boundary condition will be rarely used in this book; the main reason for discussing it here is that it proves convenient for the analysis of Lifshitz tails asymptotics in Sect. 2.3.5.

In addition, we briefly mention in Sect. 2.7.3 the Hamiltonian $H_\Lambda^{\mathrm{per},0}$ in $\Lambda = \Lambda_L(0)$ with periodic boundary conditions.

Remark 2.1.4. Before we proceed with further formal requisites, we would like to stress the following. Despite the impression that the "infinite-volume" Hamiltonian $H(\omega)$ is in the center of our attention, it is important to quote (in an adapted version) a maxim popular in several mathematical disciplines: "*Analyzing the Hamiltonian* $H(\omega)$ *on the whole lattice is a good warm-up for studying the Hamiltonian* $H_{\Lambda_L(u)}(\omega)$ *on a large but finite volume* $\Lambda_L(u)$."

Given a non-empty set $\Lambda \subsetneq \mathbb{Z}^d$, we introduce its inner boundary $\partial^-\Lambda$ and outer boundary $\partial^+\Lambda$:

$$\begin{aligned} \partial^-\Lambda &= \{y \in \Lambda : \ \mathrm{dist}_1(y, \mathbb{Z}^d \setminus \Lambda) = 1\}, \\ \partial^+\Lambda &= \{y \in \mathbb{Z}^d \setminus \Lambda : \ \mathrm{dist}_1(y, \Lambda) = 1\}. \end{aligned} \tag{2.14}$$

Finally, we set:

$$\partial\Lambda = \{(x, y) : \ x \in \partial^-\Lambda, \ y \in \partial^+\Lambda, \ |x - y|_1 = 1\}. \tag{2.15}$$

$\partial\Lambda$ is often called the edge boundary of Λ.

The discrete Laplacians Δ and $\Delta_\Lambda^{\mathrm{D},0}$ can be written as follows:

$$\Delta\phi(x) = \sum_{y\in\mathbb{Z}^d:\,|x-y|_1=1} \phi(y) - 2d\phi(x),$$

$$\Delta_\Lambda^{\mathrm{D},0}\phi(x) = \sum_{y\in\Lambda:\,|x-y|_1=1} \phi(y) - 2d\phi(x).$$

2.1.2 The MSA Induction

We mentioned in the preceding chapter that the MSA method represents a (specific) inductive procedure. The induction scheme at the core of the MSA runs through a sequence of integer values

$$2 \le L_0 < L_1 < \cdots < L_k < \cdots$$

(length scales) where

$$L_{k+1} = \lfloor L_k^\alpha \rfloor, \quad k \ge 0. \tag{2.16}$$

Here and below, $\lfloor r \rfloor$ stands for the integer part of a real number r. The initial scale L_0 and the exponent α are subject to a judicial choice: it will be assumed that L_0 is large enough (see Remark 2.1.5 below), and $1 < \alpha < 2$. In fact, we will always set $\alpha = 3/2$. Pictorially, $L_k \sim (L_0)^{\alpha^k}$. In what follows, we refer to this induction as the scale induction.

At the start, the desired properties of the operator $H_\Lambda(\omega)$ are established for $\Lambda = \Lambda_{L_0}(u)$, $u \in \mathbb{Z}^d$ (the initial step of the scale induction). Then, assuming that such properties hold for all $\Lambda = \Lambda_{L_{k-1}}(u)$, $u \in \mathbb{Z}^d$, one aims at deducing them for $\Lambda = \Lambda_{L_k}(u)$ (the inductive step).

More precisely, the MSA studies resolvents (Green's operators)

$$G_\Lambda(E,\omega) := (H_\Lambda(\omega) - E)^{-1} \tag{2.17}$$

by analyzing their norms $\|G_\Lambda(E,\omega)\|$ and matrix entries $G_\Lambda(x,y;E,\omega)$ in the lattice Dirac delta-basis δ_x, $x \in \Lambda$ (also known as the Green's functions):

$$G_\Lambda(x,y;E,\omega) := \left\langle \delta_x, \left(H_\Lambda(\omega) - E\right)^{-1}\delta_y \right\rangle, \quad x,y \in \Lambda, \ E \in \mathbb{R}. \tag{2.18}$$

We adhere to a popular tradition of writing $H_\Lambda(\omega) - E$ for the operator $H_\Lambda(\omega) - E\mathbf{I}$ where $\mathbf{I}$ is the identity operator (in this instance, in $\ell^2(\Lambda)$). The angular brackets $\langle \cdot, \cdot \rangle$ stand for the inner product in $\ell^2(\Lambda)$. By definition, the Green's functions $G_\Lambda(x,y;E,\omega)$ are well-defined for E and ω such that $E \notin \sigma(H_\Lambda(\omega))$. The value of E is referred to as an energy (at which the Green's function is considered).

Remark 2.1.5. We would like to make an *important comment* on the role of the value L_0. In the course of the scaling analysis, we will use a number of inequalities involving the length scales L_k. It is straightforward that for any $c_i, C_i \in (0, +\infty)$, $i = 1, 2, 3$, and $L \gg 1$ one has

$$C_1 \ln^{c_1} L \ll C_2 L^{c_2} \ll C_3 e^{L^{c_3}}.$$

All these inequalities hold true asymptotically, starting from some L_* which depends upon the constants C_i, c_i. Applied to the sequence $\{L_k, k \geq 0\}$, they hold for all L_k, provided that L_0 is large enough: $L_0 \geq L_*$, with L_* depending upon the respective parameters C_i, c_i. For this reason, many assertions will include a condition of the form "*...provided that* L_0 *is large enough* ...", although the assertion itself concerns the value L_k with an arbitrarily large k.

Under no circumstances such a requirement will mean that there is an infinite number of conditions on L_0, for $k = 0, 1, 2, \ldots$. It would indeed have been disastrous if one had to require that $L_0 \geq A_k$ for an inequality involving L_k to hold, where $A_k \to \infty$ as $k \to \infty$.

In certain, relatively simple situations we will be able to provide an explicit lower bound on L_0. For example, in Remark 2.4.3, Sect. 2.4.2, we require that $L_0 \geq 18^{8/3}$, and the latter explicit threshold does not depend upon k (and the value of L_k). However, in many cases an explicit condition on L_0 is both cumbersome and useless, from the purely mathematical point of view, while it might be of some importance for applications to the physics of the disordered systems.

2.1.3 Fixed-Energy and Variable-Energy MSA: Informal Discussion

The MSA focuses on decay properties of the functions $G_\Lambda(x, y; E, \omega)$ when the cube Λ is "large" and x and y are distant from each other. To this end, we introduce, complying in essence with paper [161], an important definition.

Definition 2.1.1. *Given* $E \in \mathbb{R}$, $m > 0$ *and a sample of potential* $V(\cdot\,; \omega)$, *a cube* $\Lambda_L(u)$ *is called* (E, m)*-non-singular* $\big((E, m)$-NS$\big)$ *if*

$$|\partial \Lambda_L(u)| \max_{x \in \Lambda_{L^{1/\alpha}}(u)} \max_{y \in \partial^- \Lambda_L(u)} \big|G_{\Lambda_L(u)}(x, y; E, \omega)\big| \leq e^{-\gamma(m, L)L} \tag{2.19}$$

with

$$\gamma(m, L) = m\,(1 + L^{-\tau}), \quad \tau = 1/8. \tag{2.20}$$

Otherwise $\Lambda_L(u)$ *is called* (E, m)*-singular* ($(E, m) -$ S).

The volume factor $|\partial \Lambda_L(u)|$ in the LHS of (2.19) is not traditional, but convenient.

The parameter $m > 0$ is a constant (which we will strive to make independent of E when E varies in a suitably chosen bounded interval $I \subset \mathbb{R}$), and $\gamma(m, L)$ is a factor approaching m from above as $L \to \infty$. (Here we diverge from [161]; the motivation for this change is provided below.) The value m is referred to as a "mass"; this terminology originated in the quantum field theory and is unrelated to the mass of quantum particles under consideration.

The event where Eq. (2.19) holds true involves several ingredients:

- A cube $\Lambda_L(u)$ and a random sample $\{V(x;\omega), x \in \Lambda_L(u)\}$;
- An energy E and a mass m.

It is worth noting that the bound (2.19) is not *directly* related to the norm $\|G_{\Lambda_L(u)}(E,\omega)\|$ of the Green's operator $G_{\Lambda_L(u)}(E,\omega)$ (and hence to the distance $\mathrm{dist}\big(E, \sigma(H_{\Lambda_L(u)})\big)$). Consequently, identifying singular cubes becomes an elusive task and requires an assortment of technicalities.

Another important property of the operator $H_{\Lambda_L(u)}$, which can be expressed in terms of $\mathrm{dist}\big(E, \sigma(H_{\Lambda_L(u)})\big)$, is given by the bound

$$\|G_{\Lambda_L(u)}(E,\omega)\| > \mathrm{e}^{L^\beta}, \tag{2.21}$$

for a given value of $\beta \in (0,1)$. (In this book, it will always be assumed that $\beta = 1/4$.) In this case $\Lambda_L(u)$ is called E-resonant; otherwise we say that $\Lambda_L(u)$ is E-non-resonant. (We provide a *formal definition in Sect. 2.2.3*; see Definition 2.2.3.) In essence (with an appropriate relation between the parameters m and β), E-non-resonance is a necessary, but by far not sufficient condition for (E,m)-non-singularity.

The MSA aims at proving certain probabilistic bounds on events related to the properties of E-resonance and (E,m)-singularity. The form of such bounds distinguishes between the two variants of the MSA which we are going to discuss now.

2.1.4 Fixed-Energy MSA

First, given an interval $I \subseteq \mathbb{R}$ (an energy band), one can want to assess the probability that a given cube of radius L is (E,m)-S for a given energy $E \in I$, i.e., seek a bound of the form

$$\mathbb{P}\{\, \Lambda_L(u) \text{ is } (E,m)\text{-S } \} \le f(L). \tag{2.22}$$

Recall that, by definition of the (E,m)-S property (cf. Eq. (2.19)), this reads as

$$\mathbb{P}\left\{ \max_{x\in\Lambda_{L^{1/\alpha}}(u)} \; \max_{y\in\partial^-\Lambda_L(u)} \left|G_{\Lambda_L(u)}(x,y;E,\omega)\right| > |\partial\Lambda_L(u)|^{-1}\mathrm{e}^{-\gamma(m,L)L} \right\} \le f(L) \tag{2.23}$$

where $\gamma(m, L) = m(1 + L^{-1/8}) \approx m$. In other words, if $f(L) \ll 1$, then, typically, the matrix elements of the resolvent (Green's functions) $G_{\Lambda_L(u)}(u, y; E, \omega)$ decay exponentially, with rate (at least) $m > 0$ when the arguments u and y are far apart.

Specifically, we will aim to prove the following bound, for all $E \in I$, all $k \geq 0$ and some[2] $m, p_0, \theta > 0$:

$(\mathbf{SS}.I, m, p_k, L_k)$: $\forall$ *cube* $\Lambda_{L_k}(u)$

$$\mathbb{P}\{\, \Lambda_{L_k}(u) \text{ is } (E, m)\text{-S} \,\} \leq L_k^{-p_k}, \quad \text{where } p_k = p_0(1+\theta)^k. \tag{2.24}$$

Here "SS" stands for "single singularity" (cf. "double singularity", Eq. (2.29) in the next subsection).

To understand possible implications of such a property, recall that the resolvent admits the following representation (we make abstraction from the randomness of the potential in the algebraic formula given below):

$$G_{\Lambda_L(u)}(x, y; E) = \sum_{E_i \in \sigma(H_{\Lambda_L(u)})} \frac{\psi_i(x)\, \overline{\psi_i(y)}}{E_i - E}. \tag{2.25}$$

We see that there may be two competing tendencies (or mechanisms), one enhancing the decay of Green's functions (at large distances $|x - y|$), and the other capable to destroy such a decay:

(i) If each eigenfunction ψ_i decays rapidly (e.g., exponentially) away from a restricted zone in $\Lambda_L(u)$, then the individual products $|\psi_i(x)\, \psi_i(y)|$ are small for $|x - y| \gg 1$;
(ii) If at least one denominator $E_i - E$ is extremely small, for example, much smaller than typical values of the products $|\psi_i(x)\, \psi_i(y)|$, then one can no longer guarantee the desired decay of the sum of the terms $|\psi_i(x)\, \psi_i(y)| \cdot |E_i - E|^{-1}$, $i = 1, \ldots, |\Lambda_L(u)|$.

We will refer to the events of the type (ii) as "resonances". It is clear that there is little hope to establish any reasonable decay properties of the eigenfunctions, unless one can rule out—with a sufficiently high probability—such resonances. The main technical tool for assessing the probability of resonances are the so-called eigenvalue concentration (EVC) estimates. Historically, the first fairly general estimate of such kind was proved by Wegner [163]. In our book, we will describe several EVC estimates.

While it is straightforward that small (e.g., exponentially small, in terms of $L \gg 1$) values of the products $|\psi_i(x)\, \psi_i(y)|$, combined with absence of strong resonances ($|E_i - E| \ll 1$), would normally imply the decay of the Green's functions (since the number of terms in the sum (2.25) is only polynomially large in L), the converse is much less obvious.

---

[2]See Remark 2.4.4 in Sect. 2.4.3 and Eq. (2.121).

One possible approach to the derivation of a result of the form

"exponential decay of Green's functions ⇒ decay of eigenfunctions"

in a fixed, finite cube $\Lambda_L(u)$, is using the Cauchy integral formula for the resolvent, written in the operator form as

$$G_{\Lambda_L(u)}(E) = \left(H_{\Lambda_L(u)} - E\right)^{-1} = \sum_{E_i \in \sigma\left(H_{\Lambda_L(u)}\right)} \frac{1}{E_i - E} P_{E_i}\left(H_{\Lambda_L(u)}\right) \tag{2.26}$$

where $P_{E_i}\left(H_{\Lambda_L(u)}\right)$ are the orthogonal projections on the eigenspaces of the operator $H_{\Lambda_L(u)}$. *Assuming* that all eigenvalues are simple and "sufficiently sparse", one may try to recover the projections as residues of the (operator-valued) rational function (2.26), and then transform the upper bounds on the Green's functions into those for the products $|\psi_i(x)\,\psi_i(y)|$.

We emphasized the word "assuming" in the previous paragraph, for this assumption is not as innocent as it might seem. In Sect. 2.7.3, we will briefly discuss known results concerning the accumulation of eigenvalues for Anderson-type Hamiltonians, but now it suffices to say that the above mentioned derivation is far from simple and obvious, and the popular methods of the localization analysis (MSA and FMM) use different paths. Still, this heuristic argument sheds some light on the links between the decay properties of the Green's functions (in large cubes) and those of the eigenfunctions.

Historically, for Anderson-type Hamiltonians in dimension $d > 1$, the a.s. exponential decay bounds on the Green's functions $G(x, y; E; \omega)$ on the entire lattice $\mathbb{Z}^d$ were proved first (by Fröhlich and Spencer [95]). Martinelli and Scoppola [132] proved that the a.s. exponential decay of Green's functions implies the a.s. absence of the absolutely continuous spectrum of the operator $H(\omega)$ (cf. Sect. 2.4.4); formally, this leaves the room for the singular continuous spectrum, so the spectrum might not be a.s. pure point. Later, it was shown by Simon and Wolff [153] for various classes of random operators that "suitable" probabilistic decay bounds on the Green's functions actually do imply a.s. pure point spectrum. We will discuss such results below, at an appropriate moment (cf. Sect. 2.5).

To conclude this topic, we would like to note that the choice of the function f in the upper bound of the form (2.23) plays an important role. It also marks a striking difference between the MSA and the FMM (which we only briefly discuss in Sect. 2.8).

The original techniques from [95] (later substantially simplified by Spencer [156]) allowed to prove fixed-energy bounds of the form (2.23) with $f(L) = L^{-p}$, where the exponent $p > 0$ depends upon the parameters of the model (e.g., the amplitude $|g|$ of the random potential) and should be large enough. In this chapter, we will show that one can easily improve this technique and obtain bounds with $f(L_k) = L_k^{-p(1+\theta)^k}$, for a sequence of "length scales" $L_k \sim (L_0)^{\alpha^k}$, where $L_0 > 1$, $\alpha \in (1, 2)$, $p > 0$ and $\theta > 0$. Alternatively, this can be written as

$$\mathbb{P}\{\, \Lambda_L(u) \text{ is } (E,m)\text{-S} \,\} \le C\,\mathrm{e}^{-a \ln^{1+c} L} \tag{2.27}$$

with some $C, a, c \in (0, +\infty)$.

Much stronger bounds have been obtained by Germinet and Klein [99], namely,

$$\mathbb{P}\{\, \Lambda_L(u) \text{ is } (E,m)\text{-S} \,\} \le \mathrm{e}^{-L^{\delta}} \tag{2.28}$$

with some $\delta \in (0, 1)$.

Unfortunately, the value of the exponent $\delta = 1$ in (2.28) seems so far to be out of the MSA's reach. Until now, exponential probabilistic bounds (of the general form (2.22)) have been obtained only by means of the FMM, due to deep analytic reasons which we cannot discuss in this informal introduction.

On the other hand, the MSA techniques proved to be more robust than their FMM counterpart, when a random potential field featuring the IID property (or, more generally, fast decay of correlations) is replaced by a random field with strong dependencies at large distances, e.g., a quasi-periodic potential (which still can be considered as an ergodic random field). The MSA approach is also much less sensitive to the singular nature of the marginal probability distribution (say, of an IID random potential field); see the works by Bourgain and Kenig [47], by Aizenman et al. [17] and by Germinet and Klein [100].

Now we are going to discuss another variant of the MSA, historically originated in the work by Fröhlich et al. [96] and later reformulated by von Dreifus and Klein [161].

2.1.5 Variable-Energy MSA

For all length scales $L = L_k$, we want to establish a property called (DS.I, m, p_k, L_k) where a specific bound holds for the probability that two non-overlapping cubes of radius L_k are (E,m)-S for some E from a (possibly unbounded) interval I:

(DS.I, m, p_k, L_k): *for any pair of disjoint cubes* $\Lambda_{L_k}(u)$, $\Lambda_{L_k}(v)$

$$\mathbb{P}\{\, \exists E \in I : \Lambda_{L_k}(u) \text{ and } \Lambda_{L_k}(v) \text{ are } (E,m)\text{-S} \,\} \le L_k^{-2p_k}. \tag{2.29}$$

Here $I \subseteq \mathbb{R}$ is an interval where localization is to be established and the value $p_k = p_k(p_0, \theta)$ is given by

$$p_k = p_0(1+\theta)^k \tag{2.30}$$

for some $p_0 > 6d$ and $\theta > 0$ (cf. Remark 2.1.7 in Sect. 2.1.6 and Eq. (2.186)). The event in the LHS of Eq. (2.29) will be referred to as a "double singularity".

Acknowledging the importance of inequality (2.29) we will also refer to it as the MSA bound(s), when the context would not lead to a confusion.

In fact, establishing the inequality (2.29) for all values of k can be considered as the ultimate goal of the MSA (as far as the proof of localization in spectral interval I is concerned). Physically, the importance of the bound (2.29) is related to the fact that, when there are two cubes with eigenvalues close to E (and hence to each other), this might lead to a quantum transport between these cubes (considered as parts of a larger cube or indeed of the whole lattice).

An important role in proving (2.29) is played by the so-called Wegner-type estimates (originated from [163]): for any energy $E \in I$ and any cube $\Lambda_L(u)$,

$$\mathbb{P}\{\, \Lambda_L(u) \text{ is } E\text{-resonant } \} \le L^{-Q}, \tag{2.31}$$

and $\forall$ disjoint cubes $\Lambda_L(u)$ and $\Lambda_L(v)$,

$$\mathbb{P}\{\exists\, E \in I : \Lambda_L(u) \text{ and } \Lambda_L(v) \text{ are } E\text{-resonant}\} \le L^{-Q}, \tag{2.32}$$

with a sufficiently large value of the power $Q > 0$.

The notable difference between Eqs. (2.29) and (2.31) is that the former addresses a “double” phenomenon (two cubes are involved) whereas the latter assesses the probability of a “single” incidence.

We would like to stress that the choice of the scales L_k and of the power-law decay of bounds in (2.32) and (2.29) is not arbitrary; in fact, it is well-adapted to the scale induction. In particular, it is convenient to have the ratio $L_{k+1}/L_k \sim L_k^{\alpha-1}$ growing with k. On the other hand, in the framework of the MSA it seems difficult to obtain a bound decaying exponentially in L_k (for the event in the LHS of Eq. (2.29)). Continuing a comparison with the FMM, the latter method makes it possible to obtain such a bound directly, without involving a scale induction (see Sect. 2.8). Note also that one can obtain stronger bounds of the form $e^{-L_k^{\kappa}}$ upon the probability of double singularity figuring in (2.29), with $\kappa \in (0, 1)$, using a more sophisticated method called bootstrap multi-scale analysis (cf. [99]).

The proof of (2.29) is done via the aforementioned inductive scheme. At the initial step, the bound (2.29) can be established for $k = 0$, for an arbitrarily large initial scale L_0; cf. Sects. 2.3.2 and 2.3.5. Separately (and in fact earlier), the bounds (2.31) and (2.32) are established for all $L \ge L_0$ when L_0 is large enough; see Sect. 2.2.3. We prove these bounds with a particular method proposed by Stollmann [157]; see also the presentation of this method in [158].

Then, assuming that the bound (2.29) holds for a given k, it is reproduced for $k + 1$. In other words, property $(\text{DS}.I, m, p_{k+1}, L_{k+1})$ is derived from $(\text{DS}.I, m, p_{k+1}, L_{k+1})$; we will often refer to it as the passage $k \rightsquigarrow k + 1$. See Sect. 2.6.6. Of course, this step requires certain technical means which have to be introduced beforehand. Our presentation in this chapter is adapted in such a way that its bulk is used in Chaps. 3 and 4 when we pass to multi-particle systems. (We could have made Chap. 2 shorter had we not faced the task of extending the presentation to the multi-particle case.)

Finally, by manipulating the bounds (2.32) and (2.29), the MSA proves that almost surely the spectrum of $H(\omega)$ in $\ell^2(\mathbb{Z}^d)$ is p.p. in I, and all eigenfunctions of $H(\omega)$ with eigenvalues from I decay exponentially. The connection from Eqs. (2.32) and (2.29) to exponential decay of the eigenfunctions requires an additional argument (see Sect. 2.7.1). Historically, this argument capped the first proof of localization in [96] while the bulk of work was done in [95]. In our presentation we again follow mainly the approach adopted in [161].

The main problem in proving the inductive step is that in the course of the passage $k \rightsquigarrow k+1$ in Eq. (2.29), we need to analyze all reasons why two disjoint cubes of radius L_{k+1} may become (E,m)-S. The sum of contributing probabilities has to be made small; although the means used are very basic, the analysis of possible situations leading to double singularity is rather tedious.

The principal analytical tool of the MSA is the second resolvent identity stating that if the operators A and $A+B$ are both invertible, then

$$(A+B)^{-1} = A^{-1} - A^{-1}B(A+B)^{-1}.$$

In particular, this identity applies to a situation where we want to compare the resolvents of two operators $H_{\Lambda_{L_k}(w)}(\omega)$ and $H_{\Lambda_{L_{k+1}}(u)}(\omega)$ where $\Lambda_{L_k}(w) \subset \Lambda_{L_{k+1}-1}(u)$. By using the representation

$$H_{\Lambda_{L_{k+1}}(u)}(\omega) = \left(H_{\Lambda_{L_k}(w)}(\omega) \oplus H_{\Lambda_{L_{k+1}}(u)\setminus\Lambda_{L_k}(w)}(\omega) \right) + T_{\partial\Lambda_{L_k}(w)} \tag{2.33}$$

the above identity applies with $A = H_{\Lambda_{L_k}(w)}(\omega) \oplus H_{\Lambda_{L_{k+1}}(u)\setminus\Lambda_{L_k}(w)}(\omega) - E$ and $B = T_{\partial\Lambda_{L_k}(w)}$, so that $A+B = H_{\Lambda_{L_{k+1}}(u)} - E$. See Sect. 2.3.1.

Formally, the outcome of the single-particle MSA developed in this chapter is the following pair of theorems. Here and below we adopt an agreement that values with upper stars (g^*, L^* and so on) representing various thresholds change from one assertion to another although they are denoted by the same symbol.

Remark 2.1.6. In Chap. 4 we will have to modify the "double singularity" bound.

2.1.6 Single-Particle Localization Results

The spectral localization in a single-particle Anderson model is summarized in the following Theorem 2.1.1.

Theorem 2.1.1.(A) *Suppose that Assumption* A *is fulfilled. There exists $g^* = g^*(F,d) \in (0,\infty)$ such that the following holds. Take $|g| \geq g^*$. Then $\exists\, m \geq 1$, $p_0 > 6d$, $L_0 > 1$ and $\theta > 0$ such that $\forall\, k \geq 0$ and $\forall$ pair of disjoint cubes $\Lambda_{L_k}(u)$ and $\Lambda_{L_k}(v)$, the following bound holds true:*

$$\mathbb{P}\{\exists\, E \in \mathbb{R}:\ \Lambda_{L_k}(u) \text{ and } \Lambda_{L_k}(v) \text{ are } (E,m)\text{-S}\ \} \leq L_k^{-2p_k} \tag{2.34}$$

where $p_k = p_0(1+\theta)^k$.

Consequently, if $|g| \geq g^*$ *then, with probability one,* $H(\omega)$ *has p.p. spectrum, i.e., it admits an orthogonal eigen-basis* $\{\psi_j, j = 1, 2, \ldots\}$ *in* $\ell^2(\mathbb{Z}^d)$*. Moreover, all normalized eigenfunctions* $\psi_j = \psi_j\big(H(\omega)\big)$ *obey*

$$|\psi_j(x)| \leq C_j(\omega)\, \mathrm{e}^{-m|x|}, \quad x \in \mathbb{Z}^d, \tag{2.35}$$

with random constants $C_j(\omega) \in (0, \infty)$.

(B) *Suppose that Assumptions* A *and* B *hold true. Given* $g \in \mathbb{R} \setminus \{0\}$, $\exists\, m > 0$, $p_0 > 6d$, $\theta > 0$, $E^* = E^*(g) > 0$ *and* $L_0 > 1$ *such that* $\forall\, k \geq 0$ *and disjoint cubes* $\Lambda_{L_k}(u)$ *and* $\Lambda_{L_k}(v)$*, the bound*

$$\mathbb{P}\{\exists\, E \in [0, E^*] : \; \Lambda_{L_k}(u) \text{ and } \Lambda_{L_k}(v) \text{ are } (E, m)\text{-S}\} \leq L_k^{-2p_k} \tag{2.36}$$

is satisfied, with p_k *as above.*

Consequently, with probability one, the operator $H(\omega)$ *has p.p. spectrum in* $[0, E^*]$*. Moreover, all normalized eigenfunctions* $\psi_j = \psi_j\big(H(\omega)\big)$ *with eigenvalues* $E_j\big(H(\omega)\big) \in [0, E^*]$ *satisfy inequality* (2.35) *with random constants* $c_j(\omega) \in (0, \infty)$.

Remark 2.1.7. The value of θ in Theorem 2.1.1 can be specified as $1/6 - d/p_0$; cf. Eq. (2.186).

Assertion (A) of Theorem 2.1.1 establishes complete localization (on the whole spectral axis $\mathbb{R}$) for large disorder. Assertion (B) shows localization near the lower edge of the spectrum $\sigma(H(\omega))$ (it can be shown that under Assumption B the lower edge of the spectrum is 0). To derive complete localization, the entire line is covered by a countable family of unit intervals $I_n = [n, n+1]$ and the MSA is performed in each interval I_n. Consequently, from now on we will work in a bounded interval $I \subset \mathbb{R}$.

Remark 2.1.8. We do not discuss the issue of simplicity of the spectrum of $H(\omega)$; the interested reader is referred to [119].

Remark 2.1.9. There is a subtle difference between Assertions (A) and (B) in the range for the mass m ($m \geq 1$ and $m > 0$, respectively). This is related to a difference in technical details of implementation of the MSA in the two situations: large amplitudes ($|g| \gg 1$) and extreme energies ($E \sim 0$). We comment on this in Sects. 2.6.3 and 2.6.6.

Remark 2.1.10. As was said in Remark 2.1.5, in a number of statements the initial length scale L_0 is assumed to be large enough. In the case of large disorder ($|g| \gg 1$), it is possible to give an explicit value $L_0 > 1$ which suits all technical requirements (although, to our knowledge, such a specification has never been provided in the existing literature). In fact, working with scales L, one simply needs to compare the growth of several kinds of functions of L: $\ln L$, L^a with $a > 0$, e^{L^b} with $b \in (0, 1)$ and e^{cL} with $c > 0$. While the asymptotic relations between such functions are obvious, they only hold true for L large enough. On the other hand, once a suitable value of L_0 is found, there is no need to look for larger values.

However, the situation is different in the case of weak disorder where the MSA bounds (and the localization phenomena) are established at "low energies" with the help of the Lifshitz tails argument; cf. Sect. 2.3.5. Here the initial length scale bound for the Green's functions requires a large deviations technique and, consequently, is proven only for $L_0 \geq L_0^*$ with L_0^* large enough. The lower threshold L_0^* depends upon the required value of the exponent $p > 0$ in (DS.I, m, p, L_0). Actually, this is not a technicality: a careful analysis of the Lifshitz tails argument shows that the eigenfunctions with eigenvalues in an interval of "extreme energies" are slowly varying, so that if and when they are exponentially decaying, their decay can be detected only at a sufficiently large scale.

In what follows, $\mathbf{1}(\cdot)$ will stand for the indicator function (an alternative notation employed throughout the book is $\mathbf{1}_\Lambda = \mathbf{1}(\Lambda)$). By a (slight) abuse of notation, symbol $\mathbf{1}_\Lambda$ will be also used for the natural orthoprojection $\ell^2(\mathbb{Z}^d) \to \ell^2(\Lambda)$.

The dynamical localization is featured in

Theorem 2.1.2.(A) *Suppose that Assumption* A *is fulfilled. Then* $\exists\, g^* \in (0, \infty)$ *such that if* g *satisfies* $|g| \geq g^*$ *then* $\forall\, s > 0$*, bounded interval* $I \subset \mathbb{R}$*, bounded Borel function* $\zeta : \mathbb{R} \to \mathbb{C}$ *and finite subset* $\mathbb{K} \subset \mathbb{Z}^d$*,*

$$\mathbb{E}\Big[\, \big\| M^s\, P_I\big(H(\omega)\big)\, \zeta\big(H(\omega)\big) \mathbf{1}_{\mathbb{K}} \big\| \,\Big] < \infty. \tag{2.37}$$

Here M *stands for the multiplication operator by the norm:* $M\phi(x) = |x|\phi(x)$*,* $x \in \mathbb{Z}^d$*.*

(B) *Suppose that Assumptions* A *and* B *are fulfilled. Then* $\forall\, g \in \mathbb{R} \setminus \{0\}\ \exists\, E^* > 0$ *such that* $\forall\, s > 0$*, bounded Borel function* $\zeta : \mathbb{R} \to \mathbb{R}$ *and finite subset* $\mathbb{K} \subset \mathbb{Z}^d$*,*

$$\mathbb{E}\Big[\, \big\| M^s\, P_{[0,E^*]}\big(H(\omega)\big)\, \zeta\big(H(\omega)\big) \mathbf{1}_{\mathbb{K}} \big\| \,\Big] < \infty. \tag{2.38}$$

As was mentioned earlier, the FMM, an alternative method to the MSA, is able to provide a stronger bound in Eqs. (2.37) and (2.38), replacing M^s with $\exp\big(sM\big)$. Speaking informally, the FMM, as compared to the MSA, is a single-scale method, making redundant the inductive analysis of finite-volume approximations H_Λ of the LSO $H(\omega)$. In contrast, the FMM provides, in a direct fashion, a valuable information on the eigenfunctions $\psi_j(H_\Lambda(\omega))$ of $H_\Lambda(\omega)$. More precisely, the FMM analyzes the decay of the eigenfunction (EF) correlators

$$\Upsilon_j^\Lambda(x, y, \omega) := \big|\psi_j\big(x, H_\Lambda(\omega)\big)\, \psi_j\big(y, H_\Lambda(\omega)\big)\big|, \quad 1 \leq j \leq |\Lambda|, \tag{2.39}$$

when the sites $x, y \in \Lambda$ are distant from each other.

In turn, the FMM-style analysis of the eigenfunction correlators opens a direct way to the dynamical localization and, consequently, by the RAGE theorems, to the spectral localization. The path used in the framework of the MSA is more involved: here, one begins with analyzing decay properties of the Green's functions $G_{\Lambda_{L_k}(u)}(x, y; E, \omega)$, then continues with translating the bounds (2.32) and (2.29) into decay properties of the eigenfunctions $\psi_j(H_{\Lambda_{L_k}(u)}(\omega))$ (spectral localization) and after that finishes with decay of eigenfunction correlators $\Upsilon_j^{\Lambda_{L_k}(u)}(x, y, \omega)$ (dynamical localization).

From the technical point of view, the FMM replaces bounds in probability in Eq. (2.32) and (2.29) by those in expectation (the former can, of course, be deduced from the latter by means of Chebyshev's inequality). However, the central tool of the FMM—the decoupling inequality—is more sensitive to regularity properties of the distribution of the random potential.

The most important prerequisites of the MSA are:

- A Wegner-type eigenvalue concentration bound (in our book, Stollmann's bound): for some $\beta \in (0, 1)$, any $E \in \mathbb{R}$ and some sufficiently large $Q > 0$

$$\mathbb{P}\left\{ \operatorname{dist}\left(\sigma(H_{\Lambda_L(u)}), E\right) < e^{-L^\beta} \right\} \le L^{-Q}.$$

  Cf. Lemma 2.2.2 in Sect. 2.2.3.
- A functional-analytic bound (valid under certain assumptions upon $L_0 \in \mathbb{N}$ and other key parameters):

$$\sum_{y \in \partial^- \Lambda_{L_0}(u)} |G_{\Lambda_{L_0}(u)}(u, y; E)| \le e^{-\gamma(m, L_0) L_0}.$$

  Under the assumption of strong disorder ($|g| \gg 1$), such a bound is established in Theorem 2.3.1 from Sect. 2.3.2. The proof is more complex when $|g|$ is not necessarily large (and may even be small, but positive). In such a case localization in $\mathbb{Z}^d$ (with $d > 1$) can be established only in a narrow energy band with the help of the so-called Combes–Thomas estimate (cf. Sect. 2.3.3).

For spectral localization, the MSA traditionally works with a property slightly weaker than (2.29), (2.30) (cf. also (2.34) and (2.36)). Namely, it is required that for some fixed $m > 0$ and $p > d$ large enough,

(DS.I, m, p, L_k): *For any pair of disjoint cubes $\Lambda_{L_k}(u)$, $\Lambda_{L_k}(v)$,*

$$\mathbb{P}\{\exists\, E \in I : \Lambda_{L_k}(u) \text{ and } \Lambda_{L_k}(v) \text{ are } (E, m)\text{-S}\} \le L_k^{-2p}. \tag{2.40}$$

The proof of the MSA bounds is based on an induction upon the scales L_k and works towards achieving two principal tasks[3]:

1. Prove (DS.I, m, p, L_0) assuming that L_0 is large enough.

   This task is relatively simple in the case of "large disorder", i.e., for $|g| \gg 1$. Cf. Theorem 2.3.1 in Sect. 2.3.2. On the other hand, when the amplitude factor g

---

[3]Here we provide a brief, informal description of the logical structure of the MSA. The reader may want to skip it, waiting until all formal definitions are given and principal technical results are formulated, and later return to this schematic overview of the MSA procedure.

in front of the random potential $gV(x;\omega)$ is not supposed to be large, Anderson localization (in dimension $d > 1$) can only be proven for "extreme" energies. For example, under Assumption B, localization is established in an interval $[0, E^*]$ where $E^* \in (0, \infty)$ with the help of the Combes–Thomas estimates (cf. Theorem 2.3.3 from Sect. 2.3.3) combined with a "Lifshitz tails" argument (see Theorem 2.3.5 in Sect. 2.3.5).

2. Deduce $(\mathrm{DS}.I, m, p, L_{k+1})$ from $(\mathrm{DS}.I, m, p, L_k)$ $\forall\, k \geq 0$.

   This task is achieved in Theorem 2.6.6 from Sect. 2.6.6.

   Traditionally, the value $p > 0$ is kept fixed in the course of the MSA inductive procedure. We improve the MSA bounds, using an exponentially growing sequence $p_k = p_0(1+\theta)^k$, $\theta > 0$ (cf. (2.30)). This leads to stronger localization results, including strong dynamical localization with decay of eigenfunction correlators faster than any power law.

   As was said above, the proof of Theorem 2.6.6 requires an exhaustive analysis of reasons why a cube $\Lambda_{L_{k+1}}(u)$ may become (E, m)-S. By the aforementioned Theorem 2.6.3, a cube $\Lambda = \Lambda_{L_{k+1}}(u)$ can be (E, m)-S for two reasons:

   - Λ is not E-CNR. Cf. Definition 2.2.3 from Sect. 2.2.3.
   - Λ is (m, K, I)-PS (it suffices to take $K = 3$). Cf. Definition 2.6.2 from Sect. 2.6.3.

   Further, two disjoint cubes, $\Lambda' = \Lambda_{L_{k+1}}(u)$ and $\Lambda'' = \Lambda_{L_{k+1}}(v)$, are (E, m)-S for the same $E \in I$ if at least one of the following two events occurs:

   - $\mathcal{R}$: neither Λ' nor Λ'' is E-CNR. Then $\mathbb{P}\{\mathcal{R}\}$ is assessed with the help of the two-volume eigenvalue concentration bound; cf. Lemma 2.2.3 from Sect. 2.2.3.
   - $\mathcal{B}$: one of the cubes Λ', Λ'' is (m, K, I)-PS. Then, by Lemma 2.6.4 from Sect. 2.6.5, either of these cubes is PS with small probability.

   The upper bounds on the probabilities $\mathbb{P}\{\mathcal{R}\}$ and $\mathbb{P}\{\mathcal{B}\}$, established in the above-mentioned statements suffice to derive the required property $(\mathrm{DS}.I, m, p_{k+1}, L_{k+1})$ (see Eqs. (2.29), (2.30), (2.34), and (2.36)).

Once the bound $(\mathrm{DS}.I, m, p_k, L_k)$ is proven by induction for all $k \geq 0$, the exponential spectral localization follows, owing to Theorem 2.7.2 from Sect. 2.7.1. One of the proofs of dynamical localization presented in this book requires an additional assessment of centers of localization; see Sect. 2.7.2.

2.2 Eigenvalue Concentration Bounds

2.2.1 Wegner's Bounds

An important part of the MSA is represented by eigenvalue concentration (EVC) bounds, the first example of which was the famous Wegner estimate. Various forms of Wegner-type EVC bounds have been the subject of many papers over the last several years; the reader can find an extensive bibliography in [75].

In this section, we give several formulations of Wegner-type bounds and quote original papers where detailed proofs can be found, along with relevant references. Many of the results on EVC bounds require the distribution of the random potential to be absolutely continuous. While such results are important for a rigorous study of the (limiting) density of states (see below), the assumption of existence of the marginal PDF of the random potential is somewhat restrictive in the framework of the MSA for the purpose of proving localization.

In mathematical literature, one calls Wegner estimate (or Wegner bound) the following inequality: for any finite subset $\Lambda \subset \mathbb{Z}^d$

$$\forall\, E \in \mathbb{R} \quad \mathbb{P}\left\{ \operatorname{dist}\big(\sigma\big(H_\Lambda(\omega)\big), E\big) \le \epsilon \right\} \le \text{Const}\, \|\rho\|_\infty\, |\Lambda|\, \epsilon \tag{2.41}$$

where $\rho(a) = \mathrm{d}F(a)/\mathrm{d}a$, $a \in \mathbb{R}$, is the probability density of the marginal distribution of the random external potential.

However, depending on assumptions upon the external potential, the RHS may vary (and even more so in the case of multi-particle systems; see Chap. 3). Quite often, one considers as an ultimate Wegner estimate an upper bound for the probability $\mathbb{P}\left\{ \operatorname{dist}\big(\sigma\big(H_\Lambda(\omega)\big), E\big) \le \epsilon \right\}$ which is proportional to $|\Lambda|$ (a separate issue is the presence of a factor ϵ in the RHS). We will see that for the purposes of the MSA the linear dependence in volume $|\Lambda|$ and in ϵ is not crucial.

2.2.2 Stollmann's Product-Measure Lemma

In this and subsequent subsections of Sect. 2.2, we follow Stollmann's analysis of EVC bounds; see [158]. Historically, Stollmann's bound was introduced in [157] to address the problem of extending the Wegner bound to the case where the marginal probability distribution of the random potential is singular. In the context of multi-particle Anderson models, which we discuss in Part II of this book, it proved to be a simple yet versatile and powerful tool. For this reason, our presentation makes use of this particular method.

Let $\mathcal{Q}$ be a finite set, with $1 < |\mathcal{Q}| < +\infty$. It is convenient to assume that the points of $\mathcal{Q}$ are labelled by (and identified with) numbers $j = 1, \ldots, |\mathcal{Q}|$. Consider the Euclidean space $\mathbb{R}^{\mathcal{Q}}$ with the standard basis $(e_1, \ldots, e_{|\mathcal{Q}|})$, and the positive orthant

$$\mathbb{R}_+^{\mathcal{Q}} = \left\{ q = \big(q_1, \ldots, q_{|\mathcal{Q}|}\big) \in \mathbb{R}^{\mathcal{Q}} : q_j \ge 0,\; j = 1, 2, \ldots, |\mathcal{Q}| \right\}.$$

Definition 2.2.1. *A function* $\Phi : \mathbb{R}^{\mathcal{Q}} \to \mathbb{R}$ *is called diagonally monotone* (DM, *for short) if it satisfies the following conditions:*

(I) *For any* $r \in \mathbb{R}_+^{\mathcal{Q}}$ *and any* $q \in \mathbb{R}^{\mathcal{Q}}$,

$$\Phi(q + r) \geq \Phi(q); \tag{2.42}$$

(II) *Moreover, with* $e = e_1 + \cdots + e_{|\mathcal{Q}|} \in \mathbb{R}^{\mathcal{Q}}$, $\forall\, q \in \mathbb{R}^{\mathcal{Q}}$ *and* $t > 0$,

$$\Phi(q + t \cdot e) - \Phi(q) \geq t. \tag{2.43}$$

Given a probability measure μ on $\mathbb{R}$, we denote by $\mu^{\mathcal{Q}}$ the product-measure on $\mathbb{R}^{\mathcal{Q}}$ with the marginal μ. Similarly, $\mu^{\mathcal{Q}\setminus\{j\}}$ stands for the induced product-measure on $\mathbb{R}^{\mathcal{Q}\setminus\{j\}}$, $1 \leq j \leq |\mathcal{Q}|$. Furthermore, for $\epsilon > 0$, define the following function, sometimes called the continuity modulus of the measure μ:

$$\epsilon \mapsto \varsigma(\mu, \epsilon) := \sup_{a \in \mathbb{R}} \mu([a, a + \epsilon]).$$

Lemma 2.2.1 (Stollmann). *Let $\mathcal{Q}$ and $\mu^{\mathcal{Q}}$ be as above. If the function $\Phi : \mathbb{R}^{\mathcal{Q}} \to \mathbb{R}$ is* DM, *then for any bounded interval $I \subset \mathbb{R}$ of length $|I|$ one has*

$$\mu^{\mathcal{Q}}\{ q \in \mathbb{R}^{\mathcal{Q}} : \Phi(q) \in I \} \leq |\mathcal{Q}|\, \varsigma(\mu, |I|).$$

Proof. Let $I = (a, b)$, with $b - a = \epsilon > 0$, and consider the set

$$A = \left\{ q \in \mathbb{R}^{\mathcal{Q}} : \Phi(q) \leq a \right\}.$$

Furthermore, define recursively the sets $A_j^\epsilon \subset \mathbb{R}^{|\mathcal{Q}|}$, $j = 0, 1 \ldots, |\mathcal{Q}|$, by setting

$$A_0^\epsilon = A,$$
$$A_j^\epsilon = A_{j-1}^\epsilon + [0, \epsilon] e_j := \left\{ q + t e_j : q \in A_{j-1}^\epsilon, t \in [0, \epsilon] \right\}, \quad 1 \leq j \leq |\mathcal{Q}|.$$

Obviously, the sets A_j^ϵ are increasing with j. The DM property implies

$$\left\{ q \in \mathbb{R}^{\mathcal{Q}} : \Phi(q) < b \right\} \subset A_{|\mathcal{Q}|}^\epsilon.$$

Indeed, if $\Phi(q) < b$, then for the vector $r := q - \epsilon \cdot e$ we have, by (II),

$$\Phi(r) \leq \Phi(r + \epsilon \cdot e) - \epsilon = \Phi(q) - \epsilon \leq b - \epsilon = a,$$

meaning that $r \in \{ \Phi \leq a \} = A$ and, therefore,

$$q = r + \epsilon \cdot e \in A_{|\mathcal{Q}|}^\epsilon.$$

We conclude that

$$\begin{aligned} \{ q : \Phi(q) \in I \} &= \{ q : \Phi(q) \in (a, b) \} \\ &= \{ q : \Phi(q) < b \} \setminus \{ q : \Phi(q) \leq a \} \subset A_{|\mathcal{Q}|}^\epsilon \setminus A. \end{aligned}$$

Furthermore,

$$
\begin{aligned}
\mu^{\mathcal{Q}}\{ q : \Phi(q) \in I \} &\le \mu^{\mathcal{Q}}\left(A^{\epsilon}_{|\mathcal{Q}|} \setminus A \right) \\
&= \mu^{\mathcal{Q}}\left(\bigcup_{j=1}^{|\mathcal{Q}|} \left(A^{\epsilon}_{j} \setminus A^{\epsilon}_{j-1} \right) \right) \\
&\le \sum_{j=1}^{|\mathcal{Q}|} \mu^{\mathcal{Q}}\left(A^{\epsilon}_{j} \setminus A^{\epsilon}_{j-1} \right).
\end{aligned}
$$

For $q'_{\mathcal{Q}\setminus\{1\}} = (q_2, \dots, q_{|\mathcal{Q}|}) \in \mathbb{R}^{\mathcal{Q}\setminus\{1\}}$,

$$
I_1(q'_{\mathcal{Q}\setminus\{1\}}) = \left\{ q_1 \in \mathbb{R} : (q_1, q'_{\mathcal{Q}\setminus\{1\}}) \in A^{\epsilon}_1 \setminus A \right\}.
$$

By monotonicity of Φ and the definition of the set A^{ϵ}_1, the interval $I_1(q'_{\mathcal{Q}\setminus\{1\}})$ is of length $\le \epsilon$. Since $\mu^{\mathcal{Q}}$ is a product measure, we have

$$
\mu^{\mathcal{Q}}(A^{\epsilon}_1 \setminus A) = \int \mathrm{d}\mu^{\mathcal{Q}\setminus\{1\}}(q'_{\mathcal{Q}\setminus\{1\}}) \int_{I_1} \mathrm{d}\mu(q_1) \le \varsigma(\mu, \epsilon). \tag{2.44}
$$

Similarly, for $j = 2, \dots, |\mathcal{Q}|$ we obtain

$$
\mu^{\mathcal{Q}}(A^{\epsilon}_j \setminus A^{\epsilon}_{j-1}) \le \varsigma(\mu, \epsilon),
$$

yielding

$$
\mu^{\mathcal{Q}}\{ q : \Phi(q) \in I \} \le \sum_{j=1}^{|\mathcal{Q}|} \mu^{\mathcal{Q}}(A^{\epsilon}_j \setminus A^{\epsilon}_{j-1}) \le |\mathcal{Q}| \cdot \varsigma(\mu, \epsilon). \qquad \square
$$

Next, the notion of diagonal monotonicity is extended to operators. In the following definition, we use the same notation as above.

Definition 2.2.2. *Let $\mathcal{H}$ be a finite-dimensional Hilbert space. A family of Hermitian operators $B(q) : \mathcal{H} \to \mathcal{H}$, parametrized by $q \in \mathbb{R}^{\mathcal{Q}}$, is called* DM *if, $\forall\ f \in \mathcal{H}$ with $\| f \| = 1$, the function $\Phi_f : \mathbb{R}^{\mathcal{Q}} \to \mathbb{R}$ defined by*

$$
\Phi_f(q) = (B(q)f, f)
$$

is DM.

In other words, $\forall\ f \in \mathcal{H}$, the quadratic form $(B(q)f, f)$ as a function of $q \in \mathbb{R}^{\mathcal{Q}}$ is monotone non-decreasing in the variables q_j, $1 \le j \le |\mathcal{Q}|$, and satisfies the inequality:

$$
\big(B(q + t \cdot e)f, f\big) - \big(B(q)f, f\big) \ge t \cdot \| f \|^2.
$$

Remark 2.2.1. By virtue of the min-max principle for Hermitian operators, if an operator family $\{B(q),\ q \in \mathbb{R}^Q\}$ in a finite-dimensional Hilbert space $\mathcal{H}$ is DM, and $\{E_k(B(q)),\ 1 \le k \le \dim \mathcal{H}\}$ are eigenvalues of $B(q)$ numbered in an increasing order: $E_1(B(q)) \le E_2(B(q)) \le \cdots \le E_{\dim \mathcal{H}}(B(q))$, then each eigenvalue $E_k(B(q))$ is a DM function in q.

Remark 2.2.2. If $\{B(q),\ q \in \mathbb{R}^Q\}$ is a DM operator family in a finite-dimensional Hilbert space $\mathcal{H}$, and $B_0 : \mathcal{H} \to \mathcal{H}$ is a given Hermitian operator, then the family $\{S(q) = B_0 + B(q)\}$ is also DM.

This explains why the notion of diagonal monotonicity is relevant in the spectral theory of random operators.

2.2.3 *Stollmann's EVC Bound*

Let $\Lambda_L(u) \subset \mathbb{Z}^d$ be a lattice cube (cf. (2.7)). Let $H_{\Lambda_L(u)}$ be as in (2.6), with an IID random potential $V(x;\omega)$, $x \in \Lambda_L(u)$. Let F be the marginal distribution function of $V(x;\omega)$. Given $\epsilon \in (0,1)$, set:

$$\varsigma(F,\epsilon) = \sup_{a \in \mathbb{R}} [F(a+\epsilon) - F(a)]. \tag{2.45}$$

Taking into account the above Remark 2.2.2, Lemma 2.2.1 yields immediately the following estimate which we call Stollmann's bound.

Lemma 2.2.2. $\forall\, E \in \mathbb{R}$,

$$\mathbb{P}\left\{\operatorname{dist}\Big(\sigma\big(H_{\Lambda_L(u)}(\omega)\big), E\Big) \le \epsilon\right\} \le (2L+1)^{2d} \varsigma(F, 2\epsilon). \tag{2.46}$$

Proof. Set $Q = \Lambda_L(u)$ and notice that the family of operators $B(q)$, where $B(q)$ is the operator of multiplication by the function $x \mapsto V(x;\omega)$, $x \in \Lambda_L(u)$, $q = \omega = (\omega_x)_{x \in Q} \in \mathbb{R}^Q$, in $\mathcal{H} = \ell^2(\Lambda_L(u))$ satisfies Definition 2.2.2, since $B(q + t \cdot e)$ is the operator of multiplication by the function

$$x \mapsto V(x; \omega + t \cdot e) = \omega_x + t,$$

so that $B(q + t \cdot e) = B(q) + t\mathbf{1}$. By Remark 2.2.2, with $g > 1$ and $B_0 = H^0_{\Lambda_L(u)}$, the operator family $S(q) = H_{\Lambda_L(u)}(\omega)$ is also DM, so by Remark 2.2.1, every eigenvalue $E_k(\omega)$ of $H_{\Lambda_L(u)}(\omega)$ is a DM function. Now one can apply Lemma 2.2.1 to each eigenvalue $E_k(\omega)$, with $I = [E - \epsilon, E + \epsilon]$, $Q = \Lambda_L(u)$, $\mu^Q = \mathbb{P}$, $q = \omega$ and $\Phi(q) = E_k(\omega)$, and obtain

$$\mathbb{P}\{\operatorname{dist}(E_k(\omega), E) \le \epsilon)\} \le (2L+1)^d \varsigma(F, 2\epsilon), \;\; k = 1, \ldots, |\Lambda_L(u)|,$$

yielding immediately the assertion (2.46). □

This is an analogue of Wegner bound (2.41). However, a visible distinction is the form of dependence upon the volume of $\Lambda_L(u)$: we have the factor $(2L+1)^{2d}$, i.e., $|\Lambda_L(u)|^2$, instead of $|\Lambda_L(u)|$ in Eq. (2.41).

In the next definition we engage two parameters, $\alpha \in (1,2)$ and $\beta \in (0,1)$, which will be used throughout the rest of the book. (Recall that α was introduced in Eq. (2.16) and β in (2.21).) While it can be assumed that $\alpha = 3/2$ and $\beta = 1/4$, a symbolic form of these parameters is both traditional and convenient.

Definition 2.2.3. *Given $E \in \mathbb{R}$, a cube $\Lambda_L(u)$ is said to be E-non-resonant (E-NR) if*

$$\|G_{\Lambda_L(u)}(E)\| \le \mathrm{e}^{L^\beta} \tag{2.47}$$

and E-resonant (E-R), otherwise. A cube $\Lambda_L(u)$ is said to be E-completely non-resonant (E-CNR) if it is E-NR and does not contain any cube $\Lambda_l(x)$ with $l \ge \lfloor L^{1/\alpha} \rfloor$ for which $\|G_{\Lambda_l(x)}(E)\| > \mathrm{e}^{l^\beta}$, otherwise, it is said to be E-partially resonant (E-PR).

It is readily seen that the bound (2.47) can be reformulated as follows:

$$\operatorname{dist}\Big(E, \sigma\left(H_{\Lambda_L(u)}\right)\Big) \ge \mathrm{e}^{-L^\beta}. \tag{2.48}$$

Lemma 2.2.3. *Consider two disjoint cubes, $\Lambda' = \Lambda_L(u')$ and $\Lambda'' = \Lambda_L(u'')$. Then*

$$\mathbb{P}\{\exists\, E \in \mathbb{R} : \Lambda' \text{ and } \Lambda'' \text{ are } E\text{-PR}\} \le L^2(2L+1)^{5d} \varsigma\left(F, 2\mathrm{e}^{-L^{\beta/\alpha}}\right). \tag{2.49}$$

Proof. First, note that the number of pairs of cubes $\Lambda_{l'}(x') \subset \Lambda_L(u')$ and $\Lambda_{l''}(x'') \subset \Lambda_L(u'')$ with fixed values l', l'' is bounded by $(2L+1)^{2d}$, and the number of pairs (l', l'') with $l', l'' \in [L^{1/\alpha}, L]$ is bounded by L^2.

Next, for fixed disjoint cubes $\Lambda' = \Lambda_{l'}(x')$, $\Lambda'' = \Lambda_{l''}(x'')$ with $l', l'' \in [L^{1/\alpha}, L]$, we can write

$$\begin{aligned}
&\mathbb{P}\{\exists\, E \in \mathbb{R} : \text{both } \Lambda' \text{ and } \Lambda'' \text{ are } E\text{-R}\}\\
&\le \mathbb{P}\left\{\operatorname{dist}\Big(\sigma\big(H_{\Lambda'}(\omega)\big), \sigma\big(H_{\Lambda''}(\omega)\big)\Big) \le 2\mathrm{e}^{-L^{\beta/\alpha}}\right\}\\
&= \mathbb{E}\Big[\mathbb{P}\left\{\operatorname{dist}\Big(\sigma\big(H_{\Lambda'}(\omega)\big), \sigma\big(H_{\Lambda''}(\omega)\big)\Big) \le 2\mathrm{e}^{-L^{\beta/\alpha}} \,|\, \mathfrak{B}(\Lambda'')\right\}\Big],
\end{aligned}$$

where $\mathfrak{B}(\Lambda'')$ is the sigma-algebra generated by the values $V(x;\omega)$ with $x \in \Lambda''$. When the cube Λ' is disjoint from Λ'', the operator $H_{\Lambda'}(\omega)$ (hence, its random spectrum) is independent of $\mathfrak{B}(\Lambda'')$, while the spectrum of $H_{\Lambda''}(\omega)$ is $\mathfrak{B}(\Lambda'')$-measurable. Therefore, for non-overlapping cubes Λ' and Λ'',

$$\begin{aligned}
&\mathbb{E}\Big[\mathbb{P}\Big\{\operatorname{dist}\Big(\sigma\big(H_{\Lambda'}(\omega)\big),\sigma\big(H_{\Lambda''}(\omega)\big)\Big)\le 2\mathrm{e}^{-L^{\beta/\alpha}}\mid\mathfrak{B}(\Lambda'')\Big\}\Big]\\
&\quad\le|\Lambda''|\sup_{E\in\mathbb{R}}\mathbb{P}\Big\{\operatorname{dist}\Big(\sigma\big(H_{\Lambda'}(\omega)\big),E\Big)\le 2\mathrm{e}^{-L^{\beta/\alpha}}\Big\}\\
&\quad\le|\Lambda''|\,|\Lambda'|^2\,\varsigma\big(F,2\mathrm{e}^{-L^{\beta/\alpha}}\big)\le(2L+1)^{3d}\,\varsigma\big(F,2\mathrm{e}^{-L^{\beta/\alpha}}\big).
\end{aligned}$$

Taking into account the above bound on the number of pairs of cubes $\Lambda_{l'}(x')\subset\Lambda_L(u')$, $\Lambda_{l''}(x'')\subset\Lambda_L(u'')$ by $(2L+1)^{2d}$, this completes the proof of the lemma. □

For the purposes of the MSA, the form of the RHS in (2.46) and (2.49) turns out to be very appropriate. More precisely, under Assumption A the continuity modulus $\varsigma(F,\epsilon)$ satisfies

$$\varsigma(F,\epsilon)\le\min\big[1,C\epsilon^{s}\big];\tag{2.50}$$

see (2.3). It is readily seen that, $\forall$ given $Q>0$, $\exists\,L^*\in(0,\infty)$ such that for any $L\ge L^*$ and any pair of disjoint cubes $\Lambda_L(x)$, $\Lambda_L(y)$,

$$\mathbb{P}\{\exists\,E\in\mathbb{R}:\ \Lambda_L(x)\text{ and }\Lambda_L(y)\text{ are }E\text{-PR}\}<L^{-Q}.\tag{2.51}$$

Moreover, a stronger bound holds true:

Corollary 2.2.4. *For any $Q>0$ there exists $L^*\in[2,+\infty)$ such that for all $L_0\ge L^*$ and some $\theta\in(0,1]$, for all $k\ge 0$ and any pair of disjoint cubes $\Lambda_{L_k}(x)$ and $\Lambda_{L_k}(y)$*

$$\mathbb{P}\{\exists\,E\in\mathbb{R}:\ \Lambda_{L_k}(x)\ \textit{and}\ \Lambda_{L_k}(y)\ \textit{are}\ E\text{-PR}\}<L_k^{-Q(1+\theta)^k}\tag{2.52}$$

where, again, $\{L_k,k\in\mathbb{N}\}$ follow the recursion $L_{k+1}=\lfloor L_k^{\alpha}\rfloor$.

For the proof, it suffices to notice that for any $\beta'>0$ and all L_0 large enough

$$L_k^{-Q(1+\theta)^k}=\mathrm{e}^{-Q(1+\theta)^k\ln L_k}>\mathrm{e}^{-L_k^{\beta'}}.\tag{2.53}$$

Indeed, (2.53) is equivalent to

$$\ln Q+k\ln(1+\theta)+\ln\ln L_k<\beta'\ln L_k\tag{2.54}$$

where $L_k\sim(L_0)^{\alpha^k}$, $\alpha>1$, so for some $\alpha'>1$, $\ln L_k\ge(\alpha')^k\ln L_0$. This shows that (2.54) (hence, (2.53)) is satisfied for L_0 large enough.

2.2.4 The Density of States and Higher-Order EVC Bounds

The contents of this Section are not used formally in the subsequent arguments. However, the concepts and results discussed here may become relevant for a forthcoming development of the multi-particle Anderson models. Besides, the material of this Section casts light on some important properties of single-particle theory.

We begin with the notion of integrated density of states (IDoS). The IDoS is defined as the limiting eigenvalue distribution function

$$\nu(E) = \lim_{L\to\infty} \frac{1}{|\Lambda_L(0)|} \operatorname{card}\left\{j : E_j\left(H_{\Lambda_L(0)}(\omega)\right) \leq E\right\}, \quad E \in \mathbb{R}, \tag{2.55}$$

where $E_j\left(H_{\Lambda_L(0)}(\omega)\right)$, $1 \leq j \leq |\Lambda_L(0)|$, are the eigenvalues of $H_{\Lambda_L(0)}(\omega)$ counted with multiplicities (cf. (2.10)). The derivative $\mathrm{d}\nu(E)/\mathrm{d}E$, when it exists, is called the density of states (DoS).

The limit in the RHS of (2.55) may not exist for an arbitrary potential $V(\cdot\,;\omega)$, and it is not difficult to construct examples of a function $V : \mathbb{Z}^d \to \mathbb{R}$ for which the ratio

$$\frac{1}{|\Lambda_L(0)|}\operatorname{card}\{j : E_j\left(H_{\Lambda_L(0)}(\omega)\right) \leq E\}$$

fails to converge when $L \to \infty$ for a broad range of energies E. For instance, consider a (deterministic) V of the following "intermittent" structure:

$$V(x) = \sum_{i=1}^{\infty} \mathbf{1}_{\mathbb{A}_i^{(0)}}(x), \quad x \in \mathbb{Z}^d,$$

where $\mathbb{A}_i^{(0)}$, $i \geq 1$, are the annuli

$$\mathbb{A}_i^{(0)} = \Lambda_{R_{2i}}(0) \setminus \Lambda_{R_{2i-1}}(0),$$

with appropriately chosen (viz., sufficiently rapidly growing) width $R_n - R_{n-1}$. If $R_2 \gg R_1$, then most of the eigenvalues of $H_{\Lambda_{R_2}(0)} = H^0_{\Lambda_{R_2}(0)} + gV_{\Lambda_{R_2}(0)}$ are close to those of the operator $H^0_{\Lambda_{R_2}(0)} + g\cdot\mathbf{1}$. However, if $R_3 \gg R_2$, then for the same reason a majority of eigenvalues of $H_{\Lambda_{R_3}(0)}$ are close to the eigenvalues of the Laplacian $H^0_{\Lambda_{R_3}(0)} + 0\cdot\mathbf{1}$, and so on.

On the other hand, the problem of existence of the IDoS function becomes straightforward in the case of a multiplication operator $0\cdot H^0_\Lambda + gV_\Lambda$: its eigenvalues are simply the values of $V(x;\omega)$, $x \in \Lambda$. In this case the Cesaro means in Eq. (2.55) in general do not necessarily have a limit. (Then it is natural to expect a similar behavior for the operator $\delta\cdot H^0_{\Lambda_L(0)} + gV_{\Lambda_L(0)}$ with small δ.) But the limit in question, for the diagonal operators $0 \cdot H^0_{\Lambda_L(0)} + gV_{\Lambda_L(0)}$, exists—with probability one—for the samples of an ergodic random potential $V(x;\omega)$. Moreover the limit (2.55) (for Hamiltonians $H_{\Lambda_L(0)} = H^0_{\Lambda_L(0)} + gV_{\Lambda_L(0)}$) also exists with probability one when the random potential $V(x;\omega)$ is ergodic; see, e.g., [53,77,144]. Due to the normalization factor $|\Lambda_L(0)|^{-1}$, the limiting function $\nu(E)$ defines a probability measure on $\mathbb{R}$ (sometimes called the measure of states for the random operator $H_{\Lambda_L(0)}(\omega)$). One can give examples where the DoS $\mathrm{d}\nu(E)/\mathrm{d}E$ does not exist, i.e. the measure of states is not absolutely continuous with respect to the Lebesgue measure on $\mathbb{R}$. However, if the distribution function F is differentiable and the probability density $\rho(a) = \mathrm{d}F(a)/\mathrm{d}a$ is a bounded function then the DoS exists and is bounded by the same constant; see [53].

Before going to the analysis of EVC bounds in finite volumes, we would like to quote a fairly general result concerning the limiting IDoS for a system of several particles with decaying interaction (see Chap. 3). It turns out that a finite range (or simply decaying) particle interaction cannot change the limiting IDoS. Therefore, it suffices to compute (or otherwise analyze) the IDoS for the corresponding non-interacting systems. Such a result (which is natural from the physical point of view), was established in [120] for continuous quantum systems, in a Euclidean space $\mathbb{R}^d$, $d \geq 1$, using the technique of almost analytic extensions of functions of a complex variable.

We believe (but do not prove here) that such an approach can be extended to (and simplified for) lattice systems, and that the result of Klopp–Zenk [120] admits a natural extension to lattice multi-particle systems. This would solve the problem of analysis of the limiting IDoS for multi-particle systems, but provide little information about the concentration of eigenvalues in finite cubes $\Lambda_L(u) \subset \mathbb{Z}^d$. The latter, however, is much more important for the purposes of the (multi-particle) MSA. Actually, MSA does not need *optimal* bounds on the probabilities

$$\mathbb{P}\left\{\sigma\big(H_{\Lambda_L(u)}(\omega)\big) \cap [a, a+\epsilon] \neq \varnothing\right\}.$$

Wegner-type bounds, when applicable, would give

$$\mathbb{P}\left\{\sigma\big(H_{\Lambda_L(u)}(\omega)\big) \cap [a, a+\epsilon] \neq \varnothing\right\} \leq \text{Const } |\Lambda_L(u)|\, \epsilon,$$

but usually MSA works fine even with a much weaker bound, e.g.,

$$\mathbb{P}\left\{\sigma\big(H_{\Lambda_L(u)}(\omega)\big) \cap [a, a+\epsilon] \neq \varnothing\right\} \leq \text{Const } \mathrm{e}^{L^b}\, \epsilon^{b'},$$

with appropriately chosen $b, b' \in (0,1)$, or even

$$\mathbb{P}\left\{\sigma\big(H_{\Lambda_L(u)}(\omega)\big) \cap [a, a+\epsilon] \neq \varnothing\right\} \leq c\, \frac{L^{c'}}{|\ln \epsilon|^{c''}},$$

with some $c, c' \in (0,\infty)$ and sufficiently large $c'' \in (0,\infty)$.

However, the above inequalities cannot guarantee the existence of the limiting DoS: the limiting IDoS may, formally, be singular (albeit continuous) in such cases.

Such a scenario does actually occur in one-dimensional single-particle Anderson tight-binding model with a random potential admitting finitely many values; see [54] and a discussion in [53], Sect. VIII.4.

We stress again that the precise form of an EVC bound used has, essentially, no impact on the qualitative results obtained with the help of the MSA. For that reason we do not seek in this book an optimal EVC bound, but choose the one that can be extended to interacting multi-particle systems. In our book, the role of such an EVC bound is played by Stollmann's bound, introduced in the previous subsection.

In the language of the theory of random point processes, the Wegner-type bounds establish regularity properties of the 1-point correlation measure (or moment measure of the first order)

$$\begin{aligned} B \subset \mathbb{R} \mapsto & \mathbb{E}\Big(\operatorname{card}\{k:\ E_k \in \sigma\big(H_{\Lambda_L(u)}(\omega)\big),\ E_k \in B\}\Big) \\ &= \mathbb{E}\left[\sum_{1\le k\le (2L+1)^d} \mathbf{1}\Big(E_k\left(H_{\Lambda_L(u)}(\omega)\right) \in B\Big)\right] \end{aligned}$$

for the random spectrum $\sigma(H_{\Lambda_L(u)}(\omega))$. It is then quite natural to study the regularity of the l-point correlation measure (representing higher-order energy-energy correlators, in physical terminology), for a general $l > 1$. The l-point correlation measure (or the moment measure of the lth order) for the spectrum of $H_{\Lambda_L(u)}$ is defined for sets $B \subset (\mathbb{R} \times \ldots \times \mathbb{R})_{\le}$ (with l Cartesian factors):

$$\begin{aligned} & B \subset (\mathbb{R} \times \cdots \times \mathbb{R})_{\le} \quad \mapsto \\ & \mathbb{E}\left[\sum_{1\le k_1<\cdots<k_l\le (2L+1)^d} \mathbf{1}\left(\left(E_{k_1}\left(H_{\Lambda_L(u)}(\omega)\right), \ldots, E_{k_l}\left(H_{\Lambda_L(u)}(\omega)\right)\right) \in B\right)\right]. \end{aligned}$$

Here

$$(\mathbb{R} \times \cdots \times \mathbb{R})_{\le} = \{(a_1, \ldots, a_l) \in \mathbb{R} \times \ldots \times \mathbb{R} :\ a_1 \le \cdots \le a_l\}.$$

The first results in this direction have been obtained, for $l = 2$, by Minami [133] in the analysis of local statistics of eigenvalues of Hamiltonians $H_{\Lambda_L(u)}(\omega)$. In the original paper by Minami, the 2-point correlation estimate appeared as an auxiliary (although crucial) technical argument, but the general character and the importance of this estimate—called now the Minami estimate or Minami bound—has been quickly recognized. It has been also understood that the approach from [133] can be extended to the general case of l-point eigenvalue correlators. Such an extension was done independently in [35] and [106]. Below we follow closely the presentation given in [106].

For a linear operator A in a complex (finite-dimensional) Hilbert space $\mathcal{K}$ its imaginary part is defined by

$$\mathrm{Im}A := \frac{1}{2\mathrm{i}}(A - A^*),$$

where A^* is the adjoint operator. Given an orthonormal basis $\{\phi_k\}$, one can define the matrix elements of A,

$$a_{jk} = \langle \phi_j, A\phi_k \rangle, \quad 1 \le j,k \le \dim \mathcal{K}.$$

(The equality $(\mathrm{Im}A)_{jk} = \mathrm{Im}\, a_{jk}$ holds iff A obeys $a_{jk} = a_{kj}$.)

For $l = 2$, the Minami bound is as follows:

Theorem 2.2.5. *Let $I \subset \mathbb{R}$ be a bounded interval of length $|I|$. Consider the spectral projection $P_I(H_{\Lambda_L(u)}(\omega))$ of $H_{\Lambda_L(u)}(\omega)$ to I. If the random variables $V(x;\omega)$ have a bounded probability density ρ with sup-norm $\|\rho\|_\infty$, then*

$$\mathbb{P}\left\{\mathrm{tr}\, P_I(H_{\Lambda_L(u)}(\omega)) \ge 2\right\} \le \pi^2(2L+1)^{2d}\|\rho\|_\infty^2|I|^2. \tag{2.56}$$

The proof of Theorem 2.2.5 is based on Lemma 2.2.6 combined with Lemma 2.2.7. In both lemmas, $G(\cdot,\cdot;\zeta) = G_{\Lambda_L(u)}(\cdot,\cdot;\zeta)$.

Lemma 2.2.6. *Let a bounded interval $I \subset \mathbb{R}$ be written in the form $I = (\lambda_0 - |I|/2, \lambda_0 + |I|/2)$ for some $\lambda_0 \in \mathbb{R}$ and set $\zeta = \zeta(I) = \lambda_0 + \mathrm{i}|I|$. Then*

$$\begin{aligned}\mathbb{P}\left\{\mathrm{tr}\, P_I(H_{\Lambda_L(u)}(\omega)) \ge 2\right\} &\le |I|^2(2L+1)^{2d}\\ &\times \max_{x,y\in\Lambda_L(u)} \mathbb{E}\left[\det\begin{pmatrix}\mathrm{Im}G(x,x;\zeta) & \mathrm{Im}G(x,y;\zeta)\\ \mathrm{Im}G(y,x;\zeta) & \mathrm{Im}G(y,y;\zeta)\end{pmatrix}\right].\end{aligned} \tag{2.57}$$

Lemma 2.2.7. *The determinant in the RHS of Eq. (2.57) admits the bound:*

$$\mathbb{E}\left[\det\begin{pmatrix}\mathrm{Im}G(x,x;\zeta) & \mathrm{Im}G(x,y;\zeta)\\ \mathrm{Im}G(y,x;\zeta) & \mathrm{Im}G(y,y;\zeta)\end{pmatrix}\right] \le \pi^2\|\rho\|_\infty^2.$$

In turn, Lemma 2.2.7 stems from the following result on 2×2 matrices.

Lemma 2.2.8. *Let $A = (a_{jk} : j,k = 1,2)$ be a 2×2 complex matrix with positive-definite imaginary part: $\mathrm{Im}A \ge 0$. Set: $\mathrm{diag}(v_1,v_2) = \begin{pmatrix}v_1 & 0\\ 0 & v_2\end{pmatrix}$. Then*

$$\int \mathrm{d}v_1\,\mathrm{d}v_2\,\rho(v_1)\rho(v_2)\det\left(\mathrm{Im}(\mathrm{diag}(v_1,v_2) - A)^{-1}\right) \le \pi^2\|\rho\|_\infty^2. \tag{2.58}$$

In the more general case of $l\times l$ matrices, one can prove the following inequality:

Lemma 2.2.9. *Let* $A = (a_{jk} : 1 \le j, k \le l)$ *be an* $l \times l$ *complex matrix with positive-definite imaginary part:* $\mathrm{Im}\, A \ge 0$. *Then*

$$\int \prod_{i=1}^{l} \mathrm{d}v_i\, \rho(v_i)\, \det\Big(\mathrm{Im}\big(\mathrm{diag}(v_1, \ldots, v_l) - A\big)^{-1}\Big) \le \pi^l \|\rho\|_\infty^l. \tag{2.59}$$

Following the same path as before, one can deduce from Lemma 2.2.9 a higher-order analog of the Minami estimate for eigenvalue correlators. The details can be found in [35].

Theorem 2.2.10. *Under the same assumptions as in Theorem 2.2.5,* $\forall\, l = 1, 2, \ldots$ *and bounded interval* I *of length* $|I|$,

$$\mathbb{P}\left\{\mathrm{tr}\, P_I\big(H_{\Lambda_L(u)}(\omega)\big) \ge l\right\} \le \pi^l (2L+1)^{ld}\, \|\rho\|_\infty^l\, |I|^l. \tag{2.60}$$

Concluding this subsection, we wish to say a few words about the Poisson limit for the eigenvalue distribution of $H_{\Lambda_L(u)}(\omega)$, under the assumption that the random variables $V(x;\omega)$ have a bounded and continuous probability density ρ. Recall that the operator $H_{\Lambda_L(u)}(\omega)$ has $(2L+1)^d$ eigenvalues $E_j = E_j(H_{\Lambda_L(u)}(\omega))$, $1 \le j \le (2L+1)^d$, numbered in increasing order, counting multiplicities. The same agreement is in use for the eigenvalues of the multiplication operator $V_{\Lambda_L(u)}(\omega)$ (i.e., the values $V(x;\omega)$, $x \in \Lambda_L(u)$) and the eigenvalues $E_j^0 = E_j(H^0_{\Lambda_L(u)})$ of the operator $H^0_{\Lambda_L(u)}$. See Eq. (2.10).

Notice that in absence of the kinetic energy operator $H^0_{\Lambda_L(u)}$, the Hamiltonian $H_{\Lambda_L(u)}(\omega)$ is reduced to the multiplication operator, by the function $x \in \Lambda_L(u) \mapsto V(x;\omega)$, with the eigenfunctions given by the lattice delta-functions δ_x, $x \in \Lambda_L(u)$, and the associated eigenvalues equal to $V(x;\omega)$. Standard probabilistic results imply that, given an interval $I \subset \mathbb{R}$, the random variable

$$n(\Lambda, I;\omega) = \mathrm{card}\{x \in \Lambda_L(u) : V(x;\omega) \in I\}$$

has the binomial distribution $B(N,q)$ with $N = (2L+1)^d$ and $q = \int_I \rho(a)\,\mathrm{d}a$. Furthermore, for the intervals of the form $I_{L,\eta} = \big[E, E + \eta(2L+1)^{-d}\big]$, with $\eta > 0$ fixed, the Poisson limit theorem implies the existence of the limiting probability distribution of the random variables $n(\Lambda_L(u), I_{L,\eta};\omega)$, as $L \to \infty$. The latter equals the Poisson distribution with parameter $\eta\rho(E)$. Indeed, for L large, we have

$$\begin{aligned}\mathbb{P}\left\{V(x;\omega) \in I_{L,\eta}\right\} &= \int_E^{E+\eta(2L+1)^{-d}} \rho(a)\,\mathrm{d}a \\ &= \big(\eta\rho(E) + o(\eta)\big)(2L+1)^{-d},\end{aligned}$$

which makes possible a direct application of the Poisson limit theorem.

Naturally, this simple analysis becomes more complicated in the presence of the kinetic energy operator, where the eigenvalues are implicit functions of the random potential. Nevertheless, it turns out that the presence of the kinetic energy does not change the qualitative picture which we observed in the simplest case.

Historically, the first mathematically rigorous result on the local statistics of eigenvalues was obtained in [135], in the aftermath of [102] and, particularly, [134]. However, the methods employed in [135] used specific features of the one-dimensional case ($d = 1$). Besides, paper [135] addressed a continuous model. The tight-binding model in dimension $d \geq 1$ was considered by Minami in the aforementioned paper [133]. Minami analyzed statistics of eigenvalues by using results from the localization theory (including bounds provided by the MSA and FMM). We postpone a more detailed discussion of results on the local statistics of eigenvalues until Sect. 2.7.3.

2.3 Decay of the Green's Functions, I

In this section we collect basic facts that will allow us to develop a methodology of controlling the decay of the Green's functions $G_\Lambda(x, y; E)$. Most of these facts will be useful for the analysis of multi-particle systems.

2.3.1 The Geometric Resolvent Identities

Recall the second resolvent identity for invertible operators A and C:

$$A^{-1} - C^{-1} = A^{-1}(C - A)C^{-1}.$$

With $C = A + B$, it can be rewritten as

$$(A + B)^{-1} = A^{-1} - A^{-1}B(A + B)^{-1}. \tag{2.61}$$

The Dirichlet Laplacian in a cube $\Lambda = \Lambda_L(u)$ can be written as follows:

$$H^0_\Lambda = 2d\,\mathrm{I} - \sum_{i=1}^{d} \sum_{z,y\in\Lambda:\,|z-y|_1=1} T_{z,y}$$

where I is the identity operator, and the rank-one operator $T_{z,y}$ acts by

$$\left(T_{z,y}\phi\right)(x) = \mathbf{1}_{\{z\}}(x)\phi(y), \quad x \in \Lambda_L(u).$$

Consider a smaller cube, $\tilde{\Lambda} = \Lambda_l(v) \subset \Lambda_L(u)$, and its complement $\tilde{\Lambda}^c_l(v) = \Lambda_L(u) \setminus \Lambda_l(v)$. Assume for simplicity that $\partial^+\tilde{\Lambda} \subset \Lambda$.

Then H^0_Λ admits the decomposition:

$$H^0_\Lambda = \left(H^0_{\tilde{\Lambda}} \oplus H^0_{\Lambda \setminus \tilde{\Lambda}} \right) + T_{\partial \tilde{\Lambda}},$$

with

$$T_{\partial \tilde{\Lambda}} = - \sum_{(z,y) \in \partial \tilde{\Lambda}} (T_{z,y} + T_{y,z}).$$

The potential energy operator $V_\Lambda(\omega)$ is a multiplication operator, hence, diagonal in the delta-basis. Thus, for the Hamiltonian $H_\Lambda(\omega)$ we have a similar decomposition:

$$H_\Lambda(\omega) = \left(H_{\tilde{\Lambda}}(\omega) \oplus H_{\Lambda \setminus \tilde{\Lambda}}(\omega) \right) + T_{\partial \tilde{\Lambda}}. \tag{2.62}$$

We apply the resolvent identity (2.61) to the Green's functions in $\Lambda = \Lambda_L(u)$ and $\tilde{\Lambda} = \Lambda_l(v)$ such that $\tilde{\Lambda} \cup \partial^+ \tilde{\Lambda} \subset \Lambda$ (cf. (2.18)). For $v \in \tilde{\Lambda}$ and $y \in \Lambda \setminus \tilde{\Lambda}$ (which are separated by the boundary $\partial^- \tilde{\Lambda}$) this yields:

$$G_\Lambda(v, y; E, \omega) = \sum_{(x,x') \in \partial \tilde{\Lambda}} G_{\tilde{\Lambda}}(v, x; E, \omega)\, G_\Lambda(x', y; E, \omega). \tag{2.63}$$

This equation is often called the geometric resolvent equation or geometric resolvent identity. It immediately leads to the *geometric resolvent inequality* (GRI) which we will use in one of the forms presented in Eq. (2.64) below. Assuming, as before, that $\tilde{\Lambda} = \Lambda_l(v) \subset \Lambda_L(u) = \Lambda$, $\partial^+ \tilde{\Lambda} \subset \Lambda$ and $y \in \Lambda \setminus \tilde{\Lambda}$,

$$\begin{aligned}
&|G_\Lambda(v, y; E, \omega)| \\
&\quad \le \sum_{(x,x') \in \partial \tilde{\Lambda}} \left| G_{\tilde{\Lambda}}(v, x; E, \omega) \right| \left| G_\Lambda(x', y; E, \omega) \right| \\
&\quad \le \left| \partial \tilde{\Lambda} \right| \max_{(x,x') \in \partial \tilde{\Lambda}} \left| G_{\tilde{\Lambda}}(v, x; E, \omega) \right| \left| G_\Lambda(x', y; E, \omega) \right| \\
&\quad \le \left| \partial \tilde{\Lambda} \right| \max_{x \in \partial^- \tilde{\Lambda}} \left| G_{\tilde{\Lambda}}(v, x; E, \omega) \right| \times \max_{x' \in \partial^+ \tilde{\Lambda}} \left| G_\Lambda(x', y; E, \omega) \right| \\
&\quad \le \left| \partial \tilde{\Lambda} \right| \left\| G_{\tilde{\Lambda}}(E) \right\| \max_{x' \in \partial^+ \tilde{\Lambda}} \left| G_\Lambda(x', y; E, \omega) \right| .
\end{aligned} \tag{2.64}$$

The GRI, playing an important role in the MSA, will be repeatedly used throughout our book.

Of particular interest is an identity that holds for the solutions of the equation $H(\omega)\psi = E\psi$. The decomposition

$$H(\omega) = H_\Lambda(\omega) \oplus H_{\mathbb{Z}^d \setminus \Lambda}(\omega) + T_{\partial \Lambda} \tag{2.65}$$

renders that

$$\mathbf{1}_\Lambda (H_\Lambda(\omega) - E)\psi = -\mathbf{1}_\Lambda T_{\partial\Lambda}\psi.$$

If $E \notin \sigma(H_\Lambda)$, we deduce:

$$\mathbf{1}_\Lambda \psi = -\mathbf{1}_\Lambda G_\Lambda(E,\omega) T_{\partial\Lambda}\psi,$$

and for $x \in \Lambda$,

$$\psi(x) = \sum_{(y,y')\in\partial\Lambda} G_\Lambda(x,y;E,\omega)\psi(y'). \tag{2.66}$$

It implies geometric resolvent inequalities (GRIs) for the eigenfunctions:

$$\begin{aligned} |\psi(x)| &\le \max_{y\in\partial^-\Lambda} |G_\Lambda(x,y;E,\omega)| \cdot \sum_{y'\in\partial^+\Lambda} |\psi(y')|, \\ |\psi(x)| &\le \sum_{y\in\partial^-\Lambda} |G_\Lambda(x,y;E,\omega)| \cdot \max_{y'\in\partial^+\Lambda} |\psi(y')|. \end{aligned} \tag{2.67}$$

Equations (2.67) show that the decay of a (random) solution $\psi = \psi(\omega)$ of the equation $H(\omega)\psi = E\psi$ is controlled by that of $|G_\Lambda(x,y;E,\omega)|$ (provided that we have an a priori estimate for $|\psi(y')|$).

2.3.2 Decay of the Green's Functions in Typical Cubes

Fix a positive integer L and consider the LSO $H_{\Lambda_L(u)}(\omega) = H^0_{\Lambda_L(u)} + gV_{\Lambda_L(u)}(\omega)$. If we assume that $|g|$ is large then the Laplacian H^0_Λ can be viewed as a perturbation of the multiplication operator $gV_{\Lambda_L(u)}(\omega)$ in $\ell^2(\Lambda_L(u))$. It is natural to expect that the eigenfunctions and the eigenvalues of $H_{\Lambda_L(u)}(\omega)$ can be obtained, or controlled, with the help of perturbation techniques.

In the finite-dimensional perturbation theory, it is well-known that the simplest situation is where all eigenvalues of the unperturbed operator (in our case, it is multiplication by $gV(\cdot\,;\omega)$) are simple (pairwise distinct) and sufficiently distant from each other. Under our Assumption A, with probability one,

$$\forall\, x,y \in \Lambda_L(u) \text{ with } x \ne y, \quad V(x;\omega) \ne V(y;\omega).$$

Moreover, for any (arbitrarily large) $A > 0$

$$\mathbb{P}\left\{ \min_{x,y\in\Lambda_L(u):\, x\ne y} |gV(x;\omega) - gV(y;\omega)| \le A \right\} \xrightarrow[|g|\to\infty]{} 0 \tag{2.68}$$

and for any $E \in \mathbb{R}$,

$$\mathbb{P}\left\{ \min_{x\in\Lambda_L(u)} |gV(x;\omega) - E| \le A \right\} \underset{|g|\to\infty}{\longrightarrow} 0. \tag{2.69}$$

In particular, with $A = e^{\gamma(m,L)L}$ where $m > 0$ (a choice we will follow in the sequel; cf. Eq. (2.20)), we obtain

$$\mathbb{P}\left\{ \min_{x\in\Lambda_L(u)} |gV(x;\omega) - E| \ge e^{\gamma(m,L)L} \right\} \underset{|g|\to\infty}{\longrightarrow} 1. \tag{2.70}$$

We see that if $|g|$ is large enough then, with high probability, the operator of multiplication by the potential $V(x;\omega)$ in $\ell^2(\Lambda_L(u))$ has sparse eigenvalues (of multiplicity 1), which lie at a large distance $\ge e^{\gamma(m,L)L}$ from energy E. Respectively, the operator $(V_{\Lambda_L(u)}(\omega) - E)^{-1}$ has a small norm, bounded by $e^{-\gamma(m,L)L}$. It is natural to expect that a similar estimate holds true for the perturbed operator $H^0_{\Lambda_L(u)} + V_{\Lambda_L(u)}(\omega)$, since $H^0_{\Lambda_L(u)}$ is a bounded operator, with $\left\| H^0_{\Lambda_L(u)} \right\| \le 2d$; see Theorem 2.3.1 below. In this theorem we follow Definition 2.1.1 of an (E,m)-singular cube, with mass m varying as a function of $|g|$, for $|g|$ large enough.

Theorem 2.3.1. *Suppose that the random field $V(x;\omega)$ satisfies Assumption A. Fix a positive integer L and a bounded interval $I \subset \mathbb{R}$. For any $m > 0$ and $p > 0$, there exists $g_0 = g_0(m,p,I,L)$ such that if $|g| \ge g_0$, then*

$$\mathbb{P}\{\exists\, E \in I : \Lambda_L(u) \text{ is } (E,m)\text{-S}\} \le L^{-p}. \tag{2.71}$$

We want to say that Theorem 2.3.1 asserts that, with high probability, a given cube $\Lambda_L(u)$ is (E,m)-NS for all $E \in I$. The price for such a strong claim is high: one has to take a very large amplitude $|g|$ of the random potential. Moreover, the required magnitude of g depends essentially upon the size L, so that there is no hope to reproduce such a uniform bound at larger scales. Indeed, as we will see later, a bound of the form (2.71) is to be replaced by a more sophisticated bound (2.183).

Proof. We will establish a much more general result: for all (arbitrarily large) $m > 0$, $p > 0$ and all (arbitrarily small) $\epsilon > 0$, there exists $g_0 = g_0(m,p,\epsilon,I)$ such that for all g with $|g| \ge g_0$

$$\mathbb{P}\{\exists\, E \in I : \Lambda_L(u) \text{ is } (E,m)\text{-S}\} \le \epsilon. \tag{2.72}$$

The estimate (2.71) is obtained from (2.72) by taking $\epsilon = L^{-p}$.

The proof combines two steps: analytic and probabilistic.

1. The analytic step. Recall that, by Definition 2.1.1 (cf. (2.19)), a cube $\Lambda_{L_0}(u)$ is called (E,m)-non-singular if

$$|\partial \Lambda_L(u)| \max_{x\in\Lambda_{L^{1/\alpha}}(u)} \max_{y\in\partial^-\Lambda_L(u)} \left|G_{\Lambda_L(u)}(x,y;E,\omega)\right| \leq \mathrm{e}^{-\gamma(m,L)L}.$$

Clearly,

$$\max_{x,y\in\Lambda_L(u)} |G_{\Lambda_L(u)}(x,y;E,\omega)| \leq \|G_{\Lambda_L}(E,\omega)\|.$$

Therefore,

$$\begin{aligned}&\mathbb{P}\{\exists E \in I: \ \Lambda_L(u) \text{ is } (E,m)\text{-S}\}\\&\qquad \leq \mathbb{P}\{\exists E \in I: \ \|G_{\Lambda_L}(E,\omega)\| > |\partial\Lambda_L(u)|^{-1}\,\mathrm{e}^{-\gamma(m,L)L}\}.\end{aligned}$$

Set for brevity $\eta_L := |\partial\Lambda_L(u)|\,\mathrm{e}^{+\gamma(m,L)L}$. Since the operator $H_\Lambda(\omega)$ is self-adjoint, we can assess the norm $\|G_\Lambda(E,\omega)\| = \|(H_\Lambda(\omega)-E)^{-1}\|$, for ω such that $E \notin \sigma\big(H_\Lambda(\omega)\big)$, by

$$\|G_\Lambda(E,\omega)\| = \frac{1}{\operatorname{dist}\Big[E,\sigma\big(H_\Lambda(\omega)\big)\Big]},$$

which yields the implication

$$\operatorname{dist}\Big[E,\sigma\big(H_\Lambda(\omega)\big)\Big] \geq \eta_L \implies \Lambda_L(u) \text{ is } (E,m)\text{-NS}. \qquad (2.73)$$

Therefore, the bound (2.72) will be established, if we prove that, for any fixed $m, p, \epsilon > 0$, $L \in \mathbb{N}^*$, any $\eta > 0$ and all sufficiently large $|g|$,

$$\begin{aligned}&\mathbb{P}\Big\{\exists E \in I: \operatorname{dist}\Big[E,\sigma\big(H_\Lambda(\omega)\big)\Big] < \eta\Big\}\\&\qquad \equiv \mathbb{P}\Big\{\operatorname{dist}\Big[I,\sigma\big(H_\Lambda(\omega)\big)\Big] < \eta\Big\} \leq \epsilon.\end{aligned}$$

Furthermore, it follows from the min-max principle for self-adjoint operators applied to $H_{\Lambda_L(u)}(\omega) = H^0_{\Lambda_L(u)} + gV(\cdot;\omega)$ and to the operator of multiplication by $gV(x;\omega)$ in $\ell^2(\Lambda_L(u))$ (diagonal in the delta-basis) that

$$\operatorname{dist}\left[\sigma(H_{\Lambda_L(u)}(\omega)), \{gV(x;\omega), x \in \Lambda_L(u)\}\right] \leq \|H^0_{\Lambda_L(u)}\| \leq 4d. \qquad (2.74)$$

Indeed, recall the standard argument based on the min-max principle and proving (2.74). Given $f \in \ell^2(\Lambda)$, $\|f\| = 1$, $\Lambda = \Lambda_L(u)$, we have

$$(H_\Lambda f, f) = (H_0 f, f) + (gVf, f) \leq \|H_0\| + (gVf, f),$$

so by the min-max principle, $E_j(H_\Lambda) \leq \|H_0\| + E_j(gV)$, where both $\{E_j(gV)\}$ and $\{E_j(H_\Lambda)\}$ are numbered in increasing order. Similarly, $E_j(H_\Lambda) \geq E_j(gV) - \|H_0\|$. This proves (2.74).

Consequently, it suffices to prove that, with $\eta' := \eta + 4d > 4d$,

$$\mathbb{P}\left\{ \operatorname{dist}\Big[I, \{gV(x;\omega), x \in \Lambda_L(u)\} \Big] < \eta' \right\} \le \epsilon, \tag{2.75}$$

provided that $|g|$ is large enough. To see this, let $I = [a,b]$, and use again the min-max principle. The values $\{V(x;\omega),\ x \in \Lambda_L(u)\}$ are the eigenvalues $E_j(gV)$ of the multiplication operator gV numbered in increasing order, and we already know that the eigenvalues $\{E_j(H_\Lambda)\}$ (also numbered in increasing order) satisfy

$$|E_j(H_\Lambda) - E_j(gV)| \le 4d. \tag{2.76}$$

If we have

$$\operatorname{dist}\Big[I, \{gV(x;\omega), x \in \Lambda_L(u)\} \Big] > 4d + \eta,\ \eta > 0,$$

then for any $j = 1, \dots, |\Lambda_L(u)|$, either $E_j(gV) < a - 4d - \eta$, or $E_j(gV) > b + 4d + \eta$. Fix any j; assume first the latter, then by (2.76),

$$E_j(H_\Lambda) \ge E_j(gV) - 4d > b + \eta + 4d - 4d = b + \eta.$$

Similarly, if $E_j(gV) < a - 4d - \eta$, then

$$E_j(H_\Lambda) \le E_j(gV) + 4d < a - \eta - 4d + 4d = a - \eta.$$

In any case, $\operatorname{dist}\big[I, E_j(H_\Lambda)\big] > \eta > 0$, $j = 1, \dots, |\Lambda_L(u)|$.

2. The probabilistic step. Given a cube $\Lambda = \Lambda_L(u)$, a bounded interval $I \subset [-R, +R]$ and $\eta' \in (0, +\infty)$, consider the extended interval

$$\tilde{I} = [-R - \eta', R + \eta'] =: [-R', R'].$$

The estimate (2.75) will follow from

$$\mathbb{P}\left\{ \exists x \in \Lambda_L(u) : |gV(x;\omega)| < R' \right\} \le \epsilon. \tag{2.77}$$

Observe that, under Assumption A, we have, for $g \neq 0$,

$$\begin{aligned}\mathbb{P}\left\{ \exists x \in \Lambda_L(u) : |gV(x;\omega)| < R' \right\} &\le \sum_{x \in \Lambda_L(u)} \mathbb{P}\left\{ |V(x;\omega)| < \frac{R'}{|g|} \right\} \\ &\le |\Lambda_L(u)|\, C_s \left(|g|^{-1} R' \right)^s \underset{|g| \to \infty}{\longrightarrow} 0,\end{aligned}$$

so (2.77) holds true for $|g|$ large enough. This completes the proof. □

It is clear from the proof of Theorem 2.3.1 that its assertion remains valid under a much weaker hypothesis than Assumption A: it suffices to assume that the common marginal probability distribution of the IID random variables $V(x;\omega)$ is merely continuous.

We would like to note also that the assertion of Theorem 2.3.1 can be deduced directly from the so-called Combes–Thomas estimate established in Sect. 2.3.3 (cf. Theorem 2.3.4). Moreover, a careful analysis shows that the Combes–Thomas estimate—at large disorder—gives rise to a substantially more efficient initial length scale decay bound on Green's functions (cf. Sect. 2.3.3).

Given a bounded interval $I \subset \mathbb{R}$, an initial length scale estimate of the form (DS.I, m, p_0, L_0) follows directly from Theorem 2.3.1: indeed, the event figuring in (DS.I, m, p_0, L_0) (two cubes are (E,m)-S) is smaller than that in the LHS of (2.71), where just one cube is (E,m)-S. However, a small modification is required to prove (DS.I, m, p_0, L_0) for $I = \mathbb{R}$.

Theorem 2.3.2. *Suppose that the IID random field $V(x;\omega)$ satisfies Assumption A. Fix a positive integer L. For any $m > 0$ and $p > 0$, there exists $g_0 = g_0(m,p,L)$ such that if $|g| \geq g_0$, then for any pair of disjoint cubes $\Lambda_L(u)$, $\Lambda_L(v)$*

$$\mathbb{P}\{\exists\, E \in \mathbb{R}:\ \Lambda_L(u) \text{ and } \Lambda_L(v) \text{ are } (E,m)\text{-S}\} \leq L^{-p}. \tag{2.78}$$

Proof. Again, it suffices to prove a more general estimate

$$\mathbb{P}\{\exists\, E \in \mathbb{R}:\ \Lambda_L(u) \text{ and } \Lambda_L(v) \text{ are } (E,m)\text{-S}\} \leq \epsilon, \tag{2.79}$$

this makes notations less cumbersome. Next, arguing as in the proof of Theorem 2.3.1 (cf. (2.75), (2.77)), we see that (2.79) will be established, if we prove that for any $\eta > 0$

$$\mathbb{P}\{\exists\, x \in \Lambda_L(u),\ \exists\, y \in \Lambda_L(v):\ |gV(x;\omega) - gV(y;\omega)| < \eta\} \leq \epsilon.$$

The probability in the LHS is bounded by

$$|\Lambda_L(u)| \cdot |\Lambda_L(v)| \cdot \max_{x \in \Lambda_L(u),\, y \in \Lambda_L(v)} \mathbb{P}\left\{|V(x;\omega) - V(y;\omega)| < |g|^{-1}\eta\right\}$$

so it suffices to show that for any points $x \in \Lambda_L(u)$, $y \in \Lambda_L(v)$ the probability in the last RHS tends to 0 as $|g| \to \infty$. It can be assessed as follows:

$$\begin{aligned} &\mathbb{P}\left\{|V(x;\omega) - V(y;\omega)| < |g|^{-1}\eta\right\} \\ &\quad \leq \mathbb{P}\left\{|V(x;\omega) - V(y;\omega)| < 2|g|^{-1}\eta\right\} \\ &\quad = \mathbb{E}\left[\mathbb{P}\left\{|V(x;\omega) - V(y;\omega)| < 2|g|^{-1}\eta \mid \mathfrak{B}_y\right\}\right], \end{aligned}$$

where $\mathfrak{B}_y$ is the sigma-algebra generated by the random variable $V(y;\cdot)$. Since $\Lambda_L(u) \cap \Lambda_L(v) = \varnothing$ (thus $x \neq y$) and the random field $V(\cdot;\omega)$ is IID, $V(x;\omega)$ is $\mathfrak{B}_y$-independent, while $V(y;\omega)$ is $\mathfrak{B}_y$-measurable, i.e., fixed under the above conditioning. Therefore, the last RHS is bounded by

$$\operatorname{ess\,sup} \sup_{\lambda\in\mathbb{R}} \mathbb{P}\left\{ |V(x;\omega) - \lambda| < 2|g|^{-1}\eta \;\middle|\; \mathfrak{B}_y \right\} = C_s\big(4|g|^{-1}\eta\big)^s \underset{|g|\to\infty}{\longrightarrow} 0,$$

with $s > 0$ and $C_s < \infty$ as in Assumption A. This completes the proof of (2.78). □

Again, it is clear from the proof that the assertion of the theorem remains valid for an IID random field V with any (uniformly) continuous marginal probability distribution.

2.3.3 The Combes–Thomas Estimates

As was stated in Assertion (I) of Theorem 2.1.1, for large values of $|g|$ the operator $H(\omega)$ exhibits a complete localization. The situation becomes more delicate when the amplitude g is not large, or even very small. As was mentioned before, there is a consensus in the physics community that starting from the dimension $d = 3$, for sufficiently small g the operator $H(\omega)$ must feature delocalization. In mathematical terms, this probably means presence of an absolutely continuous component in the spectrum $\sigma\big(H(\omega)\big)$. At present, such a claim remains an open—and challenging—mathematical problem—at least for Anderson models on $\mathbb{Z}^d$. On the other hand, there exists an argument, discovered a time ago by the physicist I. M. Lifshitz (and called traditionally "Lifshitz tails" argument) ensuring that even for small amplitudes g localization persists at sufficiently low energies. Cf. [126, 127]. The relevant zone of energies $E \in \mathbb{R}$ is described in terms of the DoS $\mathrm{d}\nu(E)/\mathrm{d}E$:

Anderson localization persists for energies E where $\mathrm{d}\nu(E)/\mathrm{d}E$ is sufficiently small.

In more general terms, when the DoS does not necessarily exist, this condition can be replaced by a requirement that the eigenvalue concentration, relative to cubes Λ of an appropriately chosen size L, is sufficiently low around the value E.

Lifshitz noticed that in the case where the potential V is bounded from below, so that the lower edge $E_* := \inf \sigma(H) > -\infty$, the DoS behaves at energies E near E_* like $\mathrm{e}^{-(E-E_*)^{-d/2}}$. More precisely,

$$\lim_{E\downarrow E_*} \frac{\ln\big|\ln\big(\mathrm{d}\nu(E)/\mathrm{d}E\big)\big|}{\ln\,(E - E_*)} = -\frac{d}{2}.$$

Before going further, note that, owing to a well-known argument, $E_* = \inf\{E \in \mathbb{R} :\; F(E) > 0\}$, where F is the common probability distribution of the IID random variables $V(x;\cdot)$. See, e.g., Theorem 3.9 in [113]. The main idea of the proof is to construct a Weyl sequence $\{\varphi_n^{(E)}, n \geq 1\}$ for E close to E_*, such that $\|\varphi_n^{(E)}\|_2 = 1$ and $\|(H(\omega) - E)\varphi_n^{(E)}\|_2 \underset{n\to\infty}{\longrightarrow} 0$. In fact, one can pick $\varphi_n^{(E)} = \mathbf{1}_{\Lambda_n(x_n(\omega))}$, where the cubes $\Lambda_n(x_n(\omega))$ are chosen in such a way (for a given, $\mathbb{P}$-a.e. ω) that

$$\sup_{y\in\Lambda_n(x_n(\omega))} |V(y;\omega) - E| \underset{n\to\infty}{\longrightarrow} 0,$$

See the details in [113], Sect. 3.4.

A short qualitative explanation of the Lifshitz tails phenomenon is as follows. For notational simplicity, assume that the values of the external potential $V(x;\omega)$ are non-negative and, more precisely, satisfy Assumption B; consequently, $E_* = 0$ (if we take into account the above mentioned construction of a Weyl sequence). Then for the lowest eigenvalue in a cube $\Lambda_L(u)$ to be very close to 0, not just one but many values of the potential must be very close to 0; more precisely, the average value of the sample $V(x;\omega)$, $x \in \Lambda_L(u)$ must be very small. The probability of such event is assessed with the help of the theory of large deviations; this gives rise to the Lifshitz tails asymptotics. Details can be found in [113].

This description of the DoS at low energies implies that, in a given large cube $\Lambda_L(u)$, it is very unlikely to have eigenvalues below a threshold $E_* + 2\eta$, if η is small enough. In other words, with high probability there is a gap of size $\eta > 0$ between the random spectrum $\sigma(H_\Lambda(\omega))$ and the interval $[E_*, E_* + \eta]$. In turn, this means that a quantum particle with energy $E \in [E_*, E_* + \eta]$ would be under the barrier of height $\eta > 0$ in $\Lambda_L(u)$, so that its wave function $\psi(E, x)$, $x \in \Lambda_L(u)$, must decay exponentially with $\mathrm{dist}(x, \partial\Lambda_L(u))$, i.e., as x moves deeper inside the cube. It is natural then to expect that the cube $\Lambda_L(u)$ is (E, m)-NS, with m of order of the gap width η. To make this argument mathematically valid, one has to translate decay properties of the wave-functions into upper bounds upon the resolvent $\left(H_{\Lambda_L(u)} - E\right)^{-1}$.

In the case of random potentials not bounded from below (such as an IID Gaussian), the spectrum of the operator $H(\omega)$ is also unbounded from below. In this case (not discussed in our book) negative energies E with sufficiently large $|E|$ also qualify as extreme: it is unlikely to have many values of an IID potential below a large negative threshold E. (Just as it is unlikely to have many values $V(x;\omega)$ extremely close to the infimum $\inf V(x;\omega)$ in the case where this infimum is $> -\infty$.) Naturally, the same is true for large positive energies E.

Combes and Thomas [74] proposed in 1970s an elegant argument which shows directly that if an energy E is separated by a distance $\eta > 0$ from the spectrum $\sigma\left(H_{\Lambda_L(u)}(\omega)\right)$, then the Green's functions $G_{\Lambda_L(u)}(x, y; E, \omega)$ decay exponentially as the distance $|x - y|$ grows, with an exponent of order $O(\eta)$.[4]

---

[4] Note that this phenomenon is more general than the decay of a wave function under the barrier, for E is allowed to be in a spectral gap, and not necessarily below/above the spectrum.

The original result by Combes–Thomas was designed to address a specific problem.[5] Later on, the key idea of their method has been generalized and turned in what is now called the Combes–Thomas estimate. Here we give it in a form suitable for the MSA for random Hamiltonians in cubes $\Lambda_L(u)$.

Theorem 2.3.3. *Set $\Lambda = \Lambda_L(u)$, and consider the operator $H_\Lambda(\omega)$ with fixed $\omega \in \Omega$. Suppose that for some $E \in \mathbb{R}$ and $0 < \eta \leq 1$*

$$\operatorname{dist}\Big[E, \sigma\big(H_\Lambda(\omega)\big)\Big] \geq \eta > 0.$$

Then, $\forall\, x, y \in \Lambda$ and the same ω,

$$|G_\Lambda(x, y; E, \omega)| \leq \frac{2}{\eta} \exp\left(-\frac{\eta}{5d}|x - y|\right). \tag{2.80}$$

For large values of η the term η in the exponent is to be replaced by $O(\ln \eta)$, as shown in Theorem 2.3.4, which is suitable for applications to the large disorder regime.

Theorem 2.3.4. *With the notations of Theorem 2.3.3, suppose that for some $E \in \mathbb{R}$ and $\eta > 0$, with fixed $\omega \in \Omega$,*

$$\operatorname{dist}(E, \sigma(H_\Lambda(\omega))) \geq \eta > 0.$$

Then for any points $x, y \in \Lambda$ and the same ω

$$|G_\Lambda(x, y; E, \omega)| \leq \frac{2}{\eta} \exp\left[-\frac{1}{2}\left(\ln \frac{\eta}{4d}\right)|x - y|\right].$$

Naturally, the above bound is useful only when $\eta > 4d$.

A better bound for small values of η has been proven in [30]. Roughly speaking, it was shown that the factor η in the exponent in (2.80) can be replaced by $C\eta^{1/2}$ with some $C > 0$.

2.3.4 *Proof of the Combes–Thomas Estimates*

Proof of Theorem 2.3.3. Here we follow closely the argument given in [113]. Note, however, that we provide a sharper bound on the decay exponent $(\eta/(5d)$, compared to $\eta/(12d)$ in [113]); this does not require any substantial modification of the proof.

---

[5] Curiously, the paper [74] treated multi-particle systems in the context of the scattering theory.

For simplicity, we assume that $\Lambda = \Lambda_L(0)$. In addition to the max-norm $|x| = \max |x_j|$, we will also use the ℓ^1-norm for vectors $x = (x_1, \ldots, x_d) \in \mathbb{Z}^d$:

$$|x|_1 := \sum_{j=1}^{d} |x_j|.$$

In particular, since $|x| \le |x|_1$, it suffices to prove the inequality (2.80) with $|x-y|_1$ instead of $|x-y|$.

Next, introduce in $\ell^2(\Lambda)$ the following multiplication operator $A = A_{v,a}$, labeled by $a > 0$ and $v \in \Lambda$:

$$(A\phi)(x) = \mathrm{e}^{a|x-v|_1}\phi(x),\ x \in \Lambda.$$

Fix $a > 0$ and $x \in \Lambda$. A straightforward calculation shows that for any operator $B : \ell^2(\Lambda) \to \ell^2(\Lambda)$, the matrix elements of the operator $\mathrm{Ad}_A(B) := A^{-1}BA$ in the delta-basis have the form

$$\begin{aligned}\big(\mathrm{Ad}_A(B)\big)(x,y) &= \big(A(x,x)\big)^{-1} B(x,y)A(y,y)\\ &= \mathrm{e}^{a(|y-v|_1 - |x-v|_1)} B(v,y).\end{aligned} \tag{2.81}$$

In particular, for the Green's operator $G_\Lambda(E) = (H_\Lambda - E)^{-1}$, with $E \in \mathbb{R} \setminus \sigma(H_\Lambda(\omega))$, this gives

$$\big(\mathrm{Ad}_A[G_\Lambda(E)]\big)(x,y) = \mathrm{e}^{a(|y-v|_1 - |x-v|_1)} G_\Lambda(x,y;E), \tag{2.82}$$

or, equivalently,

$$G_\Lambda(x,y;E) = \mathrm{e}^{-a(|y-v|_1 - |x-v|_1)} \big[\mathrm{Ad}_A(G_\Lambda(E))\big](x,y).$$

Hence, with $v = x$, we obtain

$$G_\Lambda(x,y;E) = \mathrm{e}^{-a|y-x|_1} \big[\mathrm{Ad}_A(G_\Lambda(E))\big](x,y). \tag{2.83}$$

Since Ad_A is an automorphism of the operator algebra in $\ell^2(\Lambda)$, we can write

$$\mathrm{Ad}_A\big[G_\Lambda(E)\big] = \mathrm{Ad}_A\left((H_\Lambda - E)^{-1}\right) = \left(\mathrm{Ad}_A(H_\Lambda) - E\right)^{-1},$$

and Eq. (2.83) implies

$$\begin{aligned}|G_\Lambda(E)(x,y)| &= \mathrm{e}^{-a|y-x|_1} \left| [\mathrm{Ad}_A(H_\Lambda) - E]^{-1}(x,y) \right|\\ &\le \mathrm{e}^{-a|y-x|_1} \left\| [\mathrm{Ad}_A(H_\Lambda) - E]^{-1} \right\|.\end{aligned}$$

For the operators $[\mathrm{Ad}_A(H_\Lambda) - E]^{-1}$ and $(H_\Lambda - E)^{-1}$ we have the resolvent identity:

$$\begin{aligned}
&[\mathrm{Ad}_A(H_\Lambda) - E]^{-1} \\
&\quad = (H_\Lambda - E)^{-1}\left[\mathbf{I} + (\mathrm{Ad}_A(H_\Lambda) - H_\Lambda)(H_\Lambda - E)^{-1}\right]^{-1}.
\end{aligned} \tag{2.84}$$

Here $\mathbf{I}$ stands for the identity operator in $\ell^2(\Lambda)$. The second factor in the RHS can be expanded in a convergent Neumann series, provided that the norm $\left\|(A^{-1}H_\Lambda A - H_\Lambda)(H_\Lambda - E)^{-1}\right\| < 1$. For any operator $B : \ell^2(\Lambda) \to \ell^2(\Lambda)$, with matrix elements $B(x, y)$ satisfying

$$\max_{x\in\Lambda} \sum_{y\in\Lambda} |B(x, y)| \le a,$$

$$\max_{y\in\Lambda} \sum_{x\in\Lambda} |B(x, y)| \le b,$$

the norm $\|B\|$ is bounded by $(ab)^{1/2}$. In particular, with $a = b$ the norm is bounded by b. For the operator $B := \mathrm{Ad}_A(H_\Lambda) - H_\Lambda$, with $H_\Lambda = H^0_\Lambda + V_\Lambda$, we have $\mathrm{Ad}_A(H_\Lambda) - H_\Lambda = \mathrm{Ad}_A(H^0_\Lambda) - H^0_\Lambda$, since the multiplication operators A and $V_\Lambda(\omega)$ commute.

Therefore, the matrix elements of $B := \mathrm{Ad}_A(H_\Lambda) - H_\Lambda$ obey

$$\begin{aligned}
\sum_{y\in\Lambda} |B(v, y)| &\le \sum_{y\in\Lambda:\, |y-v|_1=1} \left|\mathrm{e}^{a(|y-x|_1 - |v-x|_1)} - 1\right| \\
&\le 2d \sup_{|t|\le 1} |\mathrm{e}^{at} - 1| \le 2d \int_0^1 a\mathrm{e}^{as}\,\mathrm{d}s \le 2d\, a\mathrm{e}^{a} =: h(a),
\end{aligned} \tag{2.85}$$

where we used the fact that

$$\Big||y - x|_1 - |v - x|_1\Big| \le |y - v|_1.$$

A similar upper bound holds for the transpose of the matrix of B. As a result, we obtain that

$$\|\mathrm{Ad}_A(H_\Lambda) - H_\Lambda\| \le h(a).$$

Since, by assumption, $\mathrm{dist}\big(\sigma(H_\Lambda), E\big) \ge \eta$, we have:

$$\left\|(H_\Lambda - E)^{-1}\right\| \le \Big(\mathrm{dist}(\sigma(H_\Lambda), E)\Big)^{-1} \le \eta^{-1}, \tag{2.86}$$

rendering

$$\left\| \left(\mathrm{Ad}_A \left(H_\Lambda \right) - H_\Lambda \right) \left(H_\Lambda - E \right)^{-1} \right\| \le h(a)\eta^{-1}. \tag{2.87}$$

So far, we did not specify the values of $a > 0$ and $\eta > 0$. Now we take $\eta \le 1$ and $a := \frac{\eta}{5d} \le \frac{1}{5} < 1$. Then, making use of the inequality $\frac{2e^{1/5}}{5} < \frac{1}{2}$, we come to the following bound of the norm in the RHS of Eq. (2.87):

$$\left\| \left(\mathrm{Ad}_A(H_\Lambda) - H_\Lambda \right) \left(H_\Lambda - E \right)^{-1} \right\| \le h(a)\eta^{-1} \le \frac{2d\,\eta\, e^{\eta/(5d)}}{5\,d\,\eta} \le \frac{1}{2}. \tag{2.88}$$

As a result, the Eq. (2.84) can be re-written with the help of the convergent Neumann series:

$$\begin{aligned} \left\| \left(\mathrm{Ad}_A \left(H_\Lambda \right) - E \right)^{-1} \right\| &\le \left\| \left(H_\Lambda - E \right)^{-1} \right\| \\ &\times \left\| \left[\mathbf{I} + \left(\mathrm{Ad}_A \left(H_\Lambda \right) - H_\Lambda \right)^{-1} \left(H_\Lambda - E \right)^{-1} \right]^{-1} \right\| \le \frac{2}{\eta}. \end{aligned} \tag{2.89}$$

Finally, combining Eq. (2.83) with $u = x$ and Eq. (2.89), we obtain

$$\begin{aligned} & |G_\Lambda(x, y; E)| \\ & \le e^{-a|x-y|} \left\| \left(\mathrm{Ad}_A \left(H_\Lambda \right) - E \right)^{-1} \right\| \le \frac{2}{\eta} \exp\left(-\frac{\eta}{5d} |x-y|_1 \right). \end{aligned}$$

This completes the proof of Theorem 2.3.3. □

Remark 2.3.1. The assumption that $0 < \eta \le 1$ does not mean, of course, that the obtained bound for the resolvent does not apply to operators with $\mathrm{dist}(\sigma(H_\Lambda), E) > 1$; we simply deal with a bound that does not benefit of larger values of the "spectral gap". It is readily seen that $\left\| (H_\Lambda - E)^{-1} \right\| \to 0$ as $\mathrm{dist}(\sigma(H_\Lambda), E) \to \infty$, but the guaranteed exponent of decay of the Green's functions $G(x, y; E)$ for large values of $\eta := \mathrm{dist}(\sigma(H_\Lambda), E)$ is no longer proportional to η, as we show in the next section.

Now consider the case where η can be arbitrarily large. Such a generalization is rarely used; nevertheless, it can be helpful for localizing the analysis of resolvents to a relatively small energy interval I and performing a necessary cut-off of the complementary area $\mathbb{R} \setminus I$. Recall, for example, that, as was mentioned in the proof of the initial scale estimate of the Green's functions (Theorem 2.3.1), the analytic step can be replaced by a direct application of the Combes–Thomas estimate for large η.

Proof of Theorem 2.3.4. We start with the inequality in Eq. (2.88): $\forall\ \eta > 0$ and $a > 0$,

$$\left\| \left(\mathrm{Ad}_A \left(H_\Lambda \right) - H_\Lambda \right) \left(H_\Lambda - E \right)^{-1} \right\| \le h(a)\eta^{-1} = 2d\, a e^a\, \eta^{-1},$$

and notice that $ae^a < e^{2a}$ for any $a > 0$. Given $\eta > 4d$, set $a(\eta) = \dfrac{1}{2} \ln \dfrac{\eta}{4d}$, then

$$h(a)\eta^{-1} < \frac{2d\, e^{2a(\eta)}}{\eta} = \frac{2d\eta}{4d\eta} = \frac{1}{2}.$$

Using again Eq. (2.89), we conclude that

$$|G_\Lambda(x,y;E)| \le \frac{2}{\eta} \exp\left[-\frac{1}{2} \ln\left(\frac{\eta}{4d}\right) |x-y|_1 \right]. \tag{2.90}$$

□

2.3.5 Applications of the Combes–Thomas Estimate to the MSA

Preliminary Analysis

In this Section we suppose that both assumptions A and B are fulfilled (see Eqs. (2.3) and (2.4)). In particular, the random potential is a.s. non-negative. Then, ∀ cube $\Lambda = \Lambda_L(u)$ and $\delta > 0$, the event

$$\Omega_\delta(\Lambda) = \{\omega : \exists\, x \in \Lambda \text{ such that } V(x;\omega) \le \delta\} \tag{2.91}$$

has a positive probability, vanishing as $\delta \to 0$:

$$0 < \mathbb{P}\{\, \Omega_\delta(\Lambda)\,\} \le \sum_{x\in\Lambda} \mathbb{P}\{\, V(x) \le \delta\,\} = |\Lambda|\, F(\delta). \tag{2.92}$$

Consequently, $\forall\, \omega \notin \Omega_\delta(\Lambda)$, the operator $H_\Lambda = H_\Lambda(\omega)$ satisfies

$$H_\Lambda > H^0_\Lambda + \delta,$$

so that for the eigenvalues $E_j = E_j(H_\Lambda)$ and $E^0_j = E_j(H^0_\Lambda)$ of H_Λ and H^0_Λ, respectively, we have

$$E_j > E^0_j + \delta,\ \ 1 \le j \le |\Lambda|.$$

Now consider the event $\Omega^{c}_{2\delta}(\Lambda) = \{\omega : \ V(x;\omega) > 2\delta \ \ \forall\ x \in \Lambda\}$. For any $\omega \in \Omega^c_{2\delta}$ and E with $E^0_0 \le E \le E^0_0 + \delta$, we have

$$\operatorname{dist}[E, \sigma(H_\Lambda)] \ge \delta,$$

so that the Combes–Thomas estimate (2.80) gives, $\forall\ x, y \in \Lambda$,

$$|G_\Lambda(x, y; E, \omega)| \le \frac{2}{\delta} \exp\left(-\frac{\delta}{5d}|x - y|\right).$$

Remark 2.3.2. The above elementary argument, or some modifications thereof, can be useful for the MSA in the case where $F(\delta) \downarrow 0$ fast enough as $\delta \downarrow 0$; e.g., $F(\delta) \le \delta^Q$, with a value $Q \gg 1$ large enough, to suite the needs of the MSA scheme. The latter condition greatly restricts the area of applications of this method, but is used in some models (not considered in this book) where the Lifshitz tails asymptotics is difficult to establish. However, it is clear that in a more general case where, e.g., $C_1\delta \le F(\delta) \le C_2\delta$, with $C_1, C_2 \in (0, +\infty)$, the inequality (2.92) provides a very small value of the "gap" δ, viz.: $\delta \ll L^{-d}$. This is insufficient for starting the MSA inductive procedure.

Our next theorem, Theorem 2.3.5, gives a more efficient estimate of the probability that the lowest eigenvalue $E_1\big(H_{\Lambda_L(u)}(\omega)\big)$ of the Hamiltonian $H_{\Lambda_L(u)}(\omega)$ lies below a certain limit.

Theorem 2.3.5. *There exists a constant* $C \in (0, \infty)$ *and arbitrarily large integers* $L > 0$ *such that* $\forall\, u \in \mathbb{Z}^d$*, the following bound holds true for the lowest eigenvalue* $E_1\left(H_{\Lambda_L(u)}(\omega)\right)$*:*

$$\mathbb{P}\left\{ E_1\left(H_{\Lambda_L(u)}(\omega)\right) \le L^{-1/2} \right\} \le \mathrm{e}^{-CL^{d/4}}. \tag{2.93}$$

Proof. The proof of Theorem 2.3.5 follows closely the argument developed in Refs. [113, 115, 158] and requires a combination of various techniques. In particular, we will use LSOs with Neumann boundary conditions.

While it can be shown that (2.93) holds true for *all* sufficiently large lengths $L \in \mathbb{N}$, for the purposes of the MSA it suffices to have just one such value, sufficiently large to satisfy various lower bounds. So, we will prove the assertion for the values L of the form

$$L = Kl + \frac{K-1}{2}, \quad K = 2l^5 + 1, \tag{2.94}$$

with sufficiently large $l \in \mathbb{N}$. Note that (2.94) implies that $2L + 1 = K(2l + 1)$, so that the cube $\Lambda_L(u)$ is partitioned into K^d pairwise disjoint cubes of the form $\Lambda_l(u^{(s)})$ where $u^{(s)} \in \mathbb{Z}^d$.

Evidently,

$$l \sim L^{1/6}, \quad K \sim L^{5/6}, \tag{2.95}$$

up to multiplicative constants.

In each of the K^d cubes $\Lambda^{(s)} := \Lambda_l(u^{(s)})$ forming the above partition, we consider an LSO $H^{\mathrm{N}}_{\Lambda^{(s)}}(\omega) = H^{0,\mathrm{N}}_{\Lambda^{(s)}} + V_{\Lambda^{(s)}}(\omega)$ where $H^{0,\mathrm{N}}_{\Lambda^{(s)}}$ is the discrete Neumann Laplacian in $\Lambda^{(s)}$, cf. [113]. Recall, our Hamiltonian

$H_{\Lambda_L(u)}(\omega)=H^0_{\Lambda_L(u)}+V_{\Lambda_L(u)}(\omega)$ contains the Dirichlet Laplacian $H^0_{\Lambda_L(u)}$; to stress this fact we will temporarily use the notation $H^{\rm D}_{\Lambda_L(u)}(\omega)$ for $H_{\Lambda_L(u)}(\omega)$.

A well-known fact (following from Eq. (2.13) and the discussion in Sect. 2.1.1) is that for any $\omega \in \Omega$,

$$H^{\rm D}_{\Lambda_L(u)}(\omega) \geq H^{\rm N}_{\Lambda_L(u)}(\omega) \geq \bigoplus_{s=1}^{K^d} H^{\rm N}_{\Lambda^{(s)}}(\omega) \tag{2.96}$$

in the sense of quadratic forms, so that the lowest eigenvalue $E_1\left(H^{\rm D}_{\Lambda_L(u)}(\omega)\right)$ admits the following lower bound:

$$E_1\left(H^{\rm D}_{\Lambda_L(u)}(\omega)\right) \geq \min_{1\leq s\leq K^d} E_1\left(H^{\rm N}_{\Lambda^{(s)}}(\omega)\right). \tag{2.97}$$

The rest of the proof focuses on properties of the lowest eigenvalue $E_1\left(H^{\rm N}_{\Lambda^{(s)}}(\omega)\right)$ and can be summarized as follows.

- First, we prove an analytic statement saying that if

$$|\Lambda^{(s)}|^{-1} \sum_{x\in\Lambda^{(s)}} V(x;\omega) \geq \eta,$$

with sufficiently small $\eta > 0$, then the first eigenvalue $E_1\left(H^{\rm N}_{\Lambda^{(s)}}(\omega)\right)$ admits the bound

$$E_1\left(H^{\rm N}_{\Lambda^{(s)}}(\omega)\right) \geq b_1\eta^2 l^{-2} \tag{2.98}$$

with some constant $b_1 \in (0,\infty)$.

- Next, we reproduce a probabilistic bound of the form

$$\mathbb{P}\left\{ E_1\left(H^{\rm N}_{\Lambda^{(s)}}(\omega)\right) < b\eta^2 l^{-2} \right\} \leq {\rm e}^{-Cl^d}. \tag{2.99}$$

Analytic Estimates

Here we follow the strategy of Ref. [158] (cf. Sect. 2.1 in [158]). An alternative approach, based on the so-called Temple inequality, is presented in [113] (for the lattice systems) and extended to more general graphs in [60].

The Neumann Laplacian $H^{0,\rm N}_{\Lambda^{(s)}}$ has the eigenvalues $E_j\left(H^{0,\rm N}_{\Lambda^{(s)}}\right)$, $1 \leq j \leq |\Lambda^{(s)}|$, which we will assume, as always, to be numbered in the increasing order. Moreover, $H^{0,\rm N}_{\Lambda^{(s)}}$ is non-negative and has the lowest eigenvalue $E_1\left(H^{0,\rm N}_{\Lambda^{(s)}}\right) = 0$; the corresponding normalized eigenfunction $\psi_1(\,\cdot\,, H^{0,\rm N}_{\Lambda^{(s)}})$ is constant:

$$\psi_1\left(x, H^{0,\mathrm{N}}_{\Lambda^{(s)}}\right) = |\Lambda^{(s)}|^{-1/2}\mathbf{1}_{\Lambda^{(s)}}(x).$$

The eigenvalues $E_j\left(H^{0,\mathrm{N}}_{\Lambda^{(s)}}\right)$ in $\Lambda^{(s)}$ are also well-known.[6] For instance,

$$E_2\left(H^{0,\mathrm{N}}_{\Lambda^{(s)}}\right) - E_1\left(H^{0,\mathrm{N}}_{\Lambda^{(s)}}\right) = \frac{\pi^2}{(2l+1)^2}. \tag{2.100}$$

In particular, $E_1\left(H^{0,\mathrm{N}}_{\Lambda^{(s)}}\right)$ is an isolated eigenvalue. Subsequently, one can analyze the eigenvalue $E_1\left(H^{\mathrm{N}}_{\Lambda^{(s)}}(\omega)\right)$ of $H^{\mathrm{N}}_{\Lambda^{(s)}}(\omega) = H^{0,\mathrm{N}}_{\Lambda^{(s)}} + V_{\Lambda^{(s)}}(\omega)$ as a perturbation of $E_1\left(H^{0,\mathrm{N}}_{\Lambda^{(s)}}\right)$. To this end, we consider a one-parameter family of operators $H^{0,\mathrm{N}}_{\Lambda^{(s)}} + tV_{\Lambda^{(s)}}(\omega)$, $t \in \mathbb{R}$, and their respective lowest eigenvalues $E_1^{(t,\Lambda^{(s)})}(\omega)$. The first-order term in the perturbation series for $E_1^{(t,\Lambda^{(s)})}(\omega)$ is not difficult to calculate explicitly:

$$\begin{aligned}\frac{\mathrm{d}}{\mathrm{d}t}E_1^{(t,\Lambda^{(s)})}(\omega)\bigg|_{t=0} &= \left\langle \psi_1(\cdot, H^{0,\mathrm{N}}_{\Lambda^{(s)}}), V_{\Lambda^{(s)}}(\omega)\psi_1(\cdot, H^{0,\mathrm{N}}_{\Lambda^{(s)}})\right\rangle \\ &= |\Lambda^{(s)}|^{-1}\sum_{x\in\Lambda^{(s)}} V(x;\omega) =: \overline{V(\cdot\,;\omega)}_{\Lambda^{(s)}}.\end{aligned}$$

(Here and below, we employ the notation $\overline{V(\cdot\,;\omega)}_{\Lambda^{(s)}}$ for the sample mean of $\{V(x;\omega), x \in \Lambda^{(s)}\}$.) Accordingly, one can represent, for small t,

$$E_1^{(t,\Lambda^{(s)})}(\omega) = t\,\overline{V(\cdot\,;\omega)}_{\Lambda^{(s)}} + O(t^2).$$

Next, we use Eq. (2.100) rendering that the function $t \mapsto E_1^{(t,\Lambda^{(s)})}$ admits an analytic continuation into a complex disk of radius $O(l^{-2})$. As a result, we arrive at the following assertion:

Lemma 2.3.6. *There exist constants $c_0, c_1 \in (0, +\infty)$ such that $\forall\ \omega \in \Omega$ and $t \in (0, c_0 l^{-2}] \subset [0,1]$, the eigenvalues $E_1^{(t,\Lambda^{(s)})}(\omega)$ and $E_1\left(H^{\mathrm{N}}_{\Lambda^{(s)}}(\omega)\right)$ in $\Lambda^{(s)}$ satisfy*

$$\left|E_1^{(t,\Lambda^{(s)})}(\omega) - t\,\overline{V(\cdot\,;\omega)}_{\Lambda^{(s)}}\right| \le c_1 l^2 t^2.$$

and

$$E_1\left(H^{\mathrm{N}}_{\Lambda^{(s)}}(\omega)\right)\left(\equiv E_1^{(1,\Lambda^{(s)})}(\omega) \ge E_1^{(t,\Lambda^{(s)})}(\omega)\right) \ge t\,\overline{V(\cdot\,;\omega)}_{\Lambda^{(s)}} - c_1 l^2 t^2,$$

[6] See an extension of this technique to more general graphs in [60].

or, equivalently,

$$\overline{V(\cdot\,;\omega)}_{\Lambda^{(s)}} \leq t^{-1} E_1\left(H^{\mathrm{N}}_{\Lambda^{(s)}}(\omega)\right) + c_1 l^2 t.$$

We wish to emphasize that in the above inequalities one can choose any value $0 < t \leq c_0 l^{-2}$. For example, pick a number $0 < \eta \leq 2c_1c_0$ and set

$$t = \frac{\eta}{2c_1 l^2}, \quad b_1 = \frac{1}{4c_1}.$$

If $E_1\left(H^{\mathrm{N}}_{\Lambda^{(s)}}(\omega)\right) \leq b_1\eta^2 l^{-2}$, then a straightforward calculation shows that

$$\overline{V(\cdot\,;\omega)}_{\Lambda^{(s)}} \leq \frac{1}{2}\eta + \frac{1}{2}\eta = \eta.$$

Corollary 2.3.7. *Consider a cube* $\Lambda^{(s)} = \Lambda_l(u^{(s)})$, $l \geq 1$. *There exist constants* $b_2 \in (0, \infty)$, $\eta_0 \in (0, 1)$ *such that for all* $\omega \in \Omega$ *and* $\eta \in (0, \eta_0)$, *if*

$$\overline{V(\cdot\,;\omega)}_{\Lambda^{(s)}} \geq \eta, \tag{2.101}$$

then the lowest eigenvalue $E_1\left(H^{\mathrm{N}}_{\Lambda^{(s)}}(\omega)\right)$ *admits the following bound:*

$$E_1\left(H^{\mathrm{N}}_{\Lambda^{(s)}}(\omega)\right) \geq b_2\eta^2 l^{-2}. \tag{2.102}$$

Probabilistic Estimates

Let us now analyze the inequality (2.101). Under our assumptions, the sample mean $\overline{V(\cdot\,;\omega)}_{\Lambda^{(s)}}$ converges to the expectation $\mathbb{E}[\,V(0;\,\cdot\,)\,]$ as $l \to \infty$. Since the distribution function F is supported by the positive half-line, we have $\mathbb{E}[\,V(0;\cdot)\,] > 0$. Hence, for any $\eta < \mathbb{E}[\,V(0;\,\cdot\,)\,]$,

$$\mathbb{P}\left\{\overline{V(\cdot\,;\omega)}_{\Lambda^{(s)}} \leq \eta\right\} \underset{l\to\infty}{\longrightarrow} 0.$$

The asymptotical behavior of the LHS, especially in the case of IID random variables V, is well-known; it is the main subject of the theory of large deviations. More precisely, for any positive $\eta < \mathbb{E}[\,V(0;\cdot)\,]$, as $|\Lambda^{(s)}|$ increases,

$$\mathbb{P}\left\{\overline{V(\cdot\,;\omega)}_{\Lambda^{(s)}} \leq \eta\right\} \sim \exp\left[-J(\eta)|\Lambda^{(s)}|\right] \tag{2.103}$$

where J is the so-called large deviation function (a non-negative convex function with $J(t)\big|_{t=\mathbb{E}[\,V(0;\cdot)\,]} = 0$). (In Eq. (2.104) below, we define $J(\cdot)$ in a slightly different way, for the sake of brevity of the argument.)

Therefore, as $l \to \infty$,

$$\mathbb{P}\left\{ \overline{V(\cdot\,;\omega)}_{\Lambda^{(s)}} \le b\eta^2 l^{-2} \right\} \sim \exp\left[-J(b\eta^2 l^{-2})\, |\Lambda^{(s)}| \right].$$

An elementary derivation of an upper bound for the probability

$$\mathbb{P}\left\{ \overline{V(\cdot\,;\omega)}_{\Lambda^{(s)}} \le \eta^2 l^{-2} \right\}$$

can be obtained by using the following argument.

Assume that the marginal distribution of the random potential $V(x;\omega)$ is such that $\mathbb{P}\{ V(x;\omega) = 0 \} < 1$. This property is required for our argument. For a non-negative random potential (cf. Assumption B) with continuous marginal distribution (Assumption A) it is automatically fulfilled.

Define the concave positive function

$$t \mapsto \varphi(t) = -\ln \mathbb{E}\left[\mathrm{e}^{-tV(0;\,\cdot\,)} \right], \quad t \ge 0,$$

and observe that $\varphi(t) > 0$ for all $t > 0$. Indeed, since, by assumption, $\mathbb{P}\{ V(x;\omega) = 0 \} < 1$, we have $\mathbb{E}\left[\mathrm{e}^{-tV(0;\,\cdot\,)} \right] < 1$.

Then, $\forall\, t, \epsilon > 0$,

$$\mathbb{P}\left\{ \overline{V(\cdot\,;\omega)}_{\Lambda^{(s)}} \le \epsilon \right\} = \mathbb{P}\left\{ \exp\left(-t \sum_{x\in\Lambda^{(s)}} V(x;\omega) \right) \ge \mathrm{e}^{-t\epsilon|\Lambda^{(s)}|} \right\}$$

(by Chebyshev's inequality)

$$\le \mathrm{e}^{t|\Lambda^{(s)}|\epsilon}\, \mathbb{E}\left[\exp\left(-t \sum_{x\in\Lambda^{(s)}} V(x;\,\cdot\,) \right) \right]$$

(since random variables $V(x;\omega)$ are IID)

$$= \mathrm{e}^{t|\Lambda^{(s)}|\epsilon} \left(\mathbb{E}\left[\mathrm{e}^{-tV(0;\,\cdot\,)} \right] \right)^{|\Lambda^{(s)}|}$$

$$= \exp\left[-|\Lambda^{(s)}|(\varphi(t) - t\epsilon) \right].$$

For $t = 1$, we obtain

$$\mathbb{P}\left\{ \overline{V(\cdot\,;\omega)}_{\Lambda^{(s)}} \le \epsilon \right\} \le \exp\left[-J(\epsilon)|\Lambda^{(s)}| \right]$$

where, for all $\epsilon < \varphi(1)$,

$$J(\epsilon) := \varphi(1) - \epsilon > 0. \tag{2.104}$$

For our purposes, it suffices to pick, e.g., $\epsilon = \varphi(1)/2 > 0$.

Thus, we come to the following conclusion:

Lemma 2.3.8. *There exist constants* $b_3, C_1 \in (0, \infty)$ *and* $\eta_0 \in (0, 1)$ *such that for any* $\eta \in (0, \eta_0)$ *and any cube* $\Lambda^{(s)}$ *figuring in* (2.96), (2.97),

$$\mathbb{P}\left\{ E_1\left(H^{\mathrm{N}}_{\Lambda^{(s)}}(\omega)\right) \le b_3\eta^2 l^{-2} \right\} \le \mathrm{e}^{-C_1 l^d}. \tag{2.105}$$

To finish the proof of Theorem 2.3.5, we now write:

$$\begin{aligned} \mathbb{P}&\left\{ E_1\left(H^{\mathrm{D}}_{\Lambda_L(u)}(\omega)\right) \le b_3\eta^2 l^{-2} \right\} \\ &\le \mathbb{P}\left\{ E_1\left(H^{\mathrm{N}}_{\Lambda^{(s)}}(\omega)\right) \le b_3\eta^2 l^{-2} \text{ for some } s \in [1, K^d] \right\} \\ &\le K^d \mathrm{e}^{-C_1 l^d} \\ &\le \mathrm{e}^{-CL^{d/6}}, \end{aligned} \tag{2.106}$$

provided that $L \sim l^6$ is large enough. Now the assertion of Theorem 2.3.5 follows from the inequality (cf. (2.95))

$$b_3\eta^2 l^{-2} > \mathrm{Const}\,(b, \eta) L^{-1/3} > L^{-1/4}$$

again holding for sufficiently large l; in particular, there are arbitrarily large values L for which one has

$$\mathbb{P}\left\{ E_1\left(H^{\mathrm{D}}_{\Lambda_L(u)}(\omega)\right) \le L^{-1/4} \right\} \le \mathrm{e}^{-CL^{d/6}}. \qquad \square$$

As was stressed in Remark 2.1.10 (cf. Sect. 2.1.6), the initial length scale bound (DS.I, m, p, L_0) for weakly disordered systems, with $I = [0, \eta)$ and small $\eta > 0$, requires L_0 to be large enough, depending upon the required value of the exponent $p > 0$: the larger p, the larger must be L_0. Apart from the probabilistic arguments (large deviations analysis), this is explained by the fact that the lowest eigenvalues here correspond to slowly varying eigenfunctions. However, the case of random potentials unbounded from below (e.g., Gaussian) is different in several aspects:

- The proof of (DS.I, m, p, L_0) can be made much simpler, as in the case of large disorder;
- The value of L_0 does not have to be large;
- The value of the "mass" $m > 0$ can be made arbitrarily large for the energy interval $(-\infty, E^*)$ with $E^* = -\eta$ and sufficiently large $\eta \ge \eta^*(m, L_0)$.

The main idea (which we only sketch here) is as follows. Given a cube $\Lambda_{L_0}(u)$ and arbitrary numbers $m > 0$, $p > 0$, the Combes–Thomas estimate implies (DS.I, m, p, L_0), provided that

$$\mathbb{P}\left\{ \exists\, x \in \Lambda_{L_0}(u) : V(x) \le E^* + 4d\,\mathrm{e}^{2m} \right\} \le L_0^{-2p}. \tag{2.107}$$

The LHS of Eq. (2.107) obviously tends to 0 as $E^* \to -\infty$, no matter how large is $m > 0$, and eventually becomes smaller than L_0^{-2p}, no matter how large is $p > 0$. The initial scale length L_0 is to be kept fixed, but it does not have to be large.

A similar phenomenon is observed when the random potential is bounded from below by $E_* > -\infty$, but its marginal probability distribution function F decays sufficiently fast near the lower edge E_*. It is used in situations where the Lifshitz-style asymptotics is difficult to prove.

2.4 Decay of the Green's Functions, II: Fixed-Energy Analysis

The aim of this section is to introduce a machinery that will enable us to use iterations of geometric resolvent inequalities in a form applicable to multi-particle systems (considered in Chap. 4).

We will employ the following notation: given a finite set $\mathbb{O}$ and a function $f : \mathbb{O} \to \mathbb{C}$, we set:

$$\mathrm{M}(f, \mathbb{O}) := \max_{x \in \mathbb{O}} |f(x)|.$$

2.4.1 Lattice Subharmonicity

Recall that we introduced in Sect. 2.1.1 (cf. Remark 2.1.3) the notion of connectedness for subsets of a graph, e.g. of the integer lattice $\mathbb{Z}^d$ (where the bonds are formed by pairs of nearest neighbors).

The importance of the condition $\Lambda_L(u) \subsetneq \mathbb{O}$ figuring in Definition 2.4.1 below (which might look artificial) is explained in Example 2.1.

Definition 2.4.1. *Given $q \in (0, 1)$, two integers $L \geq \ell \geq 0$ and a finite connected subset $\mathbb{O} \subset \mathbb{Z}^d$, we say that a function $f : \mathbb{O} \to \mathbb{R}_+$ is (ℓ, q)-subharmonic in a cube $\Lambda_L(u) \subsetneq \mathbb{O}$ if for any $\Lambda_\ell(x) \subset \Lambda_L(u)$, the following inequality holds true:*

$$f(x) \leq q\mathrm{M}(f, \Lambda_\ell(x)). \tag{2.108}$$

Theorem 2.4.1. *Let be given two integers $L \geq \ell \geq 0$, a finite connected subset $\Lambda \subset \mathbb{Z}^d$, $\Lambda \supset \Lambda_{L+1}(x)$, and a function $f : \Lambda \to \mathbb{R}_+$ which is (ℓ, q)-subharmonic in $\Lambda_L(x)$. Then*

$$f(x) \leq q^{\lfloor \frac{L+1}{\ell+1} \rfloor}\mathrm{M}(f, \Lambda_L(x)) \leq q^{\frac{L-\ell}{\ell+1}}\mathrm{M}(f, \Lambda_{L+1}(x)). \tag{2.109}$$

Proof. Since $L \geq \ell \geq 0$, the quantity $\frac{L+1}{\ell+1} \geq 1$ is well-defined, so we can set $n+1 := \lfloor \frac{L+1}{\ell+1} \rfloor$, $n \geq 0$.

Set $\tilde{\Lambda}_j := \Lambda_{j(\ell+1)}(x)$, $0 \leq j \leq n$, and note that $\tilde{\Lambda}_{n+1} \subset \Lambda_{L+1}(x)$, since $(n+1)(\ell+1) \leq L+1$. Further, if $y \in \tilde{\Lambda}_j$ with $0 \leq j < n$ and $z \in \Lambda_{\ell+1}(y)$, then, by the triangle inequality, $z \in \tilde{\Lambda}_{j+1}$. Consider the monotone nondecreasing function $h : [0, L+1] \cap \mathbb{N} \to \mathbb{R}_+$ defined by $h(r) = \mathrm{M}(f, \tilde{\Lambda}_r(x))$. Using the (ℓ, q)-subharmonicity of f, we obtain, as long as $j \leq n$,

$$h\big(j(\ell+1)\big) \leq q \max_{y \in \tilde{\Lambda}_j} \max_{z \in \Lambda_{\ell+1}(y)} f(z) \leq q h\big((j+1)(\ell+1)\big),$$

in particular,

$$h\big(n(\ell+1)\big) \leq q h(L+1) \leq q\mathrm{M}(f, \Lambda_{L+1}(x)).$$

By using a recursion in j from n to 0 (in n steps), we obtain the bounds in Eq. (2.109), since $f(x) = h(0)$. □

Example 2.1. Take $d = 1$, $\Lambda = [-R, R] \cap \mathbb{Z}^1$, $R = L + 1 \geq 1$, $\ell = 0$, and $q \in (0, 1)$. The function

$$\begin{aligned} f : \Lambda &\to \mathbb{R}_+ \\ x &\mapsto q^{\operatorname{dist}\left(x, \{-R, R\}\right)} \end{aligned}$$

is (ℓ, q)-subharmonic in $\Lambda_L(0)$ and bounded by 1. Indeed, observe first that

$$f(0) = q f(1) = q f(-1).$$

Since the function f is even, it remains to check its subharmonicity for $1 \leq x \leq L$:

$$\begin{aligned} f(x) = q^{\operatorname{dist}\left(x, \{-R, R\}\right)} &= q^{R-x} = q \cdot q^{R-(x+1)} = q \max\left\{q^{R-(x-1)}, q^{R-(x+1)}\right\} \\ &= q \max_{y:\, |y-x| = \ell+1} f(y). \end{aligned}$$

Note also that

$$f(0) = q^{L+1} = q^{\frac{L+1}{\ell+1}},$$

which is precisely the bound given by Theorem 2.4.1. This shows that the bound (2.109) is sharp.

It is also worth noticing that the above bound cannot be extended to the exterior point $x = L + 1$ (or $x = -L - 1$), since, with $\ell = 0$,

$$1 = f(L+1) > \max_{y \in \Lambda:\, |y-(L+1)| = \ell+1} f(y) = f(L) = q.$$

This is not accidental. Indeed, if the inequality (2.108) were fulfilled for all points $x \in \Lambda$, then one would have

$$\mathrm{M}(f, \Lambda) = f(x) \le q \max_{x \in \Lambda} \max_{|y-x|=\ell+1} f(y) \le q \mathrm{M}(f, \Lambda)$$

with $0 < q < 1$, which implies $\mathrm{M}(f, \Lambda) = 0$, hence $f \equiv 0$. Therefore, such an extended condition of lattice subharmonicity would be useless for applications to Green's functions and eigenfunctions.

Recall that we have introduced the notion of an (E, m)-non-singular ((E, m)-NS) cube in Definition 2.1.1 (cf. Eq. (2.19) in Sect. 2.1.3), and the notion of an E-non-resonant (E-NR) cube in Definition 2.2.3 (cf. Eq. (2.47) in Sect. 2.2.3).

Theorem 2.4.2. *Let $\Lambda \subset \mathbb{Z}^d$ be a finite connected subset, and*

$$\begin{aligned} f : \Lambda \times \Lambda &\to \mathbb{R}_+ \\ (x, y) &\mapsto f(x, y) \end{aligned}$$

a function which is separately (ℓ, q)-subharmonic in $x \in \Lambda_{r'}(u') \subsetneq \Lambda$ and in $y \in \Lambda_{r''}(u'') \subsetneq \Lambda$, with $r', r'' \ge \ell \ge 0$ and $|u' - u''| \ge r' + r'' + 2$. Then

$$f(u', u'') \le q^{\left\lfloor \frac{r'+1}{\ell+1} \right\rfloor + \left\lfloor \frac{r''+1}{\ell+1} \right\rfloor} \mathrm{M}(f, \Lambda \times \Lambda) \le q^{\frac{r'+r''-2\ell}{\ell+1}} \mathrm{M}(f, \Lambda \times \Lambda). \qquad (2.110)$$

Proof. For each $y'' \in \Lambda_{r''+1}(u'')$ define the function $f_{y''} : x' \mapsto f(x', y'')$ in Λ. By assumption, it is (ℓ, q)-subharmonic in $\Lambda_{r'}(u')$, so Theorem 2.4.1 implies,

$$\forall\, y'' \in \Lambda_{r''+1}(u'') \quad f(u', y'') = f_{y''}(u') \le q^{\frac{r'-\ell}{\ell+1}} \mathrm{M}(f, \Lambda \times \Lambda).$$

Consider now another function, $\tilde{f}_{u'} : y'' \mapsto f(u', y'')$, $y'' \in \Lambda$. It is (ℓ, q)-subharmonic in $\Lambda_{r''}(u'')$, by hypothesis. The above inequality reads as

$$\mathrm{M}\big(\tilde{f}_{u'}, \Lambda_{r''+1}(u'')\big) \le q^{\frac{r'-\ell}{\ell+1}} \mathrm{M}(f, \Lambda \times \Lambda),$$

so another application of Theorem 2.4.1 proves the claim:

$$f(u', u'') = \tilde{f}_{u'}(u'') \le q^{\frac{r'-\ell}{\ell+1}} \mathrm{M}\big(\tilde{f}_{u'}, \Lambda_{r''+1}(u'')\big) \le q^{\frac{r'+r''-2\ell}{\ell+1}} \mathrm{M}(f, \Lambda \times \Lambda).$$

□

Theorem 2.4.3. *Let be given a number $m > 0$ and integers $L \ge l \ge 0$. Consider a finite connected subset $\Lambda \subset \mathbb{Z}^d$ and a cube $\Lambda_L(u) \subsetneq \Lambda$.*

(A) *Suppose that for some $E \in \mathbb{R} \setminus \sigma(H_\Lambda)$, all cubes of radius ℓ in $\Lambda_L(u)$ are (E, m)-NS. Then for any $y \in \Lambda \setminus \Lambda_L(u)$ the function*

$$f_y : x \mapsto |G_\Lambda(x,y;E)| \tag{2.111}$$

is (ℓ, q)*-subharmonic in* $\Lambda_L(u)$ *with*

$$q = \mathrm{e}^{-\gamma(m,l)l}. \tag{2.112}$$

(B) *Let* ψ_j *be an eigenfunction of* H_Λ *with eigenvalue* E_j*. Suppose that all cubes of radius* ℓ *in* $\Lambda_L(u)$ *are* (E_j, m)*-NS. Then for any* $y \in \Lambda \setminus \Lambda_L(u)$ *the function*

$$f_{j,y} : x \mapsto |\psi_j(x)\psi_j(y)| \tag{2.113}$$

is (ℓ, q)*-subharmonic in* $\Lambda_L(u)$ *with* $q = \mathrm{e}^{-\gamma(m,l)l}$*.*

Proof.

(A) Fix a cube $\Lambda_\ell(x) \subset \Lambda_L(u)$ and a point $y \in \Lambda \setminus \Lambda_L(u)$. Let f_y be defined by (2.111). By the GRI (2.64) combined with the definition of an (E, m)-NS cube (cf. Eq. (2.19)), one has

$$\begin{aligned} f_y(x) &= |G_\Lambda(x,y;E)| \\ &\le |\partial\Lambda_\ell(x)| \max_{(z,z')\in\partial\Lambda_\ell(x)} |G_{\Lambda_\ell(x)}(x,z;E)| \cdot |G_\Lambda(z',y;E)| \\ &\le |\partial\Lambda_\ell(x)| \cdot |\partial\Lambda_\ell(x)|^{-1} \mathrm{e}^{-\gamma(m,\ell)\ell} f(z'). \end{aligned}$$

This yields the (ℓ, q)-subharmonicity of the function f_y in $\Lambda_L(u)$.

(B) The proof is similar and uses the GRI for eigenfunctions; cf. Eqs. (2.67). □

Remark 2.4.1. Clearly, the factor $|\psi_j(y)|$ in the definition (2.113) of the function $f_{j,y}$ is constant. The reason why we introduce it is that it is the value of the product $|\psi_j(x)\psi_j(y)|$ which is crucial for the dynamical localization phenomenon which we discuss below.

Remark 2.4.2. By symmetry of the Green's functions, the assertion (A) of Theorem 2.4.3 admits a direct analog for the function $y \mapsto |G_\Lambda(x,y;E)|$.

2.4.2 Tunneling and Decay of Green's Functions

Recall that the sequence $\{L_k, k \ge 0\}$ was defined inductively: $L_{k+1} = \lfloor L_k^\alpha \rfloor$.

Definition 2.2. *Given two numbers* $E \in \mathbb{R}$, $m > 0$ *and an integer* $k \ge 0$*, a cube* $\Lambda_{L_{k+1}}(u)$ *is called* (E, m)*-tunneling* ((E, m)*-T*) *if it contains at least two disjoint* (E, m)*-S cubes of radius* L_k*, and* (E, m)*-non-tunneling* ((E, m)*-NT*)*, otherwise.*

Definition 2.3. *Given a number $m > 0$, an interval $I \subset \mathbb{R}$ and an integer $k \geq 0$, a cube $\Lambda_{L_{k+1}}(u)$ is called (I,m)-tunneling ((I,m)-T) if it is (E,m)-T for some $E \in I$, and (I,m)-non-tunneling ((I,m)-NT), otherwise.*

Lemma 2.4.4. *If a cube $\Lambda_{L_{k+1}}(x)$ is E-NR and (E,m)-NT, for some $E \in \mathbb{R}$ and $m \geq 1$, then it is (E,m)-NS, provided that L_0 is large enough.*[7]

Proof. Fix any $y \in \partial^- \Lambda_{L_{k+1}}(x)$ and $x' \in \Lambda_{L_k}(x)$. Since $\Lambda_{L_{k+1}}(u)$ is (E,m)-NT, there is a cube $\Lambda_{2L_k}(w)$ such that any L_k-cube disjoint from $\Lambda_{2L_k}(w)$ is (E,m)-NS. By the triangle inequality, there are integers $r', r'' \geq 0$ such that $r' + r'' \geq L_{k+1} - 2L_k - 2$, the cubes $\Lambda_{r'}(x')$ and $\Lambda_{r''}(y)$ are disjoint from each other and from $\Lambda_{2L_k}(w)$, so any L_k-cube inside $\Lambda_{r'}(x')$ and inside $\Lambda_{r''}(y)$ is (E,m)-NS.

Assume first that $r' \geq L_k$ and $r'' \geq L_k$ (otherwise, one of the points x', y is covered by $\Lambda_{2L_k}(w)$, so one of the radii $r', r'' \geq L_{k+1} - 3L_k$ and the same argument as below applies). By assertion (A) of Theorem 2.4.3 (cf. also Remark 2.4.2), the function $f : (v,z) \mapsto |G_{\Lambda_{L_{k+1}}}(v,z;E)|$ is (L_k, q)-subharmonic in $v \in \Lambda_{r'}(x')$ and in $z \in \Lambda_{r''}(y)$, with $q \leq e^{-m(1+L_k^{-\tau})L_k}$. Note that $|y - x'| \geq L_{k+1} - L_k$, so by Theorem 2.4.2, one can write, with the convention $-\ln 0 = +\infty$:

$$
\begin{aligned}
-\ln f(x',y) &\geq -\ln \left\{ \left(e^{-m(1+L_k^{-\tau})L_k} \right)^{\frac{(L_{k+1}-L_k)-2L_k-2}{L_k+1}} e^{L_k^\beta} \right\} \\
&\geq mL_{k+1} \left(\left(1 + L_k^{-\tau}\right) \frac{L_k}{L_k+1} \left(1 - 4L_k^{-\alpha+1}\right) - \frac{L_k^\beta}{mL_{k+1}} \right)
\end{aligned}
$$

with $m \geq 1$

$$
\geq mL_{k+1} \left(\left(1 + L_k^{-1/8}\right) \frac{1}{1 + L_k^{-1}} \left(1 - 4L_k^{1-3/2}\right) - \frac{L_k^{1/2}}{L_k^{3/2}} \right)
$$

using $\frac{1-4L_k^{-1/2}}{1+L_k^{-1}} \geq 1 - 5L_k^{-1/2}$, with $L_k \geq L_0 \geq 1$

$$
\begin{aligned}
&\geq mL_{k+1} \left((1 + L_k^{-1/8})(1 - 5L_k^{-1/2}) - L_k^{-1} \right) \\
&= mL_{k+1} \left(1 + L_k^{-1/8} - 5L_k^{-1/2} - 5L_k^{-5/8} - L_k^{-1} \right)
\end{aligned}
$$

---

[7] See Remark 2.4.3 below

with $L_k^{-1/8} \geq 22L_k^{-1/2}$, which holds true, if $L_k \geq L_0$ and L_0 is large enough,

$$\geq mL_{k+1}\left(1 + \tfrac{1}{2}L_k^{-1/8}\right) \geq \gamma(m, L_{k+1})\, L_{k+1}. \qquad \square$$

Remark 2.4.3. The argument used in the above paragraph illustrates the role of the assumptions of the form "L_0 is large enough", often used in this book as well as in a number of papers on the multi-scale analysis. The condition $L^{-1/8} \geq 22L^{-1/2}$ is equivalent to $L^{3/8} \geq 22$, so if $L_0 \geq (22)^{8/3}$, then for any $k \geq 0$, we also have $L_k \geq L_0 \geq (22)^{8/3}$, hence $L_k^{-1/8} \geq 22L_k^{-1/2}$. It is important that the required assumption on L_0 *does not depend upon* k, although the condition $L_k^{-1/8} \geq 22L_k^{-1/2}$ is formulated in terms of L_k with an arbitrary $k \geq 0$. As explained in Remark 2.1.5, we will often use asymptotic relations of the form $L^{-a} \ll L^{-b} \ll 1$ for $0 < a < b$, meaning that for any given $0 < a < b$ and any $C \in (0, +\infty)$ there is some L^* such that for all $L_0 \geq L^*$ and $\{L_k\}_{k \geq 0}$ defined by the recursion $L_{k+1} = \lfloor L_k^{\alpha} \rfloor$, one has $CL_k^{-a} \leq L^{-b} < 1$, for all $k \geq 0$.

2.4.3 Probability of Tunneling and Scale Induction

Lemma 2.4.5. *Suppose that for some $E \in \mathbb{R}$, $m > 0$, $p > 0$, $k \in \mathbb{N}$ and any cube $\Lambda_{L_k}(x) \subset \mathbb{Z}^d$ the following inequality holds true:*

$$\mathbb{P}\{\, \Lambda_{L_k}(x) \text{ is } (E,m)\text{-}S \} \leq L_k^{-p}. \tag{2.114}$$

Then for any cube $\Lambda_{L_{k+1}}(u) \subset \mathbb{Z}^d$ and the same value of E,

$$\mathbb{P}\left\{\, \Lambda_{L_{k+1}}(u) \text{ is } (E,m)\text{-}T \right\} \leq \frac{3^{2d}}{2} L_{k+1}^{-\frac{2p}{\alpha}+2d}. \tag{2.115}$$

In addition, if

$$p > \frac{2\alpha d}{2-\alpha} \tag{2.116}$$

and L_0 is large enough, then the RHS of (2.115) *is bounded by $\frac{1}{2}L_{k+1}^{-p}$.*

Proof. Take a pair of disjoint cubes $\Lambda_{L_k}(x), \Lambda_{L_k}(y) \subset \Lambda_{L_{k+1}}(u)$. Owing to the independence of the operators $H_{\Lambda_{L_k}(x)}$, $H_{\Lambda_{L_k}(y)}$, the condition (2.114) implies that

$$\mathbb{P}\{\, \Lambda_{L_k}(x) \text{ and } \Lambda_{L_k}(y) \text{ are } (E,m)\text{-S} \,\} \leq L_k^{-2p}.$$

Next, the number of pairs x, y in $\Lambda_{L_{k+1}}(u)$ is bounded by $\frac{(2L_{k+1}+1)^{2d}}{2}$ which is $\leq \frac{3^{2d}}{2} L_{k+1}^{2d}$. This yields the first assertion of the lemma.

To prove the second assertion, note that the inequality

$$\frac{2p}{\alpha} - 2d > p$$

is equivalent to $p > 2\alpha d/(2-\alpha)$. Thus, under the assumption (2.116), and for L_0 large enough (hence, with L_k large enough), one has

$$\frac{3^{2d}}{2} L_{k+1}^{-\frac{2p}{\alpha}+2d} = L_{k+1}^{-p} \cdot \left(\frac{3^{2d}}{2} L_{k+1}^{-\left(\frac{2p}{\alpha}-2d-p\right)} \right) < \frac{1}{2} L_{k+1}^{-p},$$

since $c := \frac{2p}{\alpha} - 2d - p > 0$, so it suffices that $L_0^{-c} > 3^{2d}/2$. □

Remark 2.4.4. The above calculation implies that the value of the exponent $p > 0$ can actually be improved at length scale L_{k+1}. Indeed, assume, instead of (2.114), that

$$\mathbb{P}\{ \Lambda_{L_k}(x) \text{ is } (E,m)\text{-S} \} \le L_k^{-p(1+\theta)^k}. \tag{2.117}$$

Then, arguing as in the proof of Lemma 2.4.5, we conclude that

$$\mathbb{P}\{ \Lambda_{L_{k+1}}(u) \text{ is } (E,m)\text{-T} \} \le \frac{3^{2d}}{2} L_{k+1}^{-\frac{2p(1+\theta)^k}{\alpha}+2d}.$$

For the last expression to be smaller than $\frac{1}{2} L_{k+1}^{-p(1+\theta)^{k+1}}$, i.e., in order to guarantee a bound of the form

$$\mathbb{P}\{ \Lambda_{L_{k+1}}(x) \text{ is } (E,m)\text{-T} \} \le \frac{1}{2} L_{k+1}^{-p(1+\theta)^{k+1}}, \tag{2.118}$$

it suffices that L_0 is large enough and

$$\frac{2p(1+\theta)^k}{\alpha} - 2d > p(1+\theta)^{k+1} \tag{2.119}$$

or, equivalently,

$$\frac{2-\alpha}{\alpha} - \frac{2d}{p(1+\theta)^k} > \theta. \tag{2.120}$$

For any $\theta > 0$, one has $\frac{2d}{p(1+\theta)^k} < \frac{2d}{p}$, so the LHS of Eq. (2.120) is bigger than $\frac{2-\alpha}{\alpha} - \frac{2d}{p} =: c' > 0$, since, by assumption, $p > \frac{2\alpha d}{2-\alpha}$. Therefore, it suffices to pick any number $\theta \in (0, c')$ and require that $L_0 \ge 3^{2d/c'}$.

Summarizing, the following conditions on $p > 0$ and $\theta > 0$ are sufficient for the inequality (2.119) (hence, (2.118)) to hold:

$$\begin{cases} p > \dfrac{2\alpha d}{2-\alpha}, \\ 0 < \theta < \dfrac{2-\alpha}{\alpha} - \dfrac{2d}{p}. \end{cases} \tag{2.121}$$

Now we can make the scale induction step.

Lemma 2.4.6. *Let be given $p > 0$ and $\theta > 0$ satisfying* (2.121) *Suppose that the property $(SS.I, m, p_k, L_k)$ (see* (2.24)*) holds true for some $k \geq 0$ with $p_k = p(1+\theta)^k$, and the EVC* (2.52) *(cf. Corollary 2.2.4) is established for all $L \geq L_0$. Then for all L_0 large enough, the property $(SS.I, m, p_{k+1}, L_{k+1})$ also holds true:*

$$\forall\, u \in \mathbb{Z}^d \quad \mathbb{P}\left\{ \Lambda_{L_{k+1}}(u) \text{ is } (E,m)\text{-}S \right\} \leq L_{k+1}^{-p(1+\theta)^{k+1}}. \tag{2.122}$$

Proof. Consider a cube $\Lambda_{k+1}(u)$. Introduce the following events:

$$\begin{aligned} \mathcal{S}_{k+1}(u) &= \{\Lambda_{k+1}(u) \text{ is } (E,m)\text{-S}\}, \\ \mathcal{R}_{k+1}(u) &= \{\Lambda_{k+1}(u) \text{ is } E\text{-R}\}, \\ \mathcal{T}_{k+1}(u) &= \{\Lambda_{k+1}(u) \text{ is } (E,m)\text{-T}\}. \end{aligned}$$

It follows from Lemma 2.4.4 that

$$\mathcal{S}_{k+1}(x,y) \subset \mathcal{R}_{k+1}(u) \cup \mathcal{T}_{k+1}(u).$$

Therefore, using Remark 2.4.4 we get

$$\begin{aligned} \mathbb{P}\{\mathcal{S}_{k+1}(u)\} &\leq \mathbb{P}\{\mathcal{R}_{k+1}(x,y)\} + \mathbb{P}\{\mathcal{T}_{k+1}(u)\} \\ &\leq \mathrm{e}^{-L_{k+1}^{\beta}} + \frac{1}{2} L_{k+1}^{-p(1+\theta)^{k+1}} \\ &< L_{k+1}^{-p(1+\theta)^{k+1}}, \end{aligned}$$

provided that L_0 is large enough. □

By induction, we come to the following important result.

Theorem 2.4.7. *Suppose that the property $(SS.I, m, p_0, L_0)$ holds true with $p_0 = p > \frac{2\alpha d}{2-\alpha}$, and the EVC* (2.52) *is established for all $L \geq L_0$. Then for any $\theta > 0$ satisfying* (2.121). *the property $(SS.I, m, p_k, L_k)$ is satisfied for all $k \in \mathbb{N}$, i.e., for any $E \in I$,*

$$\mathbb{P}\{\ \Lambda_{L_k}(u) \text{ is } (E,m)\text{-S}\ \} \leq L_k^{-p(1+\theta)^k}. \tag{2.123}$$

In particular, if L_0 is large enough, the bounds (2.123) *hold true for all $k \geq 0$, in each of the following cases:*

1. *The amplitude $|g|$ of the random potential field $gV(\cdot;\omega)$ is large enough, in which case one can take any interval I.*
2. *The energy E is sufficiently close to the lower edge of the spectrum,*[8] *i.e., if $I = [E_*, E_* + \eta]$ with $\eta > 0$ small enough and*

$$E_* \overset{\text{a.s.}}{:=} \inf \sigma(H(\omega)) \overset{\text{a.s.}}{=} \inf\{\lambda : F_V(\lambda) > 0\}. \tag{2.124}$$

In other words, an exponential decay of the Green's functions $G_{\Lambda_{L_k}(x)}(x, y; E)$ in arbitrarily large cubes $\Lambda_{L_k}(x)$, with respect to $|x - y| \gg 1$, occurs with high probability (decaying faster than any power-law L_k^{-p}, $p > 0$). In physics, this is considered as one of the manifestations of the Anderson localization phenomenon. In particular, this implies that the conductivity (expressed with the help of the so-called Kubo formula) vanishes; cf. the discussion in [95].

Mathematically speaking, Theorem 2.4.7 implies that the random operator $H(\omega)$ in $\ell^2(\mathbb{Z}^d)$ has a.s. no absolutely continuous spectrum in any interval $I \subset \mathbb{R}$ such that the first hypothesis of Theorem 2.4.7 holds true for every $E \in I$, as we show in the next subsection. However, to prove that the spectrum of $H(\omega)$ is a.s. pure point in an interval $I \subset \mathbb{R}$, we will need additional arguments; see Sect. 2.5.

2.4.4 Absence of A.C. Spectrum

The argument we present in this Section is due to Martinelli and Scoppola [132].

Theorem 2.4.8. *Let $I \subset \mathbb{R}$ be a bounded interval and assume that for a sequence of integers $\{L_k, k \in \mathbb{N}\}$ growing to $+\infty$ and for any $E \in I$ the following property holds true:*

$$\sum_k \mathbb{P}\{\, \Lambda_{L_k}(0) \text{ is } (E, m)\text{-S} \,\} < \infty.$$

Then, with probability one, $\sigma_{a.c.}(H(\omega)) \cap I = \varnothing$.

Proof. It is convenient to assume that $|I| = 1$, so the interval I with the Lebesgue measure mes is a probability space; otherwise, one can simply pass to the normalized length. For each $E \in I$, let Ω_E be the set of samples of the potential $V(\cdot;\omega)$ such that for some $k_0 = k_0(\omega) < \infty$ and all $k \geq k_0$, the cube $\Lambda_{L_k}(0)$ is (E, m)-NS. By the Borel–Cantelli lemma, $\mathbb{P}\{\,\Omega_E\,\} = 1$, owing to the assumption of the theorem. For each $\omega \in \Omega_E$, the equation $H(\omega)\psi = E\psi$ has no nontrivial

[8] The proof of the a.s.-equality in (2.124) can be found, e.g., in [113]; cf. Theorem 3.9 from [113].

polynomially bounded solution. Indeed, assume the opposite and let A be the degree of the bounding polynomial for a nontrivial solution ψ, and $\hat{x}$ be a point where $\psi(\hat{x}) \neq 0$.

Then there exists $k_1 \geq k_0$ such that $\hat{x} \in \Lambda_{L_{k-1}}(0)$, for all $k \geq k_1$. Then, owing to the (E, m)-NS property of $\Lambda_{L_k}(0)$, with $k \geq k_1 \geq k_0$,

$$|\psi(\hat{x})| \leq \mathrm{e}^{-mL_k} \cdot \max_{|y|=L_k+1} |\psi(y)| \leq O(L_k^A)\mathrm{e}^{-mL_k} \underset{k\to\infty}{\longrightarrow} 0,$$

in contradiction with the assumption that $\psi(\hat{x}) \neq 0$.

Now consider the probability space $(\Omega \times I, \mathbb{P} \times \mathrm{d}E)$. Set

$$\tilde{\Omega} = \{(\omega, E) \in \Omega \times I : \omega \in \Omega_E\}.$$

This set is $\mathbb{P} \times \mathrm{d}E$-measurable, since

$$\tilde{\Omega} = \bigcup_{n=1}^{\infty} \bigcap_{k=n}^{\infty} \{(\omega, E) : \Lambda_{L_k}(0) \text{ is } (E, m)\text{-NS}\}$$

and the sets in the RHS are defined in terms of measurable functions $G_{\Lambda_{L_k}}(\cdot, \cdot; E; \omega)$. For each $E \in I$, $\mathbb{P}\{\, \Omega_E \,\} = 1$, therefore, by the Fubini theorem,

$$1 = \int_I \mathrm{d}E \int_\Omega \mathrm{d}\mathbb{P}(\omega)\, \mathbf{1}_{\tilde{\Omega}}(\omega, E) = \int_\Omega \mathrm{d}\mathbb{P}(\omega) \int_I \mathrm{d}E\, \mathbf{1}_{\tilde{\Omega}}(\omega, E).$$

Consequently, with probability 1, for Lebesgue-a.e. $E \in I$, the operator $H(\omega)$ admits no polynomially bounded generalized eigenfunction with (generalized) eigenvalue E. On the other hand, for a.e. $E \in \mathbb{R}$, relative to any given spectral measure of $H(\omega)$, there exists a polynomially bounded generalized eigenfunction with (generalized) eigenvalue E (cf. [53]). As a result, the spectrum of $H(\omega)$ in I has a.s. zero Lebesgue measure. Hence, $\sigma_{a.c.}\big(H(\omega)\big) \cap I$ is empty, with probability one. □

A stronger result was obtained later by Simon and Wolff [153] who proved that, for a large class of random operators including the single-particle random LSO $H(\omega)$, a.s. square-summability of the Green's functions *at every fixed energy* E in an interval $I \subset \mathbb{R}$ implies that the operator $H(\omega)$ has a.s. pure point spectrum in I. Such a strategy of spectral reduction [to a fixed-energy localization analysis] has been employed in a number of works (cf., e.g., [7, 156, 162]). Different variants of spectral reduction have been developed, e.g., in [47, 99, 100].

In the next section, following [61], we establish a quantitative analog of the Simon–Wolff argument which will allow us to infer from the fixed-energy MSA bounds the strong dynamical (and spectral) localization. In Chap. 4, this approach will be adapted to multi-particle LSOs for which no analog of the Simon–Wolff argument was proved earlier (prior to a recent work [62]).

2.5 From Fixed-Energy to Variable-Energy Analysis

2.5.1 Eigenfunction Correlators in a Bounded Energy Interval

Throughout this subsection we will assume that the length of the interval I is finite; this assumption is sufficient for the application to the proof of strong dynamical localization for the random operator $H(\omega)$ with a.s. bounded random potential $V(\cdot;\omega)$.

Given an integer $L \geq 0$ and $x \in \mathbb{Z}^d$, set for brevity $\Lambda_L = \Lambda_L(x)$ and define the function

$$E \mapsto \mathrm{M}_x(E) = \mathrm{M}_{x,L}(E,\omega) = \max_{v \in \partial^- \Lambda_L} |G_{\Lambda_L}(x, v; E, \omega)|. \tag{2.125}$$

We also make in this section a stronger assumption on the value of the parameter p_0: given the Hölder exponent $s \in (0, 1]$ from Assumption A (cf. (2.3)), we require that

$$p_0 \geq 20ds^{-1}, \tag{2.126}$$

so that for all $k \geq 0$,

$$\frac{p_k s}{10} \geq 2d. \tag{2.127}$$

Observe that, with $s \in (0, 1]$, we have $20ds^{-1} \geq 20d > 6d$, so (2.127) is indeed stronger that the assumption $p_0 > 6d$.

Next, introduce subsets of I parameterized by $a > 0$:

$$\begin{aligned} \mathcal{E}_x(a) &= \{E \in I : \mathrm{M}_x(E,\omega) \geq a\}, \\ \mathcal{E}_{x,y}(a) &= \{E \in I : |G_{\Lambda_L}(x, y; E, \omega)| \geq a\} \subset \mathcal{E}_x(a) \end{aligned} \tag{2.128}$$

where $y \in \partial^- \Lambda_L$ (the argument L is again omitted for brevity).

Theorem 2.5.1. *Given $L \geq 0$ and $x \in \mathbb{Z}^d$, consider the operator $H_{\Lambda_L}(\omega)$ where $\Lambda_L = \Lambda_L(x)$. Let E_j, $1 \leq j \leq |\Lambda_L|$, be its eigenvalues and take a bounded interval $I \subset \mathbb{R}$. Suppose that the numbers $a, b, c, q_L > 0$ satisfy*

$$b \leq \min\left\{|\Lambda_L|^{-1} ac^2, c\right\} \tag{2.129}$$

and for all $E \in I$

$$\mathbb{P}\{\mathrm{M}_x(E) \geq a\} \leq q_L. \tag{2.130}$$

Then there is an event $\mathcal{B}_x(b) \subset \Omega$ with $\mathbb{P}\{\mathcal{B}_x(b)\} \le |I|\, b^{-1} q_L$ such that for all $\omega \notin \mathcal{B}_x(b)$, the set

$$\mathcal{E}_x(2a) = \left\{ E \in I : \mathrm{M}_x(E) \ge 2a \right\}$$

is covered by the finite union of intervals $\cup_{1\le j\le |\Lambda_L(u)|} I_j$, where $I_j := \{E : |E - E_j| \le 2c\}$, $1 \le j \le |\Lambda_L(u)|$.

Proof. Fix numbers a, b, c, q_L obeying (2.129)–(2.130) and let

$$\begin{aligned} \mathcal{S}_x(a) &= \{ (\omega, E) \in \Omega \times I : \mathrm{M}_x(E) \ge a \}, \\ \mathcal{B}_x(b) &= \{ \omega \in \Omega : \mathrm{mes}\big(\mathcal{E}_x(a)\big) \ge b \}. \end{aligned} \tag{2.131}$$

Apply Chebyshev's inequality and the Fubini theorem combined with (2.128):

$$\begin{aligned} \mathbb{P}\{\mathcal{B}_x(b)\} &\le b^{-1}\mathbb{E}\Big[\mathrm{mes}\big(\mathcal{E}_x(a)\big) \Big] \\ &= b^{-1} \int_I \mathrm{d}E \; \mathbb{E}\Big[\mathbf{1}_{\{\mathrm{M}_x(E)\ge a\}} \Big] \le |I|\, b^{-1} q_L. \end{aligned} \tag{2.132}$$

Now fix any $\omega \notin \mathcal{B}_x(b)$, so that $\mathrm{mes}\big(\mathcal{E}_x(a)\big) \le b$. Consider the eigenvalues $\{E_j\}_{j=1}^K$, $K = |\Lambda_L|$, of the operator H_{Λ_L}, and the respective normalized eigenfunctions $\{\psi_j\}_{j=1}^K$. The Green's function $E \mapsto G_{\Lambda_L}(x, y; E)$ is a rational function with poles at $\{E_j\}$,

$$G_{\Lambda_L}(x, y; E) =: \sum_{j=1}^K \frac{\kappa_j}{E - E_j}. \tag{2.133}$$

Here $\sum_j |\kappa_j| \le \sum_j |\psi_j(x)\psi_j(y)| \le |\Lambda_L|$. Set

$$\begin{aligned} \mathcal{R}(c) &= \{ E \in \mathbb{R} : \min_{1\le j\le K} |E - E_j| \ge c \}, \\ \mathcal{R}(2c) &= \{ E \in \mathbb{R} : \min_{1\le j\le K} |E - E_j| \ge 2c \}. \end{aligned}$$

If $E \in \mathcal{R}(c)$, then $\min_j |E - E_j| \ge c$ and

$$\left| \frac{\mathrm{d}}{\mathrm{d}E} G_{\Lambda_L}(x, y; E) \right| \le |\Lambda_L| c^{-2}.$$

Let us show by contraposition that, for any $\omega \notin \mathcal{B}_x(b)$,

$$\{ E : |G_{\Lambda_L}(x, y; E)| \ge 2a \} \cap \mathcal{R}(2c) = \varnothing.$$

Assume otherwise and pick any point E^* in the non-empty set in the LHS. Then for any E with $|E - E^*| \le b$ one has, owing to the assumption (2.129),

$$|G_{\Lambda_L}(x, y; E)| \ge |G_{\Lambda_L}(x, y; E^*)| - b \sup_{|E'-E^*|\le b} \left| \frac{\mathrm{d}}{\mathrm{d}E'} G_{\Lambda_L}(x, y; E') \right|$$
$$> 2a - |\Lambda_L| c^{-2} b \ge a.$$

Hence, each point $E \in J_b := [E^* - b, E^* + b]$ lies in $\mathcal{E}_{x,y}(a) \subset \mathcal{E}_x(a)$. Consequently, $\mathrm{mes}\big(\mathcal{E}_{x,y}(a)\big) \ge \mathrm{mes}(J_b) > b$, contrary to the choice of ω. Since the set $\mathcal{R}(2c)$ is independent of y, the claim follows from (2.132). □

Remark 2.5.1. One can easily verify that the condition (2.129) is fulfilled with $a = a(L_k)$, $b = b(L_k)$ and $c = c(L_k)$ where

$$a(L_k) = L_k^{-\frac{p_k}{5}}, \ b(L_k) = L_k^{-\frac{4p_k}{5}}, \ c(L_k) = 3^{\frac{d}{2}} L_k^{-\frac{1}{5}p_k}, \ q_{L_k} = L_k^{-p_k}, \quad (2.134)$$

provided that $p_k \ge 5d$. The latter inequality follows from our assumption that $p_0 > 6d$, since $p_k = p_0(1+\theta)^k > p_0$.

Theorem 2.5.2. *Fix a bounded interval $I \subset \mathbb{R}$ of length $|I|$. Given $x \in \mathbb{Z}^d$, consider the function* M_x *introduced in* (2.125)*:*

$$\mathrm{M}_x : E \in I \mapsto \max_{v \in \partial^- \Lambda_L(x)} |G_{\Lambda_L(x)}(x, v; E)|, \quad (2.135)$$

Next, assume that positive numbers a and $b \le c$ are given such that $b \le |\Lambda_L(0)|^{-1} a c^2$. Suppose that the following bound holds true for some integer $L > 0$, some $q_L \in (0, 1]$, any $u \in \mathbb{Z}^d$ and all $E \in I$:

$$\mathbb{P}\{\mathrm{M}_x(E) \ge a\} \le q_L. \quad (2.136)$$

Then under the assumption A *(cf.* (2.3)*), for any pair of disjoint cubes $\Lambda_L(x), \Lambda_L(y) \subset \mathbb{Z}^d$ the following bound holds true:*

$$\mathbb{P}\left\{\exists\, E \in I : \min\big(\mathrm{M}_x(E), \mathrm{M}_y(E)\big) > a\right\} \le C_s (2L+1)^{2d} (2c)^s + \frac{2|I|\, q_L}{b}. \quad (2.137)$$

Here the number $s \in (0, 1]$ and the constant C_s are those given in (2.3)*.*

Proof. Let $\mathcal{B}_x(b)$ and $\mathcal{B}_y(b)$ be two events defined as in (2.131). Set $\mathcal{B} := \mathcal{B}_x \cup \mathcal{B}_y$ and write that

$$\begin{aligned}\mathbb{P}\{\mathcal{E}_x(a) \cap \mathcal{E}_y(a) \ne \varnothing\} &\le \mathbb{P}\{\mathcal{B}\} + \mathbb{P}\{\{\mathcal{E}_x(a) \cap \mathcal{E}_y(a) \ne \varnothing\} \cap \mathcal{B}^c\} \\ &\le 2b^{-1} q_L\, |I| + \mathbb{E}\big[\mathbb{P}\{\{\mathcal{E}_x(a) \cap \mathcal{E}_y(a) \ne \varnothing\} \cap \mathcal{B}^c \,|\, \mathfrak{B}(\Lambda_L(y))\}\big].\end{aligned} \quad (2.138)$$

Here, as before, $\mathfrak{B}(\Lambda_L(y))$ stands for the sigma-algebra generated by the values $V(u;\omega)$ with $u \in \Lambda_L(y)$.

It remains to assess the conditional probability in the RHS of (2.138). For $\omega \notin \mathcal{B}$, each of the sets $\mathcal{E}_x(a)$, $\mathcal{E}_y(a)$ is covered by intervals of width $4c$ around the respective EVs $E_i \in \sigma(H_{\Lambda_L(x)})$ and $E'_j \in \sigma(H_{\Lambda_L(y)})$, and for disjoint cubes $\Lambda_L(x), \Lambda_L(y)$ these spectra are independent. Let I_i^x and I_j^y be such intervals of width $4c$ around the eigenvalues $E_i \in \sigma(H_{\Lambda_L}(x))$ and $E'_j \in \sigma(H_{\Lambda_L}(y))$, respectively. Then, using the Wegner–Stollmann estimate (2.46),

$$
\begin{aligned}
&\mathbb{P}\left\{ \{\mathcal{E}_x(a) \cap \mathcal{E}_y(a) \neq \varnothing\} \cap \mathcal{B}^c \,|\, \mathfrak{B}(\Lambda_L(y)) \right\} \\
&\quad \le \mathbb{P}\left\{ \exists\, i,j : I_i^x \cap I_j^y \neq \varnothing \,|\, \mathfrak{B}(\Lambda_L(y)) \right\} \\
&\quad \le |\Lambda_L(y)| \max_j \mathbb{P}\left\{ \exists\, i : I_i^x \cap I_j^y \neq \varnothing \,|\, \mathfrak{B}(\Lambda_L(y)) \right\} \\
&\quad \le |\Lambda_L(y)| \max_j \mathbb{P}\left\{ \exists\, i : |E_i - E'_j| \le 4c \,|\, \mathfrak{B}(\Lambda_L(y)) \right\} \\
&\quad \le |\Lambda_L(y)| \sup_{\lambda \in I} \mathbb{P}\{ \exists\, i : |E_i - \lambda| \le 4c \} \\
&\quad \le |\Lambda_L(y)| \cdot |\Lambda_L(x)| \cdot C_s (4c)^s.
\end{aligned}
\tag{2.139}
$$

Collecting (2.138) and (2.139), the assertion follows. □

2.5.2 *Dynamical Localization Bounds in Finite Volumes*

Now we are going to derive bounds on eigenfunction correlators in finite (but arbitrarily large) cubes, closely related to the phenomenon of strong dynamical localization on the entire lattice $\mathbb{Z}^d$. Theorem 2.5.3 below is a simplified version of an elegant (and much more general) argument from [99]. The simplification comes from replacing the infinite-space analysis by that in an (arbitrarily large) bounded subset of the configuration space ($\mathbb{Z}^d$, in our case) of the quantum particle. This allows to avoid a deep analysis of growth properties of generalized eigenfunctions (necessary in [99]) and replace it by an elementary application of the Bessel inequality.

As was noticed earlier (cf. Remark 2.1.3 in Sect. 2.1.1), the lattice $\mathbb{Z}^d$ has a natural graph structure induced by the edge set $\mathcal{E}_{\mathbb{Z}^d} = \{(x,y) : |x-y|_1 = 1\}$. Recall that we refer to connectedness of a subset $\Lambda \subset \mathbb{Z}^d$ meaning that the graph with the vertex set Λ and the edge set formed by the edges inherited from $\mathcal{E}_{\mathbb{Z}^d}$ is connected.

Given an interval $I \subseteq \mathbb{R}$, we denote by $\mathscr{B}_1(I)$ the set of all Borel functions $\phi : \mathbb{R} \to \mathbb{C}$ with $\operatorname{supp} \phi \subset I$ and $\|\phi\|_\infty \le 1$.

Theorem 2.5.3. *Fix a bounded interval $I \subset \mathbb{R}$ and consider the function*

$$\mathrm{M}_x : E \in I \mapsto \max_{v \in \partial^- \Lambda_L(x)} |G_{\Lambda_L(x)}(x, v; E)|. \tag{2.140}$$

Fix an integer $L \geq 1$ and assume that the following bound holds true for any pair of disjoint cubes $\Lambda_L(x)$, $\Lambda_L(y)$ and some $a, h(L) \in (0, \infty)$:

$$\mathbb{P}\{\exists\, E \in I : \min(\mathrm{M}_x, \mathrm{M}_y) > a\} \leq h(L).$$

Then for any $x, y \in \mathbb{Z}^d$ with $\mathrm{dist}(x, y) > 2L + 1$, any finite connected set $\Lambda \subset \mathbb{Z}^d$ with $\Lambda \supseteq \Lambda_L(x) \cup \Lambda_L(y)$ and any Borel function $\phi \in \mathscr{B}_1(I)$, the random operator $H_\Lambda = H_\Lambda(\omega)$ satisfies

$$\mathbb{E}\Big[\,|\langle \delta_x, \phi(H_\Lambda)\delta_y\rangle|\,\Big] \leq 4d(2L+1)^{d-1} \cdot a + h(L). \tag{2.141}$$

Proof. Fix $x, y \in \mathbb{Z}^d$ with $|x - y| > 2L + 1$ and let Λ and ϕ be as in the statement of the theorem. The operator H_Λ has a finite orthonormal eigenbasis $\{\psi_i\}$ with respective eigenvalues E_i. Set $\Gamma = \partial\Lambda_L(x) \cup \partial\Lambda_L(y)$ (recall: this is a set of pairs of neighboring lattice sites (u, u')) and let $|\Gamma|$ stand for the cardinality of Γ. Suppose that for some ω, for each i there is $z_i \in \{x, y\}$ such that $\mathrm{M}_z \leq a$ and let v_i be the complementary point: $\{v_i\} = \{x, y\} \setminus \{z_i\}$. Denote by $\mu_{x,y}(\phi) = \mu^{\Lambda}_{x,y}(\phi)$ the scalar product $\langle \delta_x, \phi(H_\Lambda)\delta_y\rangle$, with $|\mu_{x,y}(\phi)| \leq 1$.

By the GRI for the eigenfunctions (see (2.67)),

$$\begin{aligned}
|\mu_{x,y}(\phi)| &\leq \|\phi\|_\infty \sum_{i:E_i \in I} |\psi_i(x)\psi_i(y)| \leq \sum_{i:E_i \in I} |\psi_i(z_i)\psi_i(v_i)| \\
&\leq \sum_{i:E_i \in I} |\psi_i(v_i)| \mathrm{e}^{-mL} \sum_{(u,u') \in \partial\Lambda_L(z_i)} |\psi_i(u)| \\
&\leq a \cdot \sum_{i:E_i \in I} \sum_{(u,u') \in S} |\psi_i(u)| \big(|\psi_i(x)| + |\psi_i(y)|\big) \\
&\leq a \cdot |\Gamma| \max_{u \in \Lambda} \sum_{i:E_i \in I} \frac{1}{2}\big(2|\psi_i(u)|^2 + |\psi_i(x)|^2 + |\psi_i(y)|^2\big) \\
&\leq a \cdot \frac{|\Gamma|}{2} \max_{u \in \Lambda} \big(2\|\delta_u\|^2 + \|\delta_x\|^2 + \|\delta_y\|^2\big) = 2a \cdot |\Gamma|
\end{aligned}$$

with $|\Gamma| \leq 2d(2L+1)^{d-1}$. Consider the event

$$\mathcal{N}_L = \{\exists\, E \in I : \ \Lambda_L(x) \text{ and } \Lambda_L(y) \text{ are } (E, m)\text{-S}\}.$$

Then $\mathbb{P}\{\mathcal{N}_L\} \leq h(L)$, by assumption. Finally,

$$\mathbb{E}\big[\,|\mu_{x,y}(\phi)|\,\big] = \mathbb{E}\big[\,\mathbf{1}_{\mathcal{N}_L}|\mu_{x,y}(\phi)|\,\big] + \mathbb{E}\big[\,\mathbf{1}_{\mathcal{N}_L^c}|\mu_{x,y}(\phi)|\,\big] \leq h(L) + 2a \cdot |\Gamma|$$

where $\mathbf{1}_{\mathcal{N}_L}$ and $\mathbf{1}_{\mathcal{N}_L^c}$ are the indicators of the event $\mathcal{N}_L$ and its complement, respectively. □

2.5.3 Adaptation to Unbounded Energy Intervals

Theorem 2.5.2 can be easily extended to unbounded intervals $I \subseteq \mathbb{R}$; this can be useful in the context of unbounded random potentials, where the spectrum $\sigma\big(H(\omega)\big)$ is also a.s. unbounded.

The most obvious generalization can be obtained for a large class of IID random fields (generating the random potential) with the decay rate of the so-called "tail probabilities" which is not too slow. More precisely, introduce the following

Assumption A1. *The common marginal distribution function*

$$F(a) = \mathbb{P}\{\, V(x;\omega) \le a \,\}, \;\; x \in \mathbb{Z}^d,$$

of the IID random variables $\{V(x;\cdot), x \in \mathbb{Z}^d\}$ *satisfies the following condition: there exist constants* $A \in (0,1]$ *and* $C \in (0,+\infty)$, *such that for any* $t \ge 1$,

$$F(-t) + \big(1 - F(t)\big) \equiv \mathbb{P}\{\, |V(0;\omega)| \ge t \,\} \le C t^{-A}. \qquad (2.142)$$

Remark 2.5.2. Naturally, a decay rate faster than $t \mapsto 1/t$ is welcome here; we assume that this rate admits at least some power-low upper bound. Technically, taking $A \le 1$ proves convenient, e.g., in the formulation of Corollary 2.5.5 below.

Remark 2.5.3. Given any $g \ne 0$, (2.142) implies that

$$\mathbb{P}\{\, |gV(0;\omega)| \ge t \,\} \le C(|g|^{-1}t)^{-A} \le C'(|g|), At^{-A}.$$

Lemma 2.5.4. *Let the common marginal distribution of IID random variables* $\{V(x;\omega),\, x \in \mathbb{Z}^d\}$ *satisfy Assumption A1. Set* $\hat{t}_{L_k} := L_k^{\frac{1}{10}p_k}$. *Then for any cube* $\Lambda_{L_k}(x) \subset \mathbb{Z}^d$ *one has*

$$\mathbb{P}\left\{\sigma\left(H_{\Lambda_{L_k}(x)}(\omega)\right) \subset [-\hat{t}_{L_k} - 4d, \hat{t}_{L_k} + 4d]\right\} \ge 1 - 3^d\, C\, L_k^{-\frac{A}{10}p_k + d}. \qquad (2.143)$$

Proof. Clearly, we have

$$\sigma\left(H_{\Lambda_{L_k}(x)}(\omega)\right) \subset \Big[\, - \|H_{\Lambda_{L_k}(x)}(\omega)\|, \|H_{\Lambda_{L_k}(x)}(\omega)\| \Big],$$

where

$$\begin{aligned}\|H_{\Lambda_{L_k}(x)}(\omega)\| &\le \|\Delta_{\Lambda_{L_k}(x)}\| + \|V(\omega)\mathbf{1}_{\Lambda_{L_k}(x)}\|_\infty \\ &\le 4d + \max_{u\in\Lambda_{L_k}(x)} |V(u;\omega)|.\end{aligned}$$

Therefore, it remains to assess the following probability:

$$\begin{aligned}\mathbb{P}\left\{ \max_{u\in\Lambda_{L_k}(x)} |V(u;\omega)| \ge \hat{t}_{L_k} \right\} &\le \sum_{u\in\Lambda_{L_k}(x)} \mathbb{P}\{\,|V(u;\omega)| \ge \hat{t}_{L_k}\,\} \\ &\le (2L_k+1)^d \cdot \big(F(-\hat{t}_{L_k}) + \big(1 - F(\hat{t}_{L_k})\big)\big) \\ &\le 3^d\, C\, L_k^{-\frac{A}{10}p_k + d}. \end{aligned}$$

□

Note that, unlike several estimates used in the scale induction, here it is meaningful to accept bounds valid for $k \ge 1$ large enough. Indeed, the above lemma provides *a posteriori* analysis of localization properties, already established at any scale k. For the application to the proof of strong dynamical localization, it suffices then to have required estimates at least for L_k with k large enough.

(For k large enough, one can replace the bound $\hat{t}_{L_k} + 4d$ by a simpler (although crude) one, $2\hat{t}_{L_k}$. This explains the form of the interval $J_k := [-2\hat{t}_{L_k}, 2\hat{t}_{L_k}]$ in Corollary 2.5.5 below.)

Remark 2.5.4. It is clear from the proof that the independence of the random variables $V(x;\cdot)$ is not required, and their marginal distribution functions $F_x(t) := \mathbb{P}\{\,V(x;\omega) \le t\,\}$ need not to be identical: it suffices to assume uniform bounds on the tail probabilities:

$$\sup_{x\in\mathbb{Z}^d} \Big[F_x(-t) + \big(1 - F_x(t)\big)\Big] \le C t^{-A}.$$

Combining Theorem 2.5.1, Lemma 2.5.4, and Remark 2.5.2, we come to the following

Corollary 2.5.5. *Let the common marginal distribution function F of the random variables $\{V(x;\omega),\ x \in \mathbb{Z}^d\}$ satisfy Assumptions A and A1 (cf.* (2.142)*). For each $k \in \mathbb{N}$ and $x \in \mathbb{Z}^d$, consider the operator $H_{\Lambda_{L_k}(x)}(\omega)$. Let E_j, $1 \le j \le |\Lambda_{L_k}|$, be its eigenvalues and take the interval $J_k := [-2\hat{t}_{L_k}, 2\hat{t}_{L_k}]$. Set*

$$a(L_k) = L_k^{-\frac{p_k}{5}},\quad b(L_k) = L_k^{-\frac{4p_k}{5}},\quad c(L_k) = 3^{\frac{d}{2}} L_k^{-\frac{p_k}{5}p_k},\quad q_{L_k} = L_k^{-p_k}, \tag{2.144}$$

and assume that k is large enough, so that $p_k \ge 20d \max\{s^{-1}, A^{-1}\}$. Further, assume that for all $E \in \mathbb{R}$

$$\mathbb{P}\{\,\mathrm{M}_x(E) \ge a(L_k)\,\} \le q_{L_k}. \tag{2.145}$$

Then there is an event $\mathcal{B}_x(b(L_k)) \subset \Omega$ with

$$\mathbb{P}\{\,\mathcal{B}_x(b(L_k))\,\} \le \frac{|J_k|\, q_{L_k}}{b(L_k)} + 3^d\, L_k^{-\frac{A}{10}p_k+d} \le C(L_0, A, d)\, L_k^{-\frac{A}{20}p_k}$$

such that for all $\omega \notin \mathcal{B}_x(b(L_k))$,

$$\sigma\left(H_{\Lambda_{L_k}(x)}(\omega)\right) \subset J_k$$

and

$$\mathcal{E}_x(2a(L_k)) = \left\{E \in J_k : \mathrm{M}_x(E) \ge 2a\right\} \subset \bigcup_{j=1}^{|\Lambda_{L_k}(x)|} I_j,$$

where $I_j := \{E : |E - E_j| \le 2c(L_k)\}$, $j = 1, \dots, |\Lambda_{L_k}(x)|$.

Theorem 2.5.2 shows that, given two disjoint cubes $\Lambda_{L_k}(x)$ and $\Lambda_{L_k}(y)$, with a high probability, for *every* energy E in an interval covering the spectra of $H_{\Lambda_{L_k}(x)}(\omega)$ and $H_{\Lambda_{L_k}(y)}(\omega)$, at least one of the cubes features a fast decay of the Green's functions; we can call this cube E-good.

Note that the decay rate of the tail probabilities in Assumption A1 allows not only for such popular models as Gaussian or exponentially distributed IID random potentials, but also for Cauchy IID potentials.

Observe, however, that using measure-theoretic arguments (Fubini theorem and Chebyshev's inequality) resulted in a significantly weaker upper bound on the Green's functions in a E-good cube: it is no longer exponential (e^{-mL}), but only as strong as the probabilistic bound that one can establish for the event $\{\omega : \Lambda_L(u) \text{ is } (E, m)\text{-S}\}$.

Exponential probabilistic bounds on the Green's functions, at a fixed energy E, can be obtained under certain conditions upon the random potential V with the help of the fractional moment method (FMM), which we only briefly discuss in this book. Theorem 2.5.2 allows then to transform such bounds into exponential bounds of the form

$$\mathbb{P}\{\,\exists\, E \in I :\ \Lambda_L(x) \text{ and } \Lambda_L(y) \text{ are } (E, m)\text{-S}\,\} \le \mathrm{e}^{-\mathrm{Const}\, mL}.$$

In Sect. 2.6 we will present a different variant of the MSA, called the variable-energy MSA and going back to the works [96, 161] which allows to obtain directly a bound of the form

$$\mathbb{P}\{\,\exists\, E \in I :\ \Lambda_{L_k}(x) \text{ and } \Lambda_{L_k}(y) \text{ are } (E, m)\text{-S}\,\} \le L_k^{-2p},$$

so that even though thc probabilistic bound is far from exponential, the event in the LHS gives an exponential decay for the Green's functions in one of the two cubes, no matter how $E \in I$ is chosen. This will require more sophisticated geometric and analytic constructions.

Theorem 2.5.6. *Let the random field V satisfy Assumption* A *(cf.* (2.3)*) and Assumption A1 (cf.* (2.142)*). Suppose that the following bound holds true for some integer $L > 0$, some $q_L \in (0, 1]$, any $u \in \mathbb{Z}^d$ and all $E \in \mathbb{R}$:*

$$\mathbb{P}\{\ \Lambda_L(u) \text{ is } (E, m)\text{-S}\} \le q_L.$$

Given $x, y \in \mathbb{Z}^d$, $k \in \mathbb{N}$ and $m > 0$, consider the functions $\mathrm{M}_{x,L_k}, \mathrm{M}_{y,L_k}$ introduced in (2.125) *and assume that positive numbers $a(L_k) \ge \mathrm{e}^{-mL_k}$ and $b(L_k) \le c(L_k)$ are given such that $b(L_k) \le |\Lambda_{L_k}(0)|^{-1} a(L_k) c^2(L_k)$. Then for any pair of disjoint cubes $\Lambda_{L_k}(x)$, $\Lambda_{L_k}(y) \subset \mathbb{Z}^d$ the following bound holds true:*

$$\begin{aligned} &\mathbb{P}\{\exists\, E \in \mathbb{R} : \min\left(\mathrm{M}_{x,L_k}(E), \mathrm{M}_{y,L_k}(E)\right) > a(L_k)\} \\ &\le 3^{2d} C_s L_k^{2d} (4c(L_k))^s + \frac{\left(4L_k^{\frac{1}{10}p_k} + 8d\,\mathrm{e}^{2m}\right) q(L_k)}{b(L_k)} + L_k^{-\frac{A}{20}p_k}. \end{aligned} \tag{2.146}$$

Here the numbers $s \in (0, 1]$, $C_s < \infty$ are those given in (2.3)*, and $A > 0$ is given in* (2.142)*.*

In particular, with $a(L_k) \ge \mathrm{e}^{-mL_k}$ and $b(L_k) \le c(L_k)$ of the form (cf. (2.144)*)*

$$a(L_k) = L_k^{-\frac{p_k}{5}},\quad b(L_k) = L_k^{-\frac{4p_k}{5}},\quad c(L_k) = 3^{\frac{d}{2}} L_k^{-\frac{p_k}{5}p_k},\quad q_{L_k} = L_k^{-p_k},$$

and $p_k \ge 5d$, one has

$$\begin{aligned} &\mathbb{P}\{\exists\, E \in \mathbb{R} : \min\left(\mathrm{M}_{x,L_k}(E), \mathrm{M}_{y,L_k}(E)\right) > a(L_k)\} \\ &\le 3^{2d} C_s L_k^{2d} (4c(L_k))^s + 8L_k^{-\frac{1}{10}p_k} + L_k^{-\frac{A}{20}p_k}. \end{aligned} \tag{2.147}$$

Proof. With $m > 0$ fixed, denote $\eta = \eta(m) = 4d\,\mathrm{e}^{2m}$ and set $\hat{t}_{L_k} = L_k^{\frac{1}{10}p_k}$. Further, introduce the intervals

$$[-\hat{t}_{L_k} - 4d, \hat{t}_{L_k} + 4d] =: \hat{I}_k \subset I_k := [-\hat{t}_{L_k} - 4d - \eta, \hat{t}_{L_k} + 4d + \eta],$$

and consider the event

$$\mathcal{A}_k = \{\omega : \sigma(H_{\Lambda_{L_k}(x)}(\omega)) \subset \hat{I}_k\}.$$

By Lemma 2.5.4, $\mathbb{P}\left\{\mathcal{A}_k^{\mathrm{c}}\right\} \le L_k^{-\frac{A}{20}p_k}$ (at least for L_0 large *or* $k \ge 1$ large). Set for brevity

$$\mathcal{M}_{x,y} := \{\omega : \exists\, E \in \mathbb{R} : \min\left(\mathrm{M}_{x,L_k}(E), \mathrm{M}_{y,L_k}(E)\right) > a(L_k)\},$$

then we have

$$\begin{aligned}\mathbb{P}\{\mathcal{M}_{x,y}\} &\le \mathbb{P}\{\mathcal{M}_{x,y} \cap \mathcal{A}_k\} + \mathbb{P}\{\mathcal{A}_k^c\} \\ &\le \mathbb{P}\{\mathcal{M}_{x,y} \cap \mathcal{A}_k\} + L_k^{-\frac{A}{20}p_k}.\end{aligned} \tag{2.148}$$

Now we focus on the first probability in the last line. Within the event $\mathcal{A}_k$, for any $E \in \mathbb{R} \setminus I_k$ we have

$$\operatorname{dist}\Big(E, \sigma(H_{\Lambda_{L_k}(x)}(\omega))\Big) \ge \eta,$$

so by Combes–Thomas estimate the Green functions of the operator $H_{\Lambda_{L_k}(x)}(\omega)$, at energy E decay exponentially with rate $\ge m$, thus $\mathrm{M}_{x,L_k}(E) \le a(L_k)$ (since $a(L_k) \ge \mathrm{e}^{-mL_k}$). The latter is incompatible with the event $\mathcal{M}_{x,y}$. Therefore,

$$\mathcal{M}_{x,y} \cap \mathcal{A}_k \subset \big\{\exists\, E \in I_k : \min\big(\mathrm{M}_{x,L_k}(E), \mathrm{M}_{y,L_k}(E)\big) > a(L_k)\big\}.$$

From this point on, we can argue as in the proof of Theorem 2.5.2, with the interval I replaced by I_k, and obtain the following bound:

$$\mathbb{P}\{\mathcal{M}_{x,y} \cap \mathcal{A}_k\} \le 3^{2d} L_k^{2d}\, C_s\,(4c(L_k))^s + \frac{2|I_k|\, q_{L_k}}{b(L_k)}. \tag{2.149}$$

Collecting (2.148) and (2.149), the assertion (2.146) follows. The bound (2.147) is obtained by taking $a(L_k)$, $b(L_k)$, $c(L_k)$ of the form (2.144). □

2.5.4 Extending the Bounds to the Entire Lattice

In this Section we show the connection between the functional $\mu_{x,y}^{\Lambda_L}(\phi)$ and its infinite-volume version. Given $x, y \in \mathbb{Z}^d$ and an interval $I \subset \mathbb{R}$, one can define a signed measure (sometimes called a charge) $\mu_{x,y}(\mathrm{d}\lambda) = \mu_{x,y}^{\mathbb{Z}^d}(\mathrm{d}\lambda)$ on $\mathbb{R}$ such that for any Borel function $\phi : \mathbb{R} \to \mathbb{C}$ with supp $\phi \subset I$ one has

$$\int_I \mathrm{d}\mu_{x,y}(\lambda)\, \phi(\lambda) = \Big\langle \delta_x, \phi\big(H(\omega)\,\delta_y\big)\Big\rangle. \tag{2.150}$$

These measures are usually called the spectral measures for $H(\omega)$ (more precisely, for the operator $H(\omega)$ and the functions δ_x, δ_y). The same notion applies to the finite-volume approximants H_{Λ_L}:

$$\int_I \mu_{x,y}^{\Lambda_L}(\mathrm{d}\lambda)\, \phi(\lambda) = \langle \delta_x, \phi(H_{\Lambda_L})\,\delta_y\rangle. \tag{2.151}$$

Lemma 2.5.7. *Given $x, y \in \mathbb{Z}^d$, consider the spectral measures $\mu_{x,y}^{\Lambda_L(0)}$ and $\mu_{x,y}$ for $H_{\Lambda_L(0)}$ and $H(\omega)$, respectively. Then $\mu_{x,y}^{\Lambda_L(0)}$ converges vaguely to $\mu_{x,y}$ as $L \to \infty$.*

Before we proceed with a formal proof of Lemma 2.5.7, we would like to make the following observation. It would be straightforward to prove the vague convergence directly for an LSO $H = H_0 + V$ with bounded potential V. To cover a more general case and to show that the statement of the lemma has a much larger field of application, we refer the reader to the monograph [112] for the following well-known result.

Theorem 2.5.8. *Let $\{A_n, n \in \mathbb{N}\}$ and A be closed operators in a Hilbert space $\mathcal{H}$ and assume that there exists a core $\mathcal{D}_A$ of the operator A such that every vector $\phi \in \mathcal{D}_A$ belongs to the domains of operators A_n with sufficiently large n, and for such values of n the sequence of vectors $A_n \psi$ converges strongly in $\mathcal{H}$ as $n \to \infty$. Then the operators A_n converge to A in the strong resolvent sense.*

Proof of Lemma 2.5.7. Since the operator $H_0 = -\Delta$ is bounded, the Hamiltonian $H(\omega)$ is defined on any vector $\psi \in \ell^2(\mathbb{Z}^d)$ for which the operator of multiplication by $V(x;\omega)$ is defined. In other words, the set of functions $\psi(x)$, $x \in \mathbb{Z}^d$, with

$$\sum_{x \in \mathbb{Z}^d} |V(x;\omega)|^2 \, |\psi(x)|^2 < \infty,$$

is the domain $\mathcal{D}_H$ of $H = H(\omega)$. Hence, the linear subspace $\mathcal{H}_0$ in $\ell^2(\mathbb{Z}^d)$ composed of functions $\psi : \mathbb{Z}^d \to \mathbb{C}$ with compact support can be taken as a core for the operator H. It is clear that, for any growing sequence of cubes $\Lambda_{L_k}(0)$, any finitely supported function ϕ belongs to the domains of all operators $H_{\Lambda_{L_k}(0)}$ with k large enough. Moreover, for all L large enough (depending on $\phi \in \mathcal{H}_0$), one has

$$H_{\Lambda_L(0)} \phi \equiv H\phi.$$

Hence, the strong convergence required for Theorem 2.5.8 holds true, by stabilization. Consequently, $H_{\Lambda_L(0)}$ converge to H in the strong resolvent sense. In turn, this implies (cf., e.g., Theorem VIII.20 from [146]) that, given a bounded interval $I \subset \mathbb{R}$, for any continuous function ϕ with supp $\phi \subset I$,

$$\langle \delta_x, \phi(H_{\Lambda_L(0)}) \, \delta_y \rangle \underset{L \to \infty}{\longrightarrow} \langle \delta_x, \phi(H) \, \delta_y \rangle. \tag{2.152}$$

Note that if the random potential $V(\cdot;\omega)$ is a.s. bounded, or, more generally, the random operator $H(\omega)$ (not necessarily having the form of the lattice Schrödinger operator) is a.s. bounded in norm, then the spectrum $\sigma(H)$ is almost surely contained in some bounded, non-random interval $[-R, R]$. In this case, there exists a continuous function χ supported in $[-R-1, R+1]$ and equal to ϕ (e.g., with

$\phi(t) = \mathrm{e}^{-\mathrm{i}\lambda t}$) on $[-R, R]$, so one can safely replace $\phi(H)$ by $\chi(H)$ in (2.152), yet having the benefit of the compact support of χ.

In the general case, some additional arguments are required. Note that for the proof of the dynamical localization, we are mainly interested by functions bounded by 1 (viz., $\phi(t) = \mathrm{e}^{-\mathrm{i}\lambda t}$, $\lambda \in \mathbb{R}$). Let $\mathcal{C}_1$ be the set of continuous functions ϕ, compactly supported in I, with $\|\phi\|_\infty \leq 1$. For all continuous functions ϕ supported in I, we have by Lemma 2.5.7

$$\langle \delta_x, \phi(H_{\Lambda_L(0)})\, \delta_y \rangle = \lim_{L\to\infty} \langle \delta_x, \phi(H_{\Lambda_L(0)})\, \delta_y \rangle. \tag{2.153}$$

Therefore,

$$\sup_{\phi\in\mathcal{C}_1} \langle \delta_x, \phi(H_{\Lambda_L(0)})\, \delta_y \rangle = \sup_{\phi\in\mathcal{C}_1} \lim_{L\to\infty} \langle \delta_x, \phi(H_{\Lambda_L(0)})\, \delta_y \rangle. \tag{2.154}$$

Next, for any $L > 0$,

$$\langle \delta_x, \phi(H_{\Lambda_L(0)})\, \delta_y \rangle \leq \sup_{\phi\in\mathcal{C}_1} \langle \delta_x, \phi(H_{\Lambda_L(0)})\, \delta_y \rangle,$$

hence

$$\sup_{\phi\in\mathcal{C}_1} \langle \delta_x, \phi(H)\, \delta_y \rangle \leq \liminf_{L\to\infty} \sup_{\phi\in\mathcal{C}_1} \langle \delta_x, \phi(H_{\Lambda_L(0)})\, \delta_y \rangle,$$

To extend this result to all Borel functions of norm bounded by 1, some work has to be done, with the help of standard approximation arguments. See, e.g., [6, 13, 14] (in particular, [14], Sect. 2.5).

and using standard approximation arguments for arbitrary Borel functions $\phi \in \mathscr{B}_1$, we conclude that the spectral measures $\mu_{x,y}^{\Lambda_L}$ defined by (2.151) converge vaguely to $\mu_{x,y}$ as $L \to \infty$. □

Recall that, given a signed measure μ, one can define a positive measure $|\mu|$ by

$$|\mu|(A) := \sup_{\phi\in\mathscr{B}_1} \int_A \phi(\lambda)\, \mathrm{d}\mu(\lambda), \quad A \subset \mathbb{R},$$

where, as before, $\mathscr{B}_1$ is the set of Borel functions bounded by 1 and A is a Borel set.

The Fatou lemma implies that if a sequence of signed measures μ^{Λ_L} converges vaguely to a measure μ then

$$|\mu|(A) \leq \liminf_{L\to\infty} |\mu^{\Lambda_L}|(A). \tag{2.155}$$

This allows us to extend to the LSO $H(\omega)$ the uniform dynamical localization bounds obtained in arbitrarily large finite cubes $\Lambda_L(0)$.

2.5.5 *Strong Dynamical Localization for a Single Particle*

We are now ready to prove the main result on dynamical localization for strongly disordered single-particle systems. Following a similar argument, we will be able to prove in Sect. 4.3.6 strong dynamical localization for multi-particle systems.

We consider first the case where the IID random potential is a.s. bounded, i.e., for some $K < \infty$,

$$\mathbb{P}\{\,|V(0;\omega)| \le K\,\} = 1$$

so that there is a bounded interval I, $|I| \le \text{Const}\,(d) + 2|g|K$, such that with $\mathbb{P}$-probability one,

$$\sigma(H(\omega)) \subset I \tag{2.156}$$

and, similarly,

$$\forall\, L \ge 1\;\; \forall\, x \in \mathbb{Z}^d \qquad \sigma(H_{\Lambda_L(x)}(\omega)) \subset I. \tag{2.157}$$

Note that one can also establish first the finiteness of the expectation in the LHS of (2.158) in a reduced form, viz., for the operators $\mathbb{P}_{[n,n+1)}(H)\,\mathrm{e}^{-\mathrm{i}tH}$, where $P_{[n,n+1)}(H)$ is the spectral projection for H on the interval $[n, n+1)$, $n \in \mathbb{Z}$. If the spectrum $\sigma\big(H(\omega)\big)$ is a.s. contained in a bounded interval $I \subset [n_1, n_2]$, $n_1, n_2 \in \mathbb{Z}$, then

$$\mathrm{e}^{-\mathrm{i}tH} = \mathbb{P}_I(H)\,\mathrm{e}^{-\mathrm{i}tH} = \sum_{n=n_1}^{n_2-1} \mathbb{P}_{[n,n+1)}(H)\,\mathrm{e}^{-\mathrm{i}tH}.$$

Theorem 2.5.9. *Assume that the random potential is a.s. bounded, and consider a bounded interval $I \subset \mathbb{R}$ as in* (2.156)–(2.157)*. There exists $g^* \in (0, +\infty)$ such that for all $|g| \ge g^*$ operators $H(\omega)$ feature strong dynamical localization. That is, there exist constants $C, c, a^* \in (0, +\infty)$ such that $\forall\, x, y \in \mathbb{Z}^d$*

$$\mathbb{E}\left[\,\sup_{t\in\mathbb{R}} \left|\langle \delta_x, \mathrm{e}^{-\mathrm{i}tH}\,\delta_y\rangle\right|\,\right] \le C\,\mathrm{e}^{-a^*(\ln|x-y|)^{1+c}}. \tag{2.158}$$

Proof. As usual, we choose a length scale $\{L_k, k \ge 0\}$ (which is fixed by a choice of L_0), together with parameters m, $p_0 = p$ and θ defining a sequence of exponents $p_k = p(1+\theta)^k$.

Step 1. By Theorem 2.4.7 (cf. (2.123)), for all $k \ge 0$,

$$\mathbb{P}\{\; \Lambda_{L_k}(u) \text{ is } (E,m)\text{-S }\} \le L_k^{-p(1+\theta)^k}, \tag{2.159}$$

provided that the initial length scale bound (SS.I, m, p_0, L_0) holds true with $p_0 > 20ds^{-1} > 6d$. The latter condition is guaranteed for $|g|$ large enough by Theorem 2.3.1.

Step 2. As in Remark 2.5.1, set

$$a(L_k) = L_k^{-\frac{1}{5}p_k},\ b(L_k) = L_k^{-\frac{4}{5}p_k},\ c(L_k) = 3^{\frac{d}{2}} L_k^{-\frac{1}{5}p_k},\ q(L_k) = L_k^{-p_k}. \tag{2.160}$$

These quantities fulfill the condition

$$b(L_k) < |\Lambda_{L_k}|^{-1} a(L_k) c(L_k)^2,$$

so a direct application of Theorem 2.5.2 gives

$$\begin{aligned}\mathbb{P}\left\{ \exists\, E \in I :\ \min\left(\mathrm{M}_x, \mathrm{M}_y\right) > L_k^{-\frac{p_k}{5}} \right\} \\ &\le C L_k^{2d-\frac{1}{5}sp_k} + 2\,|I|\, L_k^{-p_k+\frac{4}{5}p_k} \\ &\le C L_k^{2d-\frac{1}{5}sp_k} + 2\,|I|\, L_k^{-\frac{1}{5}p_k}\end{aligned}$$

with $C = C(d,s)$. Since $2d \le \frac{sp_k}{10}$ (cf. (2.127)), we obtain

$$\begin{aligned}\mathbb{P}\left\{ \exists\, E \in I :\ \min\left(\mathrm{M}_{\mathbf{x}}, \mathrm{M}_{\mathbf{y}}\right) > L_k^{-\frac{p_k}{5}} \right\} &\le C L_k^{\frac{sp_k}{10}-\frac{sp_k}{5}} + 2\,|I|\, L_k^{-\frac{1}{5}p_k} \\ &\le C' L_k^{-\tilde{p}_k},\end{aligned}$$

with some $C' = C'(d, s, |I|)$ and

$$\tilde{p}_k = \tilde{p}_k(s) = \tilde{p}(1+\theta)^k,\ \ \tilde{p} = \tilde{p}(s) := \frac{sp_0}{10},\ \theta > 0.$$

Step 3. By virtue of Theorem 2.5.3, for any bounded Borel function ϕ with $\|\phi\|_\infty \le 1$, including $\phi_t : \lambda \mapsto \mathrm{e}^{-it\lambda}$ with any $t \in \mathbb{R}$,

$$\mathbb{E}\left[\left| \langle \delta_x, (\mathbf{1}_I \cdot \phi)(H_{\Lambda_{L_k}(0)})\, \delta_y \rangle \right| \right] \le C' L_k^{-\tilde{p}_k} + O(L_k^d) a(L_k) \le C'' L_k^{-\tilde{p}_k/2}. \tag{2.161}$$

Moreover, the factor $\mathbf{1}_I(\cdot)$ in the LHS can be safely removed, since by assumption, $\sigma(H_{\Lambda_{L_k}(0)}) \subset I$. Thus, we have

$$\mathbb{E}\left[\left| \langle \delta_x, \phi(H_{\Lambda_{L_k}(0)})\, \delta_y \rangle \right| \right] \le C' L_k^{-\tilde{p}_k} + O(L_k^d) L_k^{-p_k/5} \le C'' L_k^{-\tilde{p}_k/2}. \tag{2.162}$$

Step 4. By Lemma 2.5.7 (cf. (2.155)), the same bound holds true for the operator $H(\omega)$: for some $p' > 0$,

$$\mathbb{E}\left[\,|\langle \delta_x\,,\phi(H)\,\delta_y\rangle|\,\right] \le C'' L_k^{-p'(1+\theta)^k}. \tag{2.163}$$

Step 5. Assume first that $|x-y| > 3L_0$. Then there is $k_\circ \ge 0$ such that $|x-y| \in (3L_{k_\circ}, 3L_{k_\circ+1}]$. An elementary calculation shows that, with $L_{j+1} = [L_j^\alpha]$, $j = 0, 1, \ldots$, one has $\ln L_k \ge C'''\alpha^k \ln L_0$, $k \ge 0$, for some $C''' \in (0,\infty)$. Let $c = \frac{\ln(1+\theta)}{\ln \alpha} > 0$, then

$$(1+\theta)^{k_\circ}\alpha^{k_\circ} = \mathrm{e}^{k_\circ \ln \alpha \cdot \frac{\ln \alpha + \ln(1+\theta)}{\ln \alpha}} = \left(\alpha^{k_\circ}\right)^{1+c}$$

and using $L_{k_\circ} \ge \mathrm{e}^{C'''\alpha^{k_\circ} \ln L_0}$ we obtain

$$-\ln L_{k_\circ}^{-p'(1+\theta)^{k_\circ}} \ge C'''p'\big(\alpha(1+\theta)\big)^{k_\circ} \ln L_0 \;=\; \frac{C'''p'}{(\ln L_0)^c} \cdot \left(\alpha^{k_\circ} \ln L_0\right)^{1+c}$$

On the other hand, $|x-y| \le 3L_{k_\circ}^\alpha$, hence

$$\alpha^{k_\circ} \ln L_0 \ge \frac{1}{\alpha} \ln\left(\frac{|x-y|}{3}\right)$$

yielding, for some $a^* > 0$,

$$L_k^{-p'(1+\theta)^k} \le \mathrm{e}^{-a^* \ln^{1+c}|x-y|}.$$

For the remaining pairs x, y with $|x-y| \le 3L_0$, the required bound can be absorbed in a sufficiently large constant $C > 0$. Note that the RHS is uniform in functions ϕ with $\|\phi\|_\infty \le 1$. Moreover, the only dependence of our upper bounds is only through $\|\phi\|_\infty$, and, making the above notations more cumbersome, we could have been writing the estimates not for a given, individual function ϕ but for the supremum over $\phi \in \mathscr{B}_1$, thus proving the claim. □

Remark 2.5.5. It is straightforward to see that the bound (2.158) can be reformulated in the following way. As in Sect. 1.4.2, denote by M the operator of multiplication by the function $x \in \mathbb{Z}^d \mapsto |x|$. Then for any finite subset $\mathbb{K} \subset \mathbb{Z}^d$, one has

$$\mathbb{E}\left[\sup_{t\in\mathbb{R}} \left\| \mathrm{e}^{+a(\ln M)^{1+c}}\, \mathrm{e}^{-itH(\omega)}\, \mathbf{1}_{\mathbb{K}} \right\|\right] < \infty \tag{2.164}$$

(the norm under the expectation is the operator norm in the space $\mathcal{B}(\ell^2(\mathbb{Z}^d))$ of bounded operators in $\ell^2(\mathbb{Z}^d)$).

Even in the case where the random potential is unbounded, the above arguments show that the fixed-energy MSA estimates imply strong dynamical localization in any given bounded energy interval $I \subset \mathbb{R}$. The derivation of strong dynamical localization (from the fixed-energy MSA) in the entire real line $\mathbb{R}$ requires more involved arguments. In the next Section, we obtain it from the variable-energy MSA estimates.

2.6 Decay of the Green's Functions, III: Variable-Energy Analysis

In this section, we present another version of the single-particle multi-scale analysis: the variable-energy MSA, going back to the papers [96, 161]. In Chap. 4, we will adapt to the multi-particle systems both the fixed-energy and the variable-energy versions of the MSA.

As we have shown in Sect. 2.5 (cf. Theorem 2.5.2), the probabilistic bound (2.136) (or its improved version (2.159)) for Green's functions at a fixed energy E in a given interval $I \subset \mathbb{R}$ implies a stronger bound (2.137) relative to *the entire interval* I. The variable-energy MSA leads directly to a bound of the form (2.137) (hence to the spectral and strong dynamical localization). However, as we will see in this section, such an improvement requires more complex analytical techniques.

Throughout this section, we consider fixed the interval $I \subseteq \mathbb{R}$ (which may, therefore, be bounded or not). Analysis of localization properties (viz., decay of Green's functions and of eigenfunctions) can be effectively restricted to the interval I; this is a very convenient feature of the MSA which allows, in particular, to establish localization in a narrow band of "extreme" energies, making use of the "Lifshitz tails" phenomenon, even if the amplitude $|g|$ of the random potential is very small. On the other hand, under the assumption of strong disorder ($|g| \gg 1$), localization can be established in $I = \mathbb{R}$.

Situation is more complex for multi-particle systems treated in Chaps. 3 and 4. In our book, we mainly focus on strongly disordered multi-particle systems, where complete localization (in $I = \mathbb{R}$) can be proven.

Note that the initial scale bound suitable for the variable-energy MSA in $I = \mathbb{R}$ is provided by Theorem 2.3.2.

2.6.1 *Sparse Singular Cubes: An Informal Discussion*

The analysis of the Green's functions $G_{\Lambda_L(u)}(x, y; E, \omega)$ in a cube $\Lambda_L(u)$ is based on properties of Green's functions in smaller cubes inside $\Lambda_L(u)$. In this Section we analyze a particular phenomenon where the presence of sparse singular subcubes in $\Lambda_L(u)$ cannot destroy the exponential decay of the Green's function; cf. Definition 2.1.1. We focus first on principal mechanisms, postponing formal definitions, statements and proofs to Sects. 2.6.2—2.6.6.

Fix a scale L_k with $k \geq 0$ (cf. Eq. (2.16)) and suppose we have two cubes $\Lambda_{L_k}(u) \subset \Lambda_{L_{k+1}}(x)$ such that $\partial^+ \Lambda_{L_k}(u) \subset \Lambda_{L_{k+1}}(x)$ and

(i) $\Lambda_{L_k}(u)$ is (E, m)-S,
(ii) All cubes $\Lambda_{L_k}(v) \subset \Lambda_{L_{k+1}}(x)$ with $|u - v| = 2L_k + 1$ are (E, m)-NS,
(iii) The cube $\Lambda_{L_{k+1}}(x)$ is E-CNR (cf. Definition 2.2.3).

Let $y \in \Lambda_{L_{k+1}}(x) \setminus \Lambda_{L_k}(u)$. Observe that the centers v of the cubes $\Lambda_{L_k}(v)$ under consideration belong to the outer boundary $\partial^+ \Lambda_{2L_k}(u)$. Note also that the cube $\Lambda_{2L_k}(u)$ must be E-NR, since, by assumption (iii), $\Lambda_{L_{k+1}}(x)$ is E-CNR. Then, applying the geometric resolvent inequality (2.64) to the cube $\Lambda_{2L_k}(u)$, we see that

$$\begin{aligned}&\left|G_{\Lambda_{L_{k+1}}(x)}(u, y; E)\right| \\ &\qquad \le e^{L_{k+1}^{\beta}} |\partial^+ \Lambda_{2L_k}(u)| \left(\max_{v \in \partial^+ \Lambda_{2L_k}(u)} \left|G_{\Lambda_{L_{k+1}}(x)}(v, y; E)\right| \right).\end{aligned}$$

Next we apply the GRI to each cube $\Lambda_{L_k}(v)$ and use its non-singularity (assumed in condition (ii)):

$$\begin{aligned}&\left|G_{\Lambda_{L_{k+1}}(x)}(v, y; E)\right| \\ &\qquad \le e^{-\gamma(m,L_k)L_k} \left(\max_{v' \in \partial^+ \Lambda_{L_k}(v)} \left|G_{\Lambda_{L_{k+1}}(x)}(v', y; E)\right| \right).\end{aligned} \tag{2.165}$$

Combining these two observations, we come to the conclusion that

$$\begin{aligned}&\left|G_{\Lambda_{L_{k+1}}(x)}(u, y; E)\right| \\ &\qquad \le e^{-\gamma(m,L_k)L_k + L_k^{\beta}} \left(\max_{v' : |v'-u| \le 3L_k + 2} \left|G_{\Lambda_{L_{k+1}}(x)}(v', y; E)\right| \right).\end{aligned} \tag{2.166}$$

Note that the bound (2.166) is less precise than (2.165) since the maximum is taken over a larger set of points v'; however, as we will see later, this is sufficient for the purposes of the MSA (and leads to a simpler proof).

Furthermore, we have, for k large enough,

$$\gamma(m, L_k)L_k - L_k^{\beta} = mL_k + mL_k^{7/8} - L_k^{1/2} \ge m\left(1 + \frac{1}{2}L_k^{-1/8}\right)L_k.$$

As a result, (2.166) leads to the following upper bound for the cube $\Lambda_{L_{k+1}}(x)$ and any $y \in \Lambda_{L_{k+1}}(x) \setminus \Lambda_{L_k}(u)$, including $y \in \partial^- \Lambda_{L_{k+1}}(x)$:

$$\begin{aligned}&\left|G_{\Lambda_{L_{k+1}}(x)}(u, y; E)\right| \\ &\qquad \le e^{-m(1+L_k^{-1/8}/2)L_k} \left(\max_{v' : |v'-u| \le 2+3L_k} \left|G_{\Lambda_{L_{k+1}}(x)}(v', y; E)\right| \right),\end{aligned} \tag{2.167}$$

while for an (E, m)-NS cube one would have

$$\left|G_{\Lambda_{L_{k+1}}(x)}(u,y;E)\right| \le e^{-\gamma(m,L_k)L_k} \left(\max_{v':|v'-u|\le 1+L_k} \left|G_{\Lambda_{L_{k+1}}(x)}(v',y;E)\right|\right). \tag{2.168}$$

Apart from a minor variation of the decay exponent, necessary to compensate for tolerable resonances and combinatorial factors, the principal difference between (2.167) and (2.168) resides in the diameter of the reference set over which the maximum in the RHS has to be taken.

It is worth noticing that, should one of the above cubes $\Lambda_{L_k}(v)$ be also (E,m)-S (along with $\Lambda_{L_k}(u)$), we would have two disjoint (E,m)-S cubes, with the same value of $E \in \mathbb{R}$. We will see later that such an event has a small probability (cf. Eq. (2.183)).

A possible aggregation of neighboring singular cubes may make estimates more cumbersome. For example, the above condition (ii) may become invalid. Nevertheless, if this occurs then one can examine the neighbors of the cube $\Lambda_{L_k+(2L_k+1)}(u)$ instead of the neighbors of $\Lambda_{L_k}(u)$, then, if necessary, the cubes

$$\Lambda_{L_k+2(2L_k+1)}(u), \Lambda_{L_k+3(2L_k+1)}(u), \dots, \Lambda_{L_k+J(2L_k+1)}(u),$$

where J is a chosen number. In the case where each cube $\Lambda_{L_k+j(2L_k+1)}(u)$, $j = 0,\dots,J$, has at least one adjacent (E,m)-S cube, there exist at least $J+1$ disjoint (E,m)-S cubes for the same value of E. As will be shown below, such an event (starting with $J=3$) has a probability that is small enough so that the MSA will be applicable. We will see in Sect. 2.6.5 that if this unlikely event does not occur and the cube $\Lambda_{L_{k+1}}(x)$ is E-CNR, then this cube is also (E,m)-NS.

We see that, while performing the scaling step, when we pass from L_k to L_{k+1}, the exponent of the form $\gamma(m,L_k) = m(1+L_k^{-1/8})$ provides a stronger decay at scale L_k than that required in order to have $\gamma(m,L_{k+1})$ at scale L_{k+1}. This allows us to keep the parameter m intact in the course of the scaling induction. It also emphasizes the fact that the decay exponent $\gamma(m,L_k)$ at any scale is larger than m.

Technically, it will be convenient to slightly modify in Sect. 2.6.2 the geometrical constructions sketched in the above informal discussion.

2.6.2 Radial Descent Bounds

We now turn to a formal analysis of resolvents in finite cubes. The key ideas here go back to [96, 156] and [161]; however, in this book we make use of a number of modifications bearing in mind future multi-particle adaptations (see Chap. 4).

The main analytic tool of the MSA scheme, operating with finite cubes, was developed in [156, 160], further refined in [161] (where it was adapted to the variable-energy MSA) and employed since then in a number of papers on Anderson localization, both in discrete and continuous configuration spaces.

Remark 2.6.1. It has to be stressed that the inequalities derived in Sects. 2.6.2 and 2.6.3 are purely deterministic; moreover, they do not rely upon a specific structure of the potential energy. In particular, they apply to LSOs of the form $H_\Lambda(\omega)$ (single-particle Anderson tight-binding Hamiltonians studied in this chapter) and $\mathbf{H}_\Lambda(\omega)$ (multi-particle Anderson tight-binding Hamiltonians considered in Chaps. 3 and 4).

In the context of the MSA, we also need a more general notion of (ℓ, q)-subharmonic function.

Definition 2.6.1. *(See Fig. 2.1.) Suppose that $q \in (0,1)$, integers $L \geq \ell \geq 1$, a finite connected subset $\Lambda \subset \mathbb{Z}^d$ and a function $f : \Lambda \to \mathbb{R}_+$ are given. Fix a cube $\Lambda_L(u) \subsetneq \Lambda$. A point $x \in \Lambda_{L-\ell}(u)$ (i.e., with $\Lambda_\ell(x) \subset \Lambda_L(u)$) will be called (ℓ, q)-regular for the function f if*

$$f(x) \leq q\mathbf{M}(f, \Lambda_{\ell+1}(x)),$$

and (ℓ, q)-singular, otherwise.[9]

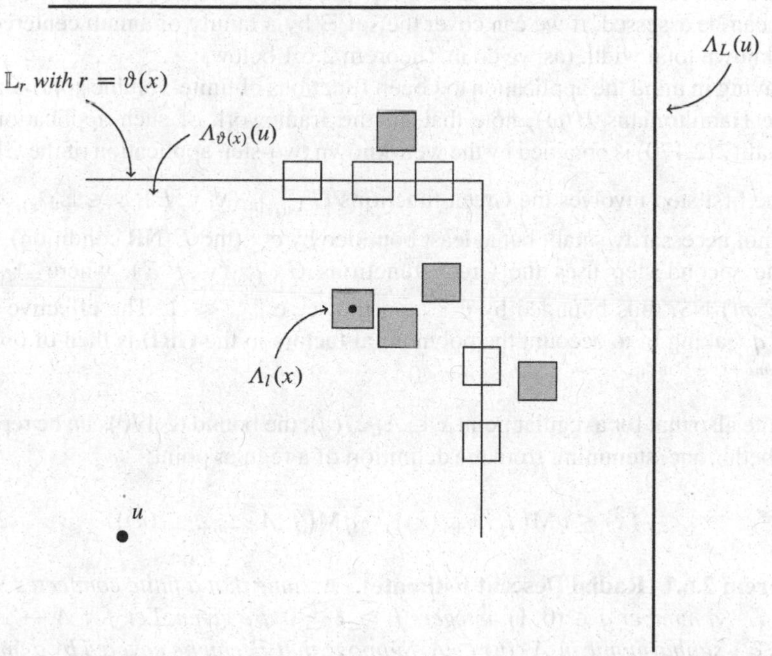

Fig. 2.1 Example for Definition 2.6.1. The set Ξ is covered by five *gray squares*, so all squares of radius l centered at $\mathbb{L}_{\vartheta(x)} = \partial^- \Lambda_{\vartheta(x)}(u)$ are regular; five such squares (*white*) are drawn by way of example

[9] Note that these notions do not apply to points x with $\Lambda_\ell(x) \not\subset \Lambda_L(u)$.

Denote by Ξ the set of all (ℓ, q)-singular points of function f in $\Lambda_L(u)$.

Next, denote by $\mathbb{L}_r(u)$ the spherical layer $\{y \in \mathbb{Z}^d : |y-u| = r\}$ centered at u of radius r, and call it regular if $\mathbb{L}_r(u) \cap var\, \Xi = \varnothing$, i.e., if it contains only regular points.

Given a point $x \in \Xi$ such that for some $|x-u|+1 \le r \le L-\ell$, the layer $\mathbb{L}_r(u)$ is regular, set:

$$\vartheta(x) = \min\{r > |x-u| : \mathbb{L}_r(u) \cap \Xi = \varnothing\}; \tag{2.169}$$

if no such regular layer exists, set formally $\vartheta(x) = +\infty$.

Given such a subset $\Xi \subset \Lambda_L(u)$, the function f will be called (ℓ, q, Ξ)-subharmonic in $\Lambda_L(u)$ if for any $x \in \Xi$ with $\vartheta(x) < \infty$, one has

$$f(x) \le q\mathrm{M}(f, \Lambda_{\vartheta(x)+\ell+1}(u)). \tag{2.170}$$

It is worth noticing that if, for some $r > 0$, the layer $\mathbb{L}_r(u)$ is regular, i.e., $\mathbb{L}_r(u) \cap \Xi = \varnothing$, then $\vartheta(x) < \infty$ for all $x \in \Lambda_{r-1}(u)$, while $\vartheta(x) = +\infty$ for all $x \in \Lambda_L(u) \setminus \Lambda_r(u)$. This means that some annulus adjacent to the boundary $\partial^- \Lambda_L(u)$ may be "wasted". On the other hand, the positive news is that the width of such a layer can be assessed, if we can cover the set Ξ by a family of annuli centered at u with known total width (as we do in Theorem 2.6.1 below).

Having in mind the application to Green functions of finite-volume approximants of the Hamiltonians $H(\omega)$, note that, in the framework of such application, the inequality (2.170) is obtained by the well-known two-step application of the GRI:

- The first step involves the Green functions $G_{\Lambda_{\vartheta(x)}(u)}(x, y; E)$, $y \in \mathbb{L}_{\vartheta(x)}$, which is not necessarily small, but at least bounded by e^{ℓ^β} (the E-NR condition), and
- The second step uses the Green functions $G_{\Lambda_\ell(y')}(y', z; E)$, where $\Lambda_{\ell(y')}$ is (E, m)-NS, thus bounded by $\mathrm{e}^{-m\ell}$, so that $\mathrm{e}^{\ell^\beta}\mathrm{e}^{-m\ell} \ll 1$. The effective value of q (taking in to account the polynomial factors in the GRI) is then of order of $\mathrm{e}^{-m\ell+\ell^\beta+C\ln \ell} \le \mathrm{e}^{-m'\ell}$, $m' \approx m > 0$.

Note also that for a regular point $x \in \Lambda_{L-\ell}(u)$, the bound (2.170) can be replaced by a better one, stemming from the definition of a regular point:

$$f(x) \le q\mathrm{M}\big(f, \Lambda_{\ell+1}(x)\big) \le q\mathrm{M}\big(f, \Lambda_{|x-u|+\ell+1}(u)\big). \tag{2.171}$$

Theorem 2.6.1 (Radial Descent Estimate). *Assume that a finite connected subset $\Lambda \subset \mathbb{Z}^d$, a number $q \in (0,1)$, integers $L \ge \ell \ge 0$ are given. Let $f : \Lambda \to \mathbb{R}_+$ be (ℓ, q, Ξ)-subharmonic in $\Lambda_L(u) \subset \Lambda$. Suppose that Ξ can be covered by a union A of concentric lattice annuli $\mathrm{A}_j \subset \Lambda_L(u)$ of width $b_j - a_j + 1$:*

$$\mathrm{A} = \bigcup_{j=1}^{J} \mathrm{A}_j \quad \textit{where} \quad \mathrm{A}_j = \Lambda_{b_j}(u) \setminus \Lambda_{a_j-1}(u), \ 1 \le j \le J.$$

Let $w(\mathrm{A}) := \sum_j (b_j - a_j + 1)$ *be the total width. If* $w(\mathrm{A}) \le L - \ell$, *then*

$$f(u) \le q^{\lfloor (L+1-w(\mathrm{A}))/(\ell+1) \rfloor} \mathrm{M}(f, \Lambda) \le q^{\frac{L-w(\mathrm{A})-\ell}{\ell+1}} \mathrm{M}(f, \Lambda). \tag{2.172}$$

Proof. The range of the radial mapping $\Phi : x \mapsto |x - u|$, defined on $\Lambda_{L+1}(u)$, is the integer interval $[[0, L+1]] := [0, L+1] \cap \mathbb{Z}$. Call an integer $r \in [[0, L-\ell]]$ regular if $\mathbb{L}_r(u) \cap \Xi = \varnothing$. Observe that every $r \in [[0, L-\ell]]$ with $\mathbb{L}_r(u) \cap \mathrm{A} = \varnothing$ is regular, so there are $\ge L - \ell - w(\mathrm{A})$ regular points in $[[0, L-\ell]]$.

By assumption, $L - w(\mathrm{A}) \ge \ell \ge 0$, so

$$\frac{L - w(\mathrm{A}) + 1}{\ell + 1} \ge \left\lfloor \frac{L - w(\mathrm{A}) + 1}{\ell + 1} \right\rfloor =: n + 1, \quad n \ge 0.$$

By decreasing j, define inductively a finite sequence of integers $\{r_j, 0 \le j \le n\}$ as follows:

$$\begin{aligned} r_n &= \max\{r \le L - \ell : \mathbb{L}_r \cap \Xi = \varnothing\} \\ r_j &= \max\{r \le r_{j+1} - \ell - 1 : \mathbb{L}_r \cap \Xi = \varnothing\}, \ j = n-1, \dots, 0. \end{aligned} \tag{2.173}$$

Set also $r_{n+1} = L + 1$ (the regularity property does not apply to r_{n+1}). Observe that one can indeed construct $n+1$ integers $r_n > \cdots > r_0 \ge 0$, since $L - w(\mathrm{A}) + 1 \ge (n+1)(\ell+1)$ and

$$L + 1 - r_0 = \sum_{j=0}^{n} (r_{j+1} - r_j) \le (n+1)(\ell+1) + w(\mathrm{A}),$$

so $r_0 \ge L + 1 - w(\mathrm{A}) - (n+1)(\ell+1) \ge 0$. Next, introduce the non-decreasing non-negative function

$$F : r \mapsto \mathrm{M}(f, \Lambda_r(u)), \ \ r = 0, 1, \dots, L+1.$$

Since for all $0 \le j \le n$, $r_j + \ell + 1 \le L + 1$ and $\mathbb{L}_{r_j}$ is regular, i.e., the points of the layer $\mathbb{L}_{r_j}$ are regular, so one has

$$\begin{aligned} F(r_n) = \mathrm{M}\big(f, \Lambda_{r_n}(u)\big) &\le q\mathrm{M}\big(f, \Lambda_{r_n+\ell+1}(u)\big) \\ &\le q\mathrm{M}(f, \Lambda) \end{aligned}$$

and for all $0 \le j \le n-1$, owing to (2.173),

$$\begin{aligned} F(r_j) = \mathrm{M}\big(f, \Lambda_{r_j}(u)\big) &\le q\mathrm{M}\big(f, \Lambda_{r_j+\ell+1}(u)\big) \\ &\le qF(r_{j+1}). \end{aligned}$$

Therefore, the backward induction in $j = n, n-1, \dots, 0$ gives rise to the following bounds:

$$
\begin{aligned}
f(u) &\le \mathrm{M}\big(f, \Lambda_{r_0}(u)\big) = F(r_0) \le q^{n+1}\mathrm{M}(f, \Lambda) \\
&\le q^{\lfloor (L+1-w(\mathrm{A}))/(\ell+1)\rfloor}\mathrm{M}(f, \Lambda) \le q^{\frac{L-w(\mathrm{A})-\ell}{\ell+1}}\mathrm{M}(f, \Lambda). \qquad \square
\end{aligned}
$$

In the context of the MSA scheme, f is simply the Green's function. From this point of view, Theorem 2.6.1 means that, on each iteration of the GRI (2.64), we earn a factor q in the bound for $|G_{\Lambda_L(u)}(x, y; E)|$ provided that we do not hit a singular cube Λ_l. On the other hand, if we hit a singular cube Λ_l we still can continue iterating the GRI, if E-R cubes of radius $\ge \ell$ are excluded; thus it will be important to assess the number of successful iterations; cf. Sect. 2.6.4.

In a nutshell, the recipe is: "Remove all singular points and see what is left." What is left is a regular, (ℓ, q)-subharmonic function.

A reader familiar with the techniques from [161] may notice that we replaced the main geometrical argument (based upon the GRI), which can be characterized as radial ascent (where one argument, y, of the Green's function $G_{\Lambda_L(x)}(x, y; E)$ moves from the center x of the cube $\Lambda_L(x)$ towards its boundary $\partial^- \Lambda_L(x)$) by radial descent. The main goal of such a modification is two-fold:

- In the course of the von Dreifus–Klein geometrical procedure, the above mentioned argument $y \in \Lambda_L(x)$ travels along a highly implicit and possibly *infinite* trajectory $(y_0, y_1, \dots)$. An astute reasoning allows, however, to conclude that a sufficiently long but finite segment of this trajectory gives rise to the desired upper bound on the Green's function. Such a formal argument does not have a direct physical meaning (and, as evidences our experience, often puzzles physicists). In our scheme, one obtains recursively upper bounds on the values

$$
f_y\big(L - n(\ell+1)\big) := \max_{x \in \Lambda_{L-n(\ell+1)}(u)} |G_{\Lambda_L(u)}(x, y; E)| \le \mathrm{e}^{-m \cdot n\ell}
$$

  for any $y \in \partial^- \Lambda_L(u)$ and $n = 1, 2, \dots, \lfloor L/(\ell+1) \rfloor$. Clearly, this geometrical procedure is essentially one-dimensional: it operates with the function of one variable

$$
f_y : r \mapsto \max_{x \in \Lambda_r(u)} |G_{\Lambda_L(u)}(x, y; E)|,
$$

  where $\Lambda_r(u)$, as usual, is the cube of radius r centered at u. In the absence of (E, m)-S cubes of radius ℓ in the annulus $\Lambda_L(u) \setminus \Lambda_r(u)$, its logarithm is roughly proportional to the width $L - r$ of the annulus. This reminds the under-barrier decay of a wave function of a quantum particle in a classically forbidden area; indeed, the latter is merely one of the possible situations[10] where the exponential decay of the Green's functions can be proved.

[10]The Combes–Thomas estimate can be considered as a far-going generalization of the under-barrier decay phenomenon, also based on the analytic continuation technique.

- In the presence of (E,m)-S cubes of radius ℓ in the annulus $\Lambda_L(u) \setminus \Lambda_r(u)$, it suffices (as our method shows) to simply exclude from $L-r$ the total width of annuli covering all singular cubes. By comparison, the Dreifus–Klein procedure requires a more elaborate "clustering" of (E,m)-S cubes. The simplification provided by our method is particularly visible (and helpful) if one allows a growing number of (E,m)-S cubes. See a recent work [61] where such an approach allowed to establish sub-exponential decay bounds on Green's functions (in single-particle models), of the form

$$\max_{y \in \partial^- \Lambda_{L_k}(u)} |G_{\Lambda_{L_k}(u)}(u,y;E)| \le e^{-L_k^{\delta}}, \ \delta \in (0,1),$$

by a fairly simple adaptation of the Dreifus–Klein method from [161], without using a more complex bootstrap MSA technique developed by Germinet and Klein [99].

2.6.3 Subharmonicity of Green's Functions in Moderately Singular Cubes

Theorem 2.6.2. *Suppose that an integer $k \ge 0$ and two numbers $m \ge L_0^{-1/4}$ and $E \in \mathbb{R}$ are given. Consider a cube $\Lambda_{L_{k+1}}(u)$ which is E-CNR.*

Let $\Xi = \Xi(E) \subset \Lambda_{L_{k+1}-L_k-1}(u)$ be a (possibly empty) subset such that any cube $\Lambda_{L_k}(x) \subset \Lambda_{L_{k+1}-L_k-1}(u) \setminus \Xi$ is (E,m)-NS.

If $L_0 \ge 16$, then for any $y \in \partial^- \Lambda_{L_{k+1}}(u)$ the function

$$f_y : x \mapsto |G_{\Lambda_{L_{k+1}}(u)}(x,y;E)|$$

is (ℓ, q, Ξ)-subharmonic in $\Lambda_{L_{k+1}-1}(u)$ with

$$q = e^{-m(1+\frac{1}{2}L_k^{-1/8})L_k}. \tag{2.174}$$

Proof. By assumption, for any cube $\Lambda_{L_k}(x) \subset \Lambda_{L_{k+1}-L_k-1}(u) \setminus \Xi$ one has, by the (E,m)-NS property,

$$f_y(x) \le e^{-m(1+L_k^{-1/8})L_k} \mathrm{M}\big(f_y, \Lambda_{L_k}(x)\big) < q\mathrm{M}\big(f_y, \Lambda_{L_k}(x)\big)$$

with $q \in (0,1)$ given by (2.174).

Next, given a point $x \in \Xi$ with $\vartheta(x) < +\infty$, apply first the GRI to the cube $\Lambda_{\vartheta(x)}(u)$, using its E-CNR property:

$$\begin{aligned}|G_{\Lambda_{L_{k+1}}(u)}(x,y;E)| &\le \|G_{\Lambda_{\vartheta(x)}(u)}(E)\| \, \mathrm{M}\big(f_y, \mathbb{L}_{\vartheta(x)+1}(u)\big) \\ &< e^{L_{k+1}^{\beta}} \mathrm{M}(f_y, \Lambda_{L_k}(x)).\end{aligned}$$

Applying the GRI to each cube $\Lambda_{L_k}(v)$, $v \in \mathbb{L}_{\vartheta(x)+1}(u)$, which has to be (E,m)-NS by definition of the radius $\vartheta(x) < \infty$, we obtain

$$\begin{aligned} f_y(v) &\le \mathrm{e}^{-m(1+L_k^{-1/8})L_k} \mathrm{M}\big(f_y, \Lambda_{L_k+1}(v)\big) \\ &\le \mathrm{e}^{-m(1+L_k^{-1/8})L_k} \mathrm{M}\big(f_y, \Lambda_{\vartheta(x)+L_k+1}(u)\big). \end{aligned}$$

Therefore,

$$\begin{aligned} |G_{\Lambda_{L_{k+1}}(u)}(x,y;E)| &< \mathrm{e}^{-m(1+L_k^{-1/8})L_k + L_{k+1}^{1/4}}\, \mathrm{M}(f_y, \Lambda_{\vartheta(x)+L_k+1}(u)) \\ &\le \mathrm{e}^{-m(1+\frac{1}{2}L_k^{-1/8})L_k}\, \mathrm{M}(f_y, \Lambda_{\vartheta(x)+L_k+1}(u)), \end{aligned} \tag{2.175}$$

provided that

$$\frac{m}{2} L_k^{7/8} - L_{k+1}^{1/4} \ge 0. \tag{2.176}$$

Analyze this condition: by hypothesis, $m \ge L_0^{-1/4} \ge L_k^{-1/4}$, thus

$$\frac{m}{2} L_k^{\frac{7}{8}} \ge \frac{1}{2} L_k^{\frac{7}{8}-\frac{1}{4}} = \frac{1}{2} L_k^{\frac{5}{8}}.$$

With $\alpha = 3/2$, one has $L_{k+1}^{1/4} = L_k^{3/8}$, so for $L_0 \ge 16$ we have

$$\frac{m}{2} L_k^{\frac{7}{8}} \ge \frac{1}{2} L_k^{\frac{5}{8}} \ge L_k^{\frac{3}{8}} = L_{k+1}^{\frac{1}{4}},$$

since $L_k^{2/8} \equiv L_k^{1/4} \ge 16^{1/4} = 2$. This justifies (2.176), hence also (2.175), and thus completes the proof of Theorem 2.6.2. □

Theorem 2.6.2 leads to the following

Definition 2.6.2. *(See Fig. 2.2.) Fix an interval $I \subseteq \mathbb{R}$, a positive number m and integers $K, k \ge 0$. A cube $\Lambda_{L_{k+1}}(u)$ is called (m, I, K)-non-partially-singular (in short: (m, I, K)-NPS) if, $\forall\ E \in I$, it does not contain any collection of $K+1$ pairwise disjoint (E,m)-S cubes of radius L_k. Consequently, all (E,m)-S cubes of radius L_k inside $\Lambda_{L_{k+1}}(u)$ can be covered by the union A of at most K disjoint concentric annuli $\mathrm{A}_i(u)$ centered at u of total width $w(\mathrm{A}) \le K \cdot 4L_k$.*

Otherwise, $\Lambda_{L_{k+1}}(u)$ is said to be (m, I, K)-partially singular.

2.6.4 Scaling of the Decay Exponent

In this subsection we estimate the effective decay parameter in a cube of radius L_{k+1}, assuming that it is E-NR, with E in a given interval I, and does not contain

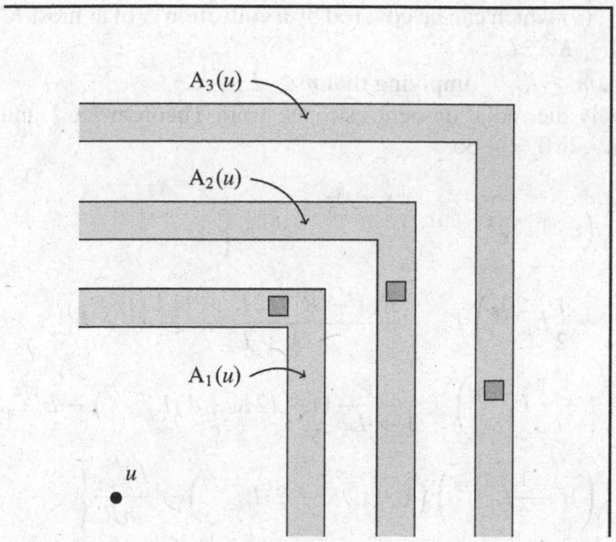

Fig. 2.2 Example for Definition 2.6.2, with $K = 3$. Three (E, m)-S squares of radius L_k (*dark gray*) are covered by three *light gray* annuli $A_i(u)$ of width $4L_k$, $i = 1, \ldots, K = 3$, so all squares of radius L_k in the remaining area (*white*) are (E, m)-NS

too many smaller (E, m)-S cubes. As was repeatedly stated earlier, we need to establish property (DS.I, m, p_k, L_k) for all $k \geq 0$. Theorem 2.6.3 below provides a key analytic argument allowing us to reproduce this property for L_{k+1}, assuming that it holds for L_k. It is stated in a manner applicable in both situations of interest: a large $|g|$ (when mass $m > 0$ can be arbitrarily large and the interval I may be unbounded) and a moderate (or small) $|g|$ when $I = (0, E^*)$ will be at the lower edge of the spectrum (and $m = m(g) > L_0^{-1/4}$).

Theorem 2.6.3. *Fix an interval $I \subset \mathbb{R}$ and an integer $K \geq 0$. There exists $L^* = L^*(K) \in \mathbb{N}$ with the following property.*

Let $L_0 \geq L^*$, $m \geq L_0^{-1/4}$ and $k \in \mathbb{N}$. If a cube $\Lambda_{L_{k+1}}(u)$ is (m, I, K)-NPS and E-CNR for some $E \in I$, then

$$|\partial^- \Lambda_{L_{k+1}}(u)| \cdot \max_{x \in \Lambda_{L_k}(u)} \max_{y \in \partial^- \Lambda_{L_{k+1}}(u)} \left| G_{\Lambda_{L_{k+1}}(u)}(x, y; E) \right| \leq e^{-\gamma(m, L_{k+1}) L_{k+1}}. \tag{2.177}$$

Proof. Fix a point $x \in \Lambda_{L_{k+1}}(u)$. Owing to Theorem 2.6.2, the function

$$f_y : x \mapsto \left| G_{\Lambda_{L_{k+1}}(u)}(x, y; E) \right|$$

is (L_k, q, Ξ)-subharmonic in $\Lambda_R(x)$ with $R = |y - x| \geq L_{k+1} - L_k - 1 > \frac{1}{2} L_{k+1}$,

$$q = e^{-m(1 + \frac{1}{2} L_k^{-1/8}) L_k},$$

and a set $\Xi \subset \Lambda_{L_{k+1}}(u)$ which can be covered by a collection A of at most K annuli of total width $w(\mathrm{A}) \le K \cdot 2L_k$.

As was assumed, $m \ge L_0^{-1/4}$ implying that $m \ge L_k^{-1/4}$.

Now we can apply the radial descent estimate from Theorem 2.6.1 and write, with the convention $-\ln 0 = +\infty$:

$$
\begin{aligned}
-\ln f_y(x) &\ge -\ln\left\{ \left(\mathrm{e}^{-m(1+\frac{1}{2}L_k^{-1/8})L_k}\right)^{\frac{R-w(\mathrm{A})-L_k}{L_k+1}} \mathrm{e}^{L_{k+1}^{\beta}} \right\} \\
&= m\left(1+\frac{1}{2}L_k^{-1/8}\right)L_k \cdot \frac{R(1-R^{-1}(2K+1)\,L_k)}{1+L_k} - L_{k+1}^{1/4} \\
&\ge Rm\left(1+\frac{1}{2}L_k^{-1/8}\right) \cdot \frac{L_k}{1+L_k}(1-(2K+1)L_k^{1-3/2}) - L_{k+1}^{1/4} \\
&\ge Rm\left\{\left(1+\frac{1}{2}L_k^{-1/8}\right)\left(1-(2K+2)L_k^{-1/2}\right) - \frac{L_{k+1}^{1/4}}{mR}\right\}
\end{aligned}
$$

(provided that $L_0 \ge (16K+16)^{8/3}$)

$$
\begin{aligned}
&\ge Rm\left(1+\frac{1}{4}L_k^{-1/8} - \frac{L_{k+1}^{1/4}}{L_k^{-1/4}\cdot\frac{1}{2}L_{k+1}}\right) \\
&\ge Rm\left(1+\frac{1}{4}L_k^{-1/8} - 2L_{k+1}^{-1/2}\right) \\
&\ge \gamma(m, L_{k+1})R + \ln|\partial\Lambda_{L_{k+1}}(u)|
\end{aligned}
\tag{2.178}
$$

provided that L_0 is large enough so that for all $k \ge 0$ (hence, $L_k \ge L_0$)

$$
\frac{1}{4}L_k^{-1/8} - 2L_{k+1}^{-1/2} \ge L_{k+1}^{-\frac{1}{8}} + \ln\left(C(d)L_k^{d-1}\right). \tag{2.179}
$$

With $\alpha = 3/2$, $L_{k+1}^{-1/8} = L_k^{-3/16}$ and $L_{k+1}^{-1/2} = L_k^{-3/4}$. Now an elementary comparison of power and logarithmic functions shows that the inequality

$$
\frac{1}{4}L^{-1/8} \ge 2L^{-3/4} + L^{-3/16} + \ln\left(C(d)L^{d-1}\right) \tag{2.180}
$$

holds true for all sufficiently large $L > 0$. Therefore, it suffices to pick L_0 large enough to guarantee the validity of (2.179) for all $k \in \mathbb{N}$. □

Observe that the statement of Theorem 2.6.3 is deterministic: the cube $\Lambda_{L_{k+1}}(u)$ *is assumed* to be (m, I, K)-NPS and E-CNR for some $E \in I$. In the next section we will show that a given cube is (m, I, K)-PS with small probability. We want to emphasize once again that one cannot guarantee that the cube $\Lambda_{L_{k+1}}(u)$ is E-CNR for all energies $E \in I$: with high probability this is false when L_{k+1} is large and I intersects the energy region where the DoS is positive. Indeed, the existence and positivity of the DoS on a given interval I implies that with high probability there are eigenvalues $E_j(\omega)$ of $H_{\Lambda_{L_{k+1}}(u)}(\omega)$ within this interval, at least for large k. Then the cube $\Lambda_{L_{k+1}}(u)$ is certainly E-R for $E = E_j\big(\Lambda_{L_{k+1}}(u)\big) \in I$. However, the MSA bound (cf. Eq. (2.29)) says that, $\forall\ E \in I$, pairs of cubes are simultaneously E-R with small probability; this is what enables us to put the whole variable-energy MSA scheme to work.

2.6.5 *Multiple Singular Cubes are Unlikely*

In this subsection we analyze the property (DS.I, m, p_k, L_k) for cubes of size L_k. For convenience, we repeat the definition (cf. (2.29)):

(DS.I, m, p_k, L_k): *For any pair of disjoint cubes $\Lambda_{L_k}(x)$, $\Lambda_{L_k}(y)$ the following bound holds true:*

$$\mathbb{P}\{\exists\, E \in I: \ \Lambda_{L_k}(x) \text{ and } \Lambda_{L_k}(y) \text{ are } (E,m)\text{-}S\ \} \le L_k^{-2p_k} \tag{2.181}$$

where $p_k = p_0(1+\theta)^k$ for some $p_0 > 6d$ and $\theta > 0$. Cf. (2.30).

Recall that (DS.I, m, p_0, L_0) with $I = \mathbb{R}$ was established in Theorem 2.3.2. This allows to take any interval $I \subseteq \mathbb{R}$ in the statements given in this subsection.

As a matter of convenience, some basic calculations will be done for a cube of radius L and for some p, where L and p are formally unrelated to the scale induction. The corresponding property is stated as follows:

(DS.I, m, p, L): *For any pair of disjoint cubes $\Lambda_L(x)$ and $\Lambda_L(y)$,*

$$\mathbb{P}\{\exists\, E \in I: \ \Lambda_L(x) \text{ and } \Lambda_L(y) \text{ are } (E,m)\text{-}S\ \} \le L^{-2p}. \tag{2.182}$$

As was repeatedly stressed before, inequalities (2.181) and (2.182) lie in the core of the MSA: the aim of the latter is to establish (DS.I, m, p, L) with $L = L_k$ and $p = p_k\ \forall\ k \ge 0$ (with an appropriate choice of L_0, p and θ). In this subsection we focus on implications, for cube $\Lambda_L(u)$, of the fact that the property holds for smaller cubes; this will allow us to justify (and then use) the form of the parameter p_k from Eq. (2.30), with $\theta > 0$ given by Eq. (2.186).

In Lemma 2.6.4 below we refer to property (DS.I, m, p, L_k) stated for cubes of radius L_k:

Lemma 2.6.4. *Suppose that for any pair of disjoint cubes $\Lambda_{L_k}(x)$ and $\Lambda_{L_k}(y)$, $k \ge 0$,*

$$\mathbb{P}\{\exists\, E \in I: \ \Lambda_{L_k}(x) \text{ and } \Lambda_{L_k}(y) \text{ are } (E,m)\text{-}S\ \} \le L_k^{-2p}. \tag{2.183}$$

Then, for L_0 large enough[11] *and $K \geq 3$,*

$$\mathbb{P}\left\{ \Lambda_{L_{k+1}}(u) \text{ is } (m, I, K)\text{-}PS \right\} \leq \frac{3^{4d}}{4!} L_{k+1}^{-\frac{4p}{\alpha}+4d}. \tag{2.184}$$

Proof. For a pair of points u', $u'' \in \mathbb{Z}^d$ we form the event

$$\mathcal{E}(u', u'') = \{\exists\, E \in I : \Lambda_{L_k}(u') \text{ and } \Lambda_{L_k}(u'') \text{ are } (E, m)\text{-S} \}.$$

The event $\left\{\Lambda_{L_{k+1}}(u) \text{ is } (m, I, K)\text{-PS}\right\}$ is contained in the larger event

$$\bigcup_{u_1,\dots,u_4 \in \Lambda_{L_{k+1}}(u)} \left(\mathcal{E}(u_1, u_2) \cap \mathcal{E}(u_3, u_4) \right).$$

Here the union $\bigcup_{u_1,\dots,u_4}$ is taken over all sequences of points $u_1, \dots, u_4 \in \Lambda_{L_{k+1}}(u)$ with distances $|u_{j'} - u_{j''}| > 2L_k + 1$ for any $1 \leq j' < j'' \leq 4$ and $\operatorname{dist}(u_j, \partial^- \Lambda_{L_{k+1}}(u)) \geq L_k$ for any $j = 1, \dots, 4$, under the agreement that each sequence is ordered according to some fixed order on $\mathbb{Z}^d$ (e.g., the lexicographic order). Equivalently, the union is over the collections of pairwise disjoint cubes $\Lambda_{L_k}(u_j) \subset \Lambda_{L_{k+1}}(u)$ taken in the specified order.

Since the cubes $\Lambda_{L_k}(u_1)$, …, $\Lambda_{L_k}(u_4)$ are pairwise disjoint, the events of the form $\mathcal{E}_i = \mathcal{E}(u_{2i-1}, u_{2i})$, $i = 1, 2$, are independent. Hence,

$$\mathbb{P}\{ \mathcal{E}_1 \cap \mathcal{E}_2 \} = \mathbb{P}\{ \mathcal{E}_1 \} \cdot \mathbb{P}\{ \mathcal{E}_2 \} \leq L_k^{-4p} \leq L_{k+1}^{-\frac{4p}{\alpha}}. \tag{2.185}$$

Next, the number of all possible collections of four pairwise disjoint cubes $\Lambda_{L_k}(u_j) \subset \Lambda_{L_{k+1}}(u)$, considered modulo their order, is bounded by

$$\frac{1}{4!}(2L_{k+1} + 1)^{4d} \leq \frac{3^{4d}}{4!} L_{k+1}^{4d}.$$

This leads to the assertion of the lemma. □

Corollary 2.6.5. *Let $\alpha = 3/2$, $p > 6d$, and let $\theta > 0$ satisfy*

$$\theta < \frac{2}{\alpha} - 1 - \frac{2d}{p} = \frac{1}{3} - \frac{2d}{p}. \tag{2.186}$$

There is $L_* \in (0, +\infty)$ such that if $L_0 \geq L_*$, then the following implication holds for any $k \in \mathbb{N}$.

---

[11] See the remarks made in the end of proof of Theorem 2.6.3 concerning the lower threshold for L_0.

Assume that for any pair of disjoint cubes $\Lambda_{L_k}(x)$ and $\Lambda_{L_k}(y)$,

$$\mathbb{P}\{\exists\, E \in I : \Lambda_{L_k}(x) \text{ and } \Lambda_{L_k}(y) \text{ are } (E,m)\text{-S}\ \} \le L_k{}^{-2p(1+\theta)^k}.$$

Then for $K \ge 3$

$$\mathbb{P}\left\{ \Lambda_{L_{k+1}}(u) \text{ is } (m, I, K)\text{-PS} \right\} \le \frac{1}{4} L_{k+1}^{-2p(1+\theta)^{k+1}}. \tag{2.187}$$

Proof. Owing to inequality (2.184), we have:

$$\mathbb{P}\left\{ \Lambda_{L_{k+1}}(u) \text{ is } (m, I, K)\text{-PS} \right\} \le \frac{3^{4d}}{4!} L_{k+1}^{-\frac{4p(1+\theta)^k}{\alpha}+4d}. \tag{2.188}$$

Observe that the inequality

$$\frac{4p(1+\theta)^k}{\alpha} - 4d > 2p(1+\theta)^{k+1} > 0 \tag{2.189}$$

is equivalent to

$$0 < \theta < \frac{2}{\alpha} - 1 - \frac{2d}{p(1+\theta)^k}.$$

Furthermore, it would follow from the stronger inequality (recall that $\alpha = 3/2$)

$$0 < \theta < \frac{2-\alpha}{\alpha} - \frac{2d}{p} \equiv \frac{1}{3} - \frac{2d}{p},$$

and the latter holds true, by hypothesis.

Moreover, if (2.189) holds true, then there is some $c > 0$ such that

$$\frac{4p(1+\theta)^k}{\alpha} - 4d = 2p(1+\theta)^{k+1} + c, \quad c > 0,$$

and therefore,

$$\frac{3^{4d}}{4!} L_{k+1}^{-\frac{4p(1+\theta)^k}{\alpha}+4d} = \frac{3^{4d}}{4!} L_{k+1}^{-c} \cdot L_{k+1}^{-2p(1+\theta)^{k+1}d} \le \frac{3^{4d}}{4!} L_0^{-c} \cdot L_{k+1}^{-2p(1+\theta)^{k+1}d}.$$

With $c > 0$ and L_0 large enough, this proves the assertion. □

2.6.6 The Inductive Scaling Step

In this subsection we complete the scale-induction transition $k \rightsquigarrow k+1$. Theorem 2.6.6 below is the culmination point of the variable-energy single-particle

MSA. Like Theorem 2.6.3 providing the key analytic component of the scale induction, it is stated in a manner allowing to address both the case of large disorder ($|g| \gg 1$, with $m(g) \ge 1$) and that of moderate or weak disorder. In the latter case, the MSA bounds (and the phenomenon of localization) in dimension $d > 1$ can be established only for a particular interval of energies, usually with the help of the Lifshitz tails argument, with mass $m \ge L_0^{-1/4}$; the latter inequality implies that $m \ge L_k^{-1/4}$ for all $k \ge 0$, since $L_k \ge L_0$.

Theorem 2.6.6. *Let L_0 be given and $\{L_k\}$ be the associated length scale. Let some parameters $p > 6d$, $\theta > 0$ satisfying* (2.186) *and $m \ge L_0^{-1/4}$ be given. We denote $p_k = p(1+\theta)^k$, $k \in \mathbb{N}$. If L_0 is large enough, and if for some $k \ge 0$ the property* (DS.I, m, p_k, L_k) *holds true, then* (DS.I, m, p_{k+1}, L_{k+1}) *is also satisfied, i.e., for any pair of disjoint cubes $\Lambda_{L_{k+1}}(u')$, $\Lambda_{L_{k+1}}(u'')$,*

$$\mathbb{P}\left\{\exists\, E \in I :\ \Lambda_{L_{k+1}}(u') \text{ and } \Lambda_{L_{k+1}}(u'') \text{ are } (E,m)\text{-S}\ \right\} \le L_{k+1}^{-2p_{k+1}}.$$

Proof. Let $\Lambda' \doteq \Lambda_{L_{k+1}}(x')$, $\Lambda'' = \Lambda_{L_{k+1}}(x'')$. Consider the events

$$\begin{aligned}\mathcal{B} &= \{\exists\, E \in I :\ \Lambda' \text{ and } \Lambda'' \text{ are } (E,m)\text{-S}\,\},\\ \mathcal{R} &= \{\exists\, E \in I :\ \Lambda' \text{ and } \Lambda'' \text{ are } E\text{-PR}\,\},\\ \mathcal{K} &= \{\,\Lambda' \text{ or } \Lambda'' \text{ is } (m,I,K)\text{-PS}\,\}.\end{aligned}$$

It follows from Theorem 2.6.3 that $\mathcal{B} \subset \mathcal{R} \cup \mathcal{K}$. Indeed, if at least one of the two cubes is CNR, then either it contains $K+1 = 4$ or more (E,m)-S cubes (so that $\mathcal{K}$ occurs), or it is (E,m)-NS (which is incompatible with $\mathcal{B}$).

By the EVC bound for pairs of cubes (cf. Corollary 2.2.4), with $Q > 2p_0$ and sufficiently large[12] L_0, we have that for any $\theta \in (0,1]$

$$\mathbb{P}\{\mathcal{R}\} \le \frac{1}{2} L_{k+1}^{-2p(1+\theta)^{k+1}} = \frac{1}{2} L_{k+1}^{-2p_k(1+\theta)}.$$

At the same time,

$$\begin{aligned}\mathbb{P}\{\mathcal{K}\} &\le \mathbb{P}\left\{\,\Lambda' \text{ is } (m,I,K)\text{-PS}\right\} + \mathbb{P}\left\{\,\Lambda'' \text{ is } (m,I,K)\text{-PS}\right\}\\ &\le \frac{1}{4} L_{k+1}^{-2p_k(1+\theta)} + \frac{1}{4} L_{k+1}^{-2p_k(1+\theta)} = \frac{1}{2} L_{k+1}^{-2p_k(1+\theta)},\end{aligned}$$

with $\theta > 0$ satisfying (2.186), by virtue of Corollary 2.6.5. Therefore,

$$\begin{aligned}\mathbb{P}\{\mathcal{B}\} \le \mathbb{P}\{\mathcal{R}\} + \mathbb{P}\{\mathcal{K}\} &\le \frac{1}{2} L_{k+1}^{-2p_k(1+\theta)} + \frac{1}{2} L_{k+1}^{-2p_k(1+\theta)}\\ &= L_{k+1}^{-2p_k(1+\theta)},\end{aligned}$$

[12]For the role of L_0, see the formulation of Corollary 2.2.4 and (2.53)–(2.54).

which is the assertion of the theorem. □

We now come to the central result of the MSA: the deduction of property (DS.I, m, p_k, L_k) for any k from the property with $k = 0$:

Theorem 2.6.7. *Assume that the random potential $V(\cdot;\omega)$ satisfies Assumption* A *(cf. Eq.* (2.3)*). There exists a positive integer L^* with the following property. Suppose that $L_0 \geq L^*$ and the inequality* (DS.I, m, p_0, L_0) *(see* (2.182)*) is satisfied for some $m \geq 1$ and $p_0 > 6d$. Then property* (DS.I, m, p_k, L_k) *holds true for all $k \in \mathbb{N}$ with $p_k = p_0(1+\theta)^k$ and $\theta = \frac{1}{6} - \frac{d}{p}$ (which is a particular case of* (2.186)*).*

Proof. The assertion follows by induction in k from Theorem 2.6.6. □

Recall that the property (DS.I, m, p_0, L_0), with $I \subseteq \mathbb{R}$, arbitrarily large $L_0 \in \mathbb{N}$, $p_0 > 0$ and $m \geq 1$, also can be inferred from Assumption A, provided that the amplitude $|g|$ of the random potential gV is large enough: $|g| \geq g^*(m, p_0, L_0)$; see Theorem 2.3.2.

On the other hand, Assumptions A and B imply (DS.I, m, p_0, L_0) for an arbitrarily large $p_0 > 0$ and sufficiently large $L_0 \geq L^*(p_0)$, with some $m \geq L_0^{-1/4}$, but only in a sufficiently narrow interval $I(p_0) = [E_*, E_* + \eta(p_0)]$, where $E_* = \inf\{\lambda : F(\lambda) > 0\}$.

2.7 From Green's Functions to Eigenfunctions and Eigenvalues

2.7.1 The MSA Bound Implies Exponential Spectral Localization

In this subsection we deduce from Theorem 2.6.7 the exponential decay of all polynomially bounded solutions of the equation $H(\omega)\psi = E\psi$ with E from a given interval $I \subset \mathbb{R}$ where the MSA bounds can be established. The approach used here goes back to the works [96, 161], and is based on a fact well-known in the spectral theory of Schrödinger operators, as well as their discrete counterparts: *spectrally almost all eigenfunctions of such operators are polynomially bounded.* The first result in this direction was obtained by Shnol [150]; later it has been extended to a large class of operators; cf. [36, 151].

The main result of this section is Theorem 2.7.2 below which asserts that with probability one, for spectrally a.e. $E \in \mathbb{R}$ solutions to the equation $H(\omega)\psi = E\psi$ are indeed decaying exponentially fast at infinity. Consequently, with probability one, $H(\omega)$ admits an orthogonal basis of exponentially fast decaying eigenfunctions. Our proof follows closely the argument given in [161], but is slightly more straightforward in the derivation of the decay exponent for the localized eigenfunctions.

For the purposes of this subsection, it would suffice to consider any unit interval $I = I_n = [n, n+1]$ and prove that, with probability one, the spectrum of $H(\omega)$ in I_n is pure point for each $n \in \mathbb{Z}$. This would imply that the entire spectrum of $H(\omega)$ is a.s. pure point.

As usual, we call a function $\phi : \mathbb{Z}^d \to \mathbb{R}$ polynomially bounded if, for some $C, s \in (0, +\infty)$ and all $x \in \mathbb{Z}^d$,

$$|\phi(x)| \leq C\,(1 + |x|)^s. \tag{2.190}$$

We start with the following simple statement.

Lemma 2.7.1. *Let ψ be a polynomially bounded solution of the equation $H\psi = E\psi$ such that $\psi(\hat{x}) \neq 0$ for some $\hat{x} \in \mathbb{Z}^d$. Fix $m > 0$ and an increasing sequence of integers $\{r_k, k \geq 1\}$ such that $r_k \to \infty$ as $k \to \infty$. Then there exists $k^* \geq 0$ such that every cube $\Lambda_{r_k}(0)$ with $k \geq k^*$ is (E, m)-singular.*

Proof. Since $r_k \to \infty$, there exists $k_\circ \geq 0$ such that for all $k \geq k_\circ$, $\hat{x} \in \Lambda_{r_k^{1/\alpha}}(0)$. We thus restrict our analysis to $k \geq k_\circ$.

Given an integer $k \geq k_\circ$, it suffices to consider the case where $E \notin \sigma\big(H_{\Lambda_{r_k}(0)}\big)$; otherwise $\Lambda_{r_k}(0)$ is automatically (E, m)-S. Owing to Eq. (2.66) for eigenfunctions, the values of ψ inside cube $\Lambda_{r_k^{1/\alpha}}(0) \ni \hat{x}$ can be recovered from its boundary values; in particular,:

$$\psi(\hat{x}) = \sum_{(v,v') \in \partial\Lambda_{r_k}(0)} G_{\Lambda_{r_k}(0)}(\hat{x}, v; E)\, \psi(v'). \tag{2.191}$$

Since $\Lambda_{r_k}(0)$ is (E, m)-NS, for some $C, t \in (0, +\infty)$

$$\begin{aligned} |\psi(\hat{x})| &\leq \sum_{(v,v') \in \partial\Lambda_{r_k}(0)} \left| G_{\Lambda_{r_k}(0)}(\hat{x}, v; E) \right| \, |\psi(v')| \\ &\leq \mathrm{e}^{-m r_k}\, C(1 + r_k)^t \to 0, \text{ as } k \to \infty. \end{aligned}$$

Thus, if $\Lambda_{r_k}(0)$ is (E, m)-NS for arbitrarily large values of r_k, then $\psi(\hat{x}) = 0$, which contradicts the hypothesis of the lemma. □

We would like to repeat that the form of the parameter $p_k = p_0(1 + \theta)^k$ in Eq. (2.181) is not essential for our proof of Theorem 2.7.2: it suffices to have a weaker property (DS.I, m, p, L_k) (see (2.182)) with $p > \alpha d$. However, we note that the RHS of (2.181) behaves as

$$L_k^{-2p_k} = L_k^{-2p_0(1+\theta)^k} \leq \exp\big(-a \ln^{1+c} L_k \big), \text{ with } a, c \in (0, \infty).$$

This will allow us to establish in Sect. 2.7.2 strong dynamical localization with eigenfunction correlators decaying faster than polynomially; a more traditional approach (cf., e.g., [78, 98]) gives rise only to a power-law decay of eigenfunction correlators.

The main idea of Theorem thm:MSA.implies.AL, inferring exponential spectral localization from the key probabilistic bound of the variable-energy MSA, goes back to the pioneering works [96, 161]; it has been adapted to a large number of Anderson-type models, including those where the random potential field $V(x;\omega)$ satisfies the condition called "Independence At Distance" (IAD, in short). In these (and some other) models, the upper bounds on the probabilities of the form

$$\mathbb{P}\{\exists\, E \in I : \Lambda_{L_k}(u) \text{ and } \Lambda_{L_k}(v) \text{ are } (E,m)\text{-S }\}$$

can be obtained only for pairs of cubes with $\operatorname{dist}\big(\Lambda_{L_k}(u), \Lambda_{L_k}(v)\big) \geq R$, where $R > 0$ depends upon the structure of the respective random Hamiltonian. The derivation of exponential localization can be easily adapted to this case. In fact, one can even have $R = R(L) = O(L)$. Such a situation arises in the analysis of multi-particle models studied in Part II of our book. For this reason, we formulate Theorem 2.7.2 below in such a way that it can be applied to a larger class of models, including multi-particle lattice systems.

Theorem 2.7.2. *Let $I \subset \mathbb{R}$ be an interval (possibly unbounded), $m > 0$, and $\mathbb{N} \ni C \geq 1$. Let the length scale $\{L_k, k \in \mathbb{N}\}$ be fixed and suppose that the following bound holds: for all $k \geq 0$ and any pair of cubes $\Lambda_{L_k}(u)$ and $\Lambda_{L_k}(v)$ with $|u-v| \geq 2CL+1$*

$$\mathbb{P}\{\exists\, E \in I : \Lambda_{L_k}(u) \text{ and } \Lambda_{L_k}(v) \text{ are } (E,m)\text{-S }\} \leq L_k^{-2p}, \tag{2.192}$$

with $p > \alpha d$. Then the exists a set $\Omega^* = \Omega^*(I)$ of probability $\mathbb{P}\{\Omega^*\} = 1$ such that, $\forall\, \omega \in \Omega^*$ and $E \in I$, if $\psi = \psi(\cdot;\omega)$ is a polynomially bounded non-zero solution to the equation $H(\omega)\psi = E\psi$ then

$$\limsup_{|x|\to\infty} \frac{\ln|\psi(x;\omega)|}{|x|} \leq -m. \tag{2.193}$$

Consequently, for $\omega \in \Omega^*$ operator $H(\omega)$ has p.p. spectrum in I, and any normalized eigenfunction $\psi_j = \psi_j\big(H(\omega)\big)$ of the operator $H(\omega)$ with eigenvalue $E_j(H(\omega)) \in I$ satisfies

$$\forall\, x \in \mathbb{Z}^d \quad |\psi_j(x)| \leq C_j(\omega)\, \mathrm{e}^{-m|x|} \tag{2.194}$$

with some random constant $C_j(\omega) \in (0,\infty)$.

Proof. (See Fig. 2.3.)
We will call two cubes $\Lambda_L(u), \Lambda_L(v)$ distant if $|u-v| \geq 2CL+1$; this simply makes shorter references to the geometrical condition figuring in the hypothesis of the theorem.

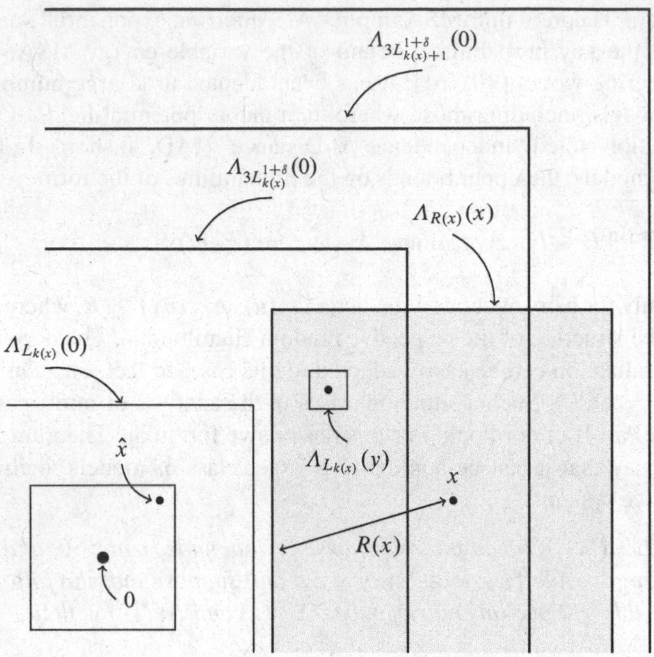

Fig. 2.3 An example for the proof of Theorem 2.7.2, with $C = 1$

As the function ψ solving $H(\omega)\psi = E\psi$ is non-zero, there exists a point $\hat{x} \in \mathbb{Z}^d$ such that $\psi(\hat{x}) \neq 0$. By virtue of Lemma 2.7.1, there exists a positive integer $k_1 = k_1(E, \hat{x}; \omega)$ such that $\forall\, k \geq k_1$ the cube $\Lambda_{L_k}(0)$ is (m, E)-S.

Next, since $2p > Nd\alpha$ with $\alpha \in (1, 2)$, there is $\delta > 0$ such that $\alpha' = (1+\delta)\alpha \in (\alpha, 2)$ and $Nd\alpha' < 2p$. Fix such numbers δ and α'.

Further, for every $k \geq 0$, introduce the annulus

$$\mathbb{A}_k^{(1)} = \Lambda_{6CL_{k+1}^{1+\delta}}(0) \setminus \Lambda_{3CL_k}(0) = \Lambda_{6CL_k^{\alpha'}}(0) \setminus \Lambda_{3CL_k}(0)$$

and the event

$$\Omega_k = \{\exists\, E \in I\ \exists\, y \in \mathbb{A}_k^{(1)} :\ \Lambda_{L_k}(y) \text{ and } \Lambda_{L_k}(0) \text{ are } (E, m) - \text{S}\}.$$

For any $y \in \mathbb{A}_k^{(1)}$, $|y| > 3CL_k > 2CL_k + 1$ thus $\Lambda_{L_k}(y)$ and $\Lambda_{L_k}(0)$ are distant. Therefore, by the assumed property (2.192), we have

$$\mathbb{P}\{\,\Omega_k\,\} \leq \left(12CL_k^{\alpha'} + 1\right)^d L_k^{-2p} \leq (13C)^d L_k^{-2p+d\alpha'}.$$

Since $2p > d\alpha'$, the series

$$\sum_{k=0}^{\infty} \mathbb{P}\{\,\Omega_k\,\} \le \sum_{k=0}^{\infty} \text{Const}\,(d, C)\, L_0^{-\alpha^k(2p-d\alpha')}$$

converges, and by Borel–Cantelli lemma, the event

$$\Omega_{<\infty} = \{\omega \in \Omega : \Omega_k \text{ occurs finitely many times}\}$$

has probability one. So, it suffices to pick a sample of the random potential $V(y;\omega)$, $y \in \mathbb{Z}^d$, with $\omega \in \Omega_{<\infty}$ and prove exponential decay of any given generalized eigenfunction ψ for the specified $\omega \in \Omega_{<\infty}$.

For any $\omega \in \Omega_{<\infty}$, there exists an integer $k_2 = k_2(\omega)$ such that for all $k \ge k_2$, the event Ω_k does *not* occur. Set $k_3 = \max\{k_1, k_2\}$. Then

$$\forall\, k \ge k_3 :\ \forall\, y \in \mathbb{A}^{(1)}_{k+1} \text{ the cube } \Lambda_{L_k}(y) \text{ is } (E, m)\text{-NS}\,.$$

From this point on, the argument becomes deterministic, and we drop the argument ω.

Introduce modified annuli

$$\begin{aligned}\mathbb{A}^{(2)}_k &= \Lambda_{3CL_k^{\alpha'}}(0) \setminus \Lambda_{3CL_k^{1+\delta}}(0)\\ &= \Lambda_{3CL_{k+1}^{1+\delta}}(0) \setminus \Lambda_{3CL_k^{1+\delta}}(0) \subset \mathbb{A}^{(1)}_k,\ k \ge 0,\end{aligned}$$

and note that for every x with $|x| > 3CL_{k_3}^{1+\delta}$ there is an integer $k(x) \ge k_3$ such that $x \in \mathbb{A}^{(2)}_{k(x)}$. Indeed,

$$\bigcup_{k\ge 0} \mathbb{A}^{(2)}_k = \bigcup_{k\ge 0}\left(\Lambda_{3CL_{k+1}^{1+\delta}}(0) \setminus \Lambda_{3CL_k^{1+\delta}}(0)\right) = \mathbb{Z}^d \setminus \Lambda_{3CL_0^{1+\delta}}(0).$$

Further, for $x \in \mathbb{A}^{(2)}_{k(x)}$ one has

$$3CL_{k(x)}^{1+\delta} < |x| \le 3CL_{k(x)+1}^{1+\delta},$$

and using the RHS inequality, we obtain

$$\begin{aligned}R(x) &:= \operatorname{dist}\big(x,\ \partial^-\Lambda_{3CL_{k(x)}}(0) \cup \partial^-\Lambda_{6CL^{\alpha'}_{k(x)}}(0)\big)\\ &= \min\left\{|x| - 3CL_{k(x)},\ 6CL_{k(x)+1}^{1+\delta} - |x|\right\} \qquad (2.195)\\ &= |x| - 3CL_{k(x)}.\end{aligned}$$

Owing to the inequality $|x| > L_{k(x)}^{1+\delta}$ with $\delta > 1$, we have that

$$\frac{R(x)}{|x|} = \frac{|x| - 3CL_{k(x)}}{|x|} \underset{|x|\to\infty}{\longrightarrow} 1. \tag{2.196}$$

Since all cubes $\Lambda_{L_k}(y) \subset \mathbb{A}^{(1)}_{k(x)}$ are (E, m)-NS, the function $\psi : \Lambda_{R(x)}(x) \to \mathbb{C}$ is (L_k, q)-subharmonic with $q \leq e^{-\gamma(m, L_k)L_k}$. By Theorem 2.4.1, for some $C' < \infty$ we have

$$|\psi(x)| \leq q^{\frac{R(x)}{L_k+1}} \mathcal{M}\big(\psi, \Lambda_{R(x)}(x)\big) \leq e^{-mR(x) - C' \ln L_k}.$$

Taking into account Eq. (2.196), we conclude that

$$\limsup_{|x|\to\infty} \frac{\ln|\psi(x)|}{|x|} \leq -m. \qquad \square$$

Remark 2.7.1. Observe that in the proof of Theorem 2.7.2, we used only the assumption that the bound (2.192) holds true with $p > \alpha d$. Recall that the proof of (2.192) actually required that a stronger lower bound: $p > 2 \cdot \alpha d$ (with $\alpha = 3/2$, the latter takes the form $p > 6d$). Therefore, the bound $p > \alpha d$ is automatically satisfied under the assumptions of Theorem 2.6.6.

Remark 2.7.2. It is straightforward that the case where the key bound (2.193) holds true for any pair of *disjoint* cubes of radius L is obtained by setting $C = 1$. In fact, replacing the condition $|u - v| \geq 2CL + 1$ for pairs of cubes of radius L by $|u - v| \geq CL^{\delta}$, with sufficiently small $\delta > 0$, would require only minor adaptations of the proof. Such a modification is useful, e.g., in the multi-scale analysis of Anderson-type models with correlated, but strongly mixing random potential, featuring a sufficiently fast decay of correlations.

We finish this subsection with the output statement produced by the MSA about spectral localization for a single-particle tight-binding model on $\mathbb{Z}^d$. Recall, we suppose that the random external potential $V(\cdot\,;\omega)$ is IID.

Corollary 2.7.3. (A) *Suppose that the marginal distribution function F satisfies Assumption* A. *Then $\exists\, g^* \in (0, +\infty)$ and a set Ω^* of probability $\mathbb{P}\{\,\Omega^*\,\} = 1$ such that $\forall\ \omega \in \Omega^*$ and g with $|g| \geq g^*$ the operator $H(\omega)$ has p.p. spectrum, and any of its normalized eigenfunctions $\psi = \psi(H(\omega))$ satisfies* (2.194) *with some $m = m(g) > 0$ and with a random constant $C\big(\psi(\omega)\big) \in (0, \infty)$.*

(B) *Suppose that the marginal distribution function F satisfies Assumptions* A *and* B. *Then, $\forall$ nonzero $g \in \mathbb{R}$, $\exists\, E^* > 0$ and a set Ω^* with $\mathbb{P}\{\,\Omega^*\,\} = 1$ such that $\forall\ \omega \in \Omega^*$ the operator $H(\omega)$ has p.p. spectrum in $(0, E^*)$ and $\forall$ normalized eigenfunction $\psi = \psi(\omega)$ with eigenvalue from $(0, E^*)$, the bound* (2.194) *holds, with some $C(\psi(\omega)) \in (0, \infty)$ and $m = m(g) > 0$.*

2.7.2 *From the MSA Bound to Dynamical Localization*

In this subsection, we give a direct derivation of the dynamical localization for the operator $H(\omega)$ on the entire lattice, bypassing the stage of finite-volume bounds on EF correlators. We follow the original scheme developed by Germinet and De Bièvre [98] and improved by Damanik and Stollmann [78], to obtain *strong* dynamical localization.

Note that the original Germinet–Klein argument from [99], unlike the adaptation presented in Sect. 2.5.2, also gives rise to the strong dynamical localization on the entire lattice; this would require, however, a functional-analytic complement on Hilbert–Schmidt norms of spectral projections of $H(\omega)$ in properly chosen weighted Hilbert spaces.

We believe that the scheme from [78,98], despite its greater complexity compared to that of Germinet–Klein, is fairly instructive, as it emphasizes the role of the so-called centers of localization (of square-summable eigenfunctions ψ of $H(\omega)$). This notion appeared already in the work [79] which has influenced further studies of the phenomenon of dynamical localization.

Let M denote the operator of multiplication by the max-norm:

$$(M\phi)(x) = |x|\,\phi(x),\ x \in \mathbb{Z}^d.$$

The main result of this subsection is the following

Theorem 2.7.4. *Suppose that Assumption* A *is satisfied. Let $I \subset \mathbb{R}$ be an interval and m, s be positive numbers. Then there exists an integer $L^* = L^*(m,s,I) \in (0,\infty)$ with the following property. Take $L_0 > L^*$ and suppose that the following bound holds: $\forall\ k \geq 0$ and $\forall$ pair of disjoint cubes $\Lambda_{L_k}(u)$ and $\Lambda_{L_k}(v)$,*

$$\mathbb{P}\{\exists\, E \in I : \Lambda_{L_k}(u) \text{ and } \Lambda_{L_k}(v) \text{ are } (E,m)\text{-S}\} \leq L_k^{-2p}, \tag{2.197}$$

with $p > (3d\alpha + \alpha s)/2$. Then $\forall$ finite subset $\mathbb{K} \subset \mathbb{Z}^d$ and bounded Borel function $\zeta : \mathbb{R} \to \mathbb{R}$, the operator $H = H(\omega)$ satisfies

$$\mathbb{E}\Big[\big\| M^s\, P_I(H)\, \zeta(H) \mathbf{1}_{\mathbb{K}} \big\|\Big] < \infty.$$

Proof. Before going to the proof, recall that property (DS.I, m, p_k, L_k) has the form

$$\mathbb{P}\big\{\exists\, E \in I :\ \Lambda_{L_k}(x') \text{ and } \Lambda_{L_k}(x'') \text{ are } (E,m)\text{-S}\ \big\} \leq L_k^{-2p_k} \tag{2.198}$$

where

$$p_k = p_0(1+\theta)^k \underset{k\to\infty}{\longrightarrow} +\infty. \tag{2.199}$$

Therefore, given an arbitrarily large $p' > 0$, we can change the numbering of scales and set $\tilde{L}_i = L_{i+k}$ with sufficiently large k. Then we can infer from (2.198) that

$$\mathbb{P}\left\{\exists\, E \in I : \Lambda_{\tilde{L}_i}(x') \text{ and } \Lambda_{\tilde{L}_i}(x'') \text{ are } (E,m)\text{-S}\right\} \le \tilde{L}_i^{-2p'}. \tag{2.200}$$

Now fix an arbitrary $s > 0$ and assume, without loss of generality, that the scales L_k are already chosen in such a way that the inequality

$$\mathbb{P}\left\{\exists\, E \in I : \Lambda_{L_k}(x') \text{ and } \Lambda_{L_k}(x'') \text{ are } (E,m)\text{-S}\right\} \le L_k^{-2p} \tag{2.201}$$

holds true for all integers $k \ge 0$, with the exponent p in the RHS satisfying

$$2p > 3d\alpha + \alpha s. \tag{2.202}$$

Bearing in mind this observation, we see that the validity of the hypothesis of Theorem 2.7.4 follows from our previous analysis, provided that $|g|$ is large enough. For the sake of brevity, we keep the parameter $p > \alpha(3d + s)/2$ fixed in the course of the proof. The lower bound (2.202) on p is always assumed in the rest of this subsection.

Theorem 2.7.4 will imply dynamical localization:

Theorem 2.7.5. (A) *Suppose that the distribution function F obeys Assumption* A. *Then $\exists\, g^* \in (0, +\infty)$ with the following property. $\forall\, g$ with $|g| \ge g^*$, $s > 0$, interval $I \subset \mathbb{R}$, bounded Borel function $\zeta : \mathbb{R} \to \mathbb{C}$ and finite subset $\mathbb{K} \subset \mathbb{Z}^d$,*

$$\mathbb{E}\left[\left\| M^s\, P_I(H)\, \zeta(H) \mathbf{1}_{\mathbb{K}} \right\|\right] < \infty. \tag{2.203}$$

(B) *Let F satisfy Assumptions* A *and* B. *Then $\forall\, g \in \mathbb{R} \setminus \{0\}\ \exists\, E^* > 0$ with the following property. $\forall\, s > 0$, bounded Borel function $\zeta : \mathbb{R} \to \mathbb{C}$ and finite subset $\mathbb{K} \subset \mathbb{Z}^d$, the inequality* (2.203) *holds for $I = [0, E^*]$.*

The rest of this Section is devoted to the proof of Theorem 2.7.4. We assume that I, m, s and p are given, as required in the theorem. We begin with the description of "singular" events and an assessment of their probabilities.

Given $j \ge 1$, consider the events $\mathcal{S}_j = \mathcal{S}_j(I, m)$ defined as follows:

$$\begin{aligned} \mathcal{S}_j = \{\exists\, E \in I\ \exists\, y, z \in \Lambda_{3L_{j+1}}(0) &\text{ such that } |y - z| > 2L_j \\ &\text{and cubes } \Lambda_{L_j}(y) \text{ and } \Lambda_{L_j}(z) \text{ are } (m, E)\text{-S}\}. \end{aligned}$$

Further, for $k \ge 1$ introduce the event

$$\Omega_k^{(\mathrm{S})} = \bigcup_{j \ge k} \mathcal{S}_j. \tag{2.204}$$

Proposition 2.7.6. *There exists a constant $c^* = c^*(p) \in (0, \infty)$ such that $\forall\, k \ge 1$,*

$$\mathbb{P}\left\{ \Omega_k^{(\mathrm{S})} \right\} \le c^* L_k^{-(2p-2d\alpha)}. \tag{2.205}$$

Proof. The total number of pairs of points $y, z \in \Lambda_{3L_{j+1}}(0)$ figuring in the definition of the event $\mathcal{S}_j$ does not exceed $(6L_j^\alpha + 1)^{2d}/2$. Combining this fact with (2.197), we obtain:

$$\begin{aligned}
\mathbb{P}\left\{ \Omega_k^{(\mathrm{S})} \right\} &\le \sum_{j\ge k} \mathbb{P}\{\mathcal{S}_j\} \le \sum_{j\ge k} L_j^{-2p} (6L_j^\alpha + 1)^{2d}/2 \\
&\le \sum_{j\ge k} (7L_j)^{2d\alpha} L_j^{-2p} \\
&\le 7^{2d\alpha} L_k^{-(2p-2d\alpha)} \left(1 + \sum_{j>k} L_k^{-(2p-2d\alpha)\alpha^{j-k}} \right).
\end{aligned}$$

The assertion of Proposition 2.7.6 follows. □

Now we introduce an important notion of localization center for an ℓ^2-function; cf. [78, 98] and also [79]. In a way, it is this notion which allows to transform MSA bounds into the proof of dynamical localization.

Fix $\omega \in \Omega^*$, where Ω^* is the set from Theorem 2.7.2. Let $\psi = \psi\big(H(\omega)\big)$ be an ℓ^2-eigenfunction of $H(\omega)$ (i.e., with $\|\psi\|_2 < \infty$) with an eigenvalue from I. We say that a point $x^* \in \mathbb{Z}^d$ is a *center of localization* for $\psi(\omega)$ if

$$|\psi(x^*)| = \max_{y\in\mathbb{Z}^d} |\psi(y)|. \tag{2.206}$$

As ψ belongs to $\ell^2(\mathbb{Z}^d)$, the collection of localization centers

$$\mathbb{X}(\psi) = \left\{ x^* \in \mathbb{Z}^d : |\psi(x^*)| = \max_{y\in\mathbb{Z}^d} |\psi(y)| \right\} \tag{2.207}$$

is non-empty. Moreover, since ψ with $\|\psi\|_2 < \infty$, the cardinality $j^*(\psi) = |\mathbb{X}(\psi)|$ is finite.

Proposition 2.7.7. *Given a site $x \in \mathbb{Z}^d$, $\exists\, k_0 = k_0(I, m)$ such that $\forall\, \omega \in \Omega^*$ and $k \ge k_0$, if a center of localization for an ℓ^2-eigenfunction ψ with eigenvalue E from I belongs to $\Lambda_{L_k}(x)$, then the cube $\Lambda_{L_{k+1}}(x)$ is (m, E)-S.*

Proof. Assume that a center of localization $x^* = x^*(\psi)$ for an ℓ^2-eigenfunction ψ lies in $\Lambda_{L_k}(x)$. Then the GRI (2.67) for the eigenfunction ψ and cube $\Lambda_{L_{k+1}}(x)$, combined with the non-singularity condition (2.19), implies that

$$|\psi(x^*)| \le 2d(2L_{k+1} + 1)^{d-1} \mathrm{e}^{-\gamma(m, L_{k+1})L_{k+1}} \max_{y\in\partial^+\Lambda_{L_{k+1}}(x)} |\psi(y)|$$

which, for large k, contradicts the definition of a center of localization. □

We see that, within the event Ω^*, pairs of localization centers should inevitably give rise to pairs of (E, m)-S cubes; the probability of such pairs can, therefore, be estimated with the help of Eq. (2.192).

Observe also that, in Proposition 2.7.7, a particular choice of a center of localization is irrelevant: the presence of at least one center of localization in the cube $\Lambda_{L_k}(x)$ suffices for the assertion to hold. Until the end of this section we will assume that, for each given eigenfunction $\psi = \psi\big(H(\omega)\big)$, its center of localization $x^*(\psi)$ is chosen in a well-defined way. For example, one can introduce a lexicographic order $\prec$ in $\mathbb{Z}^d$ and adopt an agreement that points of the set $\mathbb{X}(\psi)$ are listed in this order:

$$x_1^*(\psi) \prec x_2^*(\psi) \prec \cdots \prec x_{j^*(\psi)}^*(\psi);$$

cf. [139]. Point $x_1^*(\psi)$ then could be chosen for representing a canonical localization center and denoted by $x^*(\psi)$.

Fix k_0 as in Proposition 2.7.7 and for $k \geq k_0$ set

$$\Omega_k^{(\mathrm{NS})} = \Omega^* \setminus \Omega_k^{(\mathrm{S})}.$$

Next, consider a sequence of annuli

$$\mathbb{A}_k^{(3)} = \Lambda_{3L_{k+1}}(0) \setminus \Lambda_{3L_k}(0). \tag{2.208}$$

Proposition 2.7.8. *$\exists\, k_1 = k_1(m) \geq k_0$ with the following property. Take $k \geq k_1$ and $\omega \in \Omega_k^{(\mathrm{NS})}$ and suppose that $\psi = \psi(H(\omega))$ is a normalized ℓ^2-eigenfunction of $H(\omega)$ with an eigenvalue $E = E(\psi) \in I$ and a localization center $x^*(\psi) \in \Lambda_{L_{k+1}}(0)$. Then*

$$\left\| \mathbf{1}_{\mathbb{Z}^d \setminus \Lambda_{3L_{k+2}}(0)}\, \psi \right\| \leq \frac{1}{4}.$$

Proof. Using the annuli $\mathbb{A}_k^{(3)}$, we can write

$$\|\mathbf{1}_{\mathbb{Z}^d \setminus \Lambda_{3L_{k+2}}(0)}\, \psi\|^2 = \sum_{i \geq k+2} \|\mathbf{1}_{\mathbb{A}_i^{(4)}}\psi\|^2 = \sum_{i \geq k+2} \sum_{y \in \mathbb{A}_i^{(3)}} |\psi(y)|^2$$

Now take $i \geq k+2$. By assumption, the cube $\Lambda_{L_{i-1}}(0) \supseteq \Lambda_{L_{k+1}}(0)$ contains the center $x^*(\psi)$. By virtue of Proposition 2.7.7, the cube $\Lambda_{L_i}(0)$ is (E, m)-S. Furthermore, for $y \in \mathbb{A}_i^{(3)}$ we have

$$|y| > 3L_i > 2L_i,$$

implying that $\Lambda_{L_i}(0) \cap \Lambda_{L_i}(y) = \varnothing$. By construction of the event $\Omega_i^{(\mathrm{NS})}$, the cube $\Lambda_{L_i}(y)$ must be (E_j, m)-NS. In turn, this implies by the GRI for eigenfunctions that

$$|\psi(y)|^2 \le \left[(2d)(2L_i+1)^{d-1}\right]^2 \mathrm{e}^{-2\gamma(m,L_i)L_i}.$$

Now the claim of Proposition 2.7.8 follows from a polynomial (in L_i) bound on the number of terms in the sum $\sum\limits_{y\in\mathbb{A}_i^{(4)}}$. □

The next step in the proof of Theorem 2.7.4 is to provide bounds for eigenfunction correlators.

Proposition 2.7.9. *With k_1 as in Proposition 2.7.8, $\exists\ c_1^* = c_1^*(I,m)$ such that for $\omega \in \Omega_k^{(\mathrm{NS})}$ and $i \ge k \ge k_1$ the following bound holds for the number of the eigenfunctions $\psi_j = \psi_j\big(H(\omega)\big)$ of $H(\omega)$ with the localization centers $x_j^* = x^*(\psi_j)$ in a given cube:*

$$\mathrm{card}\Big\{\psi_j : x_j^* \in \Lambda_{L_{i+1}}(0)\Big\} \le c_1^*\, L_{i+1}^{\alpha d}. \tag{2.209}$$

Proof. First, observe that, with $\psi_j = \psi_j(H(\omega))$ running over the normalized eigenfunctions of $H(\omega)$ and $x_j^* = x^*(\psi_j)$ standing for the localization center, we have:

$$\begin{aligned}
\sum_{\psi_j:\, x_j^*\in\Lambda_{L_{i+1}}(0)} & \Big\langle \mathbf{1}_{\Lambda_{L_{i+2}}(0)} P_I(H) \mathbf{1}_{\Lambda_{L_{i+2}}(0)}\psi_j,\, \psi_j\Big\rangle \\
&\le \sum_{\psi_j}\Big\langle \mathbf{1}_{\Lambda_{L_{i+2}}(0)} P_I(H) \mathbf{1}_{\Lambda_{L_{i+2}}(0)}\psi_j,\, \psi_j\Big\rangle \\
&= \mathrm{tr}\left[\mathbf{1}_{\Lambda_{L_{i+2}}(0)} P_I(H)\right].
\end{aligned} \tag{2.210}$$

Next, the trace

$$\begin{aligned}
\mathrm{tr}\left[\mathbf{1}_{\Lambda_{L_{i+2}}(0)} P_I\big(H(\omega)\big)\right] &\le \|P_I\big(H(\omega)\big)\|\,\mathrm{tr}\left[\mathbf{1}_{\Lambda_{L_{i+2}}(0)}\right] \\
&\le 1\cdot\left|\Lambda_{L_{i+2}}(0)\right| \\
&\le \left(2L_{i+1}^{\alpha}+1\right)^d,
\end{aligned}$$

trivially. Therefore, for the statement of the lemma it is enough to show that each term in the sum $\sum\limits_{\psi_j:\, x_j^*\in\Lambda_{L_{i+1}}(0)}$ in the LHS is not smaller than $1/2$. Indeed,

$$\begin{aligned}
\Big\langle \mathbf{1}_{\Lambda_{L_{i+2}}(0)} P_I\big(H(\omega)\big)\mathbf{1}_{\Lambda_{L_{i+2}}(0)}\,\psi_j,\,\psi_j\Big\rangle &= \Big\langle \mathbf{1}_{\Lambda_{L_{i+2}}(0)} P_I\big(H(\omega)\big)\,\psi_j,\,\psi_j\Big\rangle \\
&\quad - \Big\langle \mathbf{1}_{\Lambda_{L_{i+2}}(0)} P_I\big(H(\omega)\big)\,\mathbf{1}_{\mathbb{Z}^d\setminus\Lambda_{L_{i+2}}(0)}\,\psi_j,\,\psi_j\Big\rangle \\
&\ge \Big\langle \mathbf{1}_{\Lambda_{L_{i+2}}(0)}\,\psi_j,\,\psi_j\Big\rangle - 1/4
\end{aligned}$$

where at the last step we made use of Proposition 2.7.8 and the fact that $j+2>k$.

Moreover, using Proposition 2.7.8 again:

$$\begin{aligned}&\left\langle \mathbf{1}_{\Lambda_{L_{i+2}}(0)}\,\psi_j\,,\,\psi_j\right\rangle - 1/4\\&\quad= \langle\psi_j\,,\,\psi_j\rangle - \left\langle \mathbf{1}_{\mathbb{Z}^d\setminus\Lambda_{L_{i+2}}(0)}\,\psi_j\,,\,\psi_j\right\rangle - 1/4 \geq 1/2.\end{aligned}$$

□

In Proposition 2.7.10 we refer to annuli $\mathbb{A}_k^{(4)}(0)$; cf. Eq. (2.208).

Proposition 2.7.10. *Let k_1 be as in Proposition 2.7.8. There exists an integer $k_2 = k_2(I,m) \geq k_1$ such that, for all L_0 large enough, all $k \geq k_2$, $\omega \in \Omega_k^{(\mathrm{NS})}$, $x \in \mathbb{A}_k^{(4)}(0)$ and $y \in \Lambda_{L_k}(0)$, the matrix elements of the operator $P_I\big(H(\omega)\big)\zeta(H(\omega))$ in the delta-basis obey:*

$$\left|\left[P_I\big(H(\omega)\big)\zeta\big(H(\omega)\big)\right](x,y)\right| \leq \mathrm{e}^{-\gamma(m,L_k)L_k/2}\|\zeta\|_\infty. \tag{2.211}$$

Proof. Without loss of generality, assume that $\|\zeta\|_\infty \leq 1$. Also, for definiteness, take $y=0$; the modifications for a general $y \in \Lambda_{L_k}(0)$ will be straightforward. Writing, as before, ψ_j and E_j for the eigenfunctions and eigenvalues of $H(\omega)$ and x_j^* for the centers of localization for ψ_j, we have:

$$\begin{aligned}\left|\left[P_I\big(H(\omega)\big)\zeta\big(H(\omega)\big)\right](x,0)\right| &\leq \sum_{(\psi_j,E_j):\,E_j\in I} |\zeta(E_j)|\,|\psi_j(x)|\,|\psi_j(0)|\\&\leq \left(\sum_{\substack{(\psi_j,E_j):\,E_j\in I\\ x_j^*\in\Lambda_{3L_k}(0)}} + \sum_{i>k}\ \sum_{\substack{(\psi_j,E_j):\,E_j\in I\\ x_j^*\in\mathbb{A}_i^{(3)}(0)}}\right)|\psi_j(x)|\,|\psi_j(0)|.\end{aligned} \tag{2.212}$$

For $k \geq k_1$ large enough and L_0 sufficiently large, the first sum in the RHS can be bounded as follows:

$$\sum_{\substack{j:\,E_j\in I\\ x_j^*\in\Lambda_{3L_k}(0)}} |\psi_j(x)|\,|\psi_j(0)| \leq c_2^* L_{k+1}^{\alpha d}\,\mathrm{e}^{-\gamma(m,L_k)L_k} \leq \frac{1}{2}\mathrm{e}^{-\gamma(m,L_k)L_k/2}, \tag{2.213}$$

where $c_2^* = c_2^*(m) \in (0,+\infty)$ is a constant. Indeed, for $x \in \mathbb{A}_k^{(4)}$, the norm $|x| \geq 3L_k > 2L_k$. Thus, $\Lambda_{L_k}(x) \cap \Lambda_{L_k}(0) = \varnothing$, and hence one of the cubes $\Lambda_{L_k}(x)$, $\Lambda_{L_k}(0)$ must be (E_j,m)-NS. Then the GRI for the eigenfunctions (see Eq. (2.67)) implies that

$$\min\left\{|\psi_j(x)|,\,|\psi_j(0)|\right\} \leq 2d(2L_k+1)^{d-1}\mathrm{e}^{-\gamma(m,L_k)L_k}.$$

Applying Proposition 2.7.9 yields that the number of summands in the sum, figuring in the LHS of Eq. (2.213), is upper-bounded by $(6L_k^\alpha + 1)^d$. This renders inequality (2.213).

Next, we pass to the sum $\sum_{i>k}$ in the RHS of Eq. (2.212). Fix $i \geq k+1$ and consider the corresponding sum

$$\sum_{\substack{(\psi_j, E_j):\, E_j \in I \\ x_j^* \in \mathbb{A}_i^{(4)}(0)}} |\psi_j(x)|\, |\psi_j(0)|.$$

The cubes $\Lambda_{L_i}(0)$ and $\Lambda_{L_i}(x_j^*)$ are again disjoint, and, by Proposition 2.7.7, cube $\Lambda_{L_i}(x_j^*)$ is (E_j, m)-S. Consequently, for $\omega \in \Omega_k^{(\mathrm{NS})}$ cube $\Lambda_{L_i}(0)$ must be (E_j, m)-NS. Therefore, again by virtue of the GRI for the eigenfunctions, for k large enough and $i > k$,

$$|\psi_j(0)| \leq c_3^*\, L_i^{d-1}\, \mathrm{e}^{-\gamma(m, L_i) L_i}$$

where $c_3^* = c_3^*(m) \in (0, \infty)$. Using again Proposition 2.7.9, we get, for k large enough,

$$\begin{aligned}\sum_{i>k} \sum_{\substack{n:\, E_j \in I \\ x_{n,1} \in \mathbb{A}_i^{(4)}(0)}} |\psi_j(x)| |\psi_j(0)| &\leq c_4^* \sum_{i>k} \mathrm{e}^{-\gamma(m, L_i) L_i}\, L_i^{\alpha d + d - 1} \\ &\leq \frac{1}{2} \mathrm{e}^{-\gamma(m, L_k) L_k / 2}.\end{aligned} \tag{2.214}$$

This leads to the bound (2.211). □

Let us conclude the proof. For any finite set $\mathbb{K} \subset \mathbb{Z}^d$ we find $k \geq k_1$ such that $\mathbb{K} \subset \Lambda_{L_k - 1}(0)$. Then write:

$$\begin{aligned}\mathbb{E}\Big[\| M^s\, P_I(H) \zeta(H) \mathbf{1}_{\mathbb{K}} \| \Big] &\leq \mathbb{E}\Big[\| M^s\, \mathbf{1}_{\Lambda_{3L_k}(0)} P_I(H) \zeta(H) \mathbf{1}_K \| \Big] \\ &\quad + \sum_{i \geq k} \mathbb{E}\Big[\Big\| M^s\, \mathbf{1}_{\mathbb{A}_i^{(4)}} P_I(H) \zeta(H) \mathbf{1}_K \Big\| \Big].\end{aligned} \tag{2.215}$$

For the first term in the RHS we have:

$$\mathbb{E}\Big[\Big\| M^s\, \mathbf{1}_{\Lambda_{3L_k}(0)} P_I(H) \zeta(H) \mathbf{1}_K \Big\| \Big] \leq (3L_k)^s\, \|\zeta\|_\infty, \tag{2.216}$$

evidently. Next, for the second summand:

$$\begin{aligned}\mathbb{E}\Big[\,\big\|M^s\,\mathbf{1}_{\mathbb{A}_i^{(4)}}P_I(H)\zeta(H)\mathbf{1}_{\mathbb{K}}\big\|\,\Big]\\ \le (3L_{i+1})^s \sum_{x\in\mathbb{A}_i^{(4)}} \mathbb{E}\Big[\,\big\|\mathbf{1}_{\{x\}}P_I(H)\zeta(H)\mathbf{1}_{\mathbb{K}}\big\|\,\Big].\end{aligned} \tag{2.217}$$

Further, the expression $\mathbb{E}\Big[\,\big\|\mathbf{1}_{\{x\}}P_I(H)\zeta(H)\mathbf{1}_{\mathbb{K}}\big\|\,\Big]$ is split into the sum

$$\mathbb{E}\left[\,\mathbf{1}_{\Omega_k^{(\mathrm{S})}}\,\big\|\mathbf{1}_{\{x\}}P_I(H)\zeta(H)\,\mathbf{1}_{\mathbb{K}}\big\|\,\right]+\mathbb{E}\left[\,\mathbf{1}_{\Omega_k^{(\mathrm{NS})}}\,\big\|\mathbf{1}_{\{x\}}P_I(H)\zeta(H)\,\mathbf{1}_{\mathbb{K}}\big\|\,\right].$$

Next, owing to Proposition 2.7.6,

$$\begin{aligned}\mathbb{E}\left[\,\mathbf{1}_{\Omega_k^{(\mathrm{S})}}\,\big\|\mathbf{1}_{\{x\}}P_I(H)\zeta(H)\,\mathbf{1}_{\mathbb{K}}\big\|\,\right] &\le \|\zeta\|_\infty\mathbb{P}\left\{\Omega_k^{(\mathrm{S})}\right\}\\ &= \|\zeta\|_\infty\, c^*\, L_k^{-(2p-2d\alpha)}.\end{aligned} \tag{2.218}$$

Moreover, $\forall\ \omega\in\Omega_k^{(\mathrm{NS})}$, $i\ge k$ and $x\in\mathbb{A}_i^{(4)}$, by virtue of Proposition 2.7.10

$$\begin{aligned}&\|\mathbf{1}_{\{x\}}P_I\big(H(\omega)\big)\zeta(H(\omega))\,\mathbf{1}_{\mathbb{K}}\|\\ &\quad\le |\mathbb{K}|\max\big|\big[P_I\big(H(\omega)\big)\zeta\big(H(\omega)\big)\big](x,y)\big|\\ &\quad\le \|\zeta\|_\infty \mathrm{e}^{-\gamma(m,L_i)L_i/2}.\end{aligned} \tag{2.219}$$

After substituting Eqs. (2.216)–(2.219) in (2.215), we obtain that

$$\mathbb{E}\Big[\,\big\|M^s\,P_I(H)\zeta(H)\mathbf{1}_{\mathbb{K}}\big\|\,\Big]<\infty.$$

This completes the proof of Theorem 2.7.4. □

To deduce Theorem 2.7.5, observe that we have established the following facts.

(I) Suppose that Assumption A is fulfilled and a bounded interval I, $m>0$ and $p_0>6d$ are given. Then, assuming that $|g|$ is large enough, for some $\theta>0$ and L_0 large enough, $\forall\ k\ge 0$ and $\forall$ pair of disjoint cubes $\Lambda_{L_k}(u)$ and $\Lambda_{L_k}(v)$,

$$\mathbb{P}\{\,\exists\, E\in I:\ \ \Lambda_{L_k}(u)\text{ and }\Lambda_{L_k}(v)\text{ are }(E,m)\text{-S}\,\}\le L_k^{-2p_k}. \tag{2.220}$$

(II) Suppose that Assumptions A and B are fulfilled. Then given g, $\exists\ m>0$ and $E^*>0$ such that, for $I=[0,E^*]$ and L_0 large enough, $\exists\ p_0>6d$ and $\theta>0$ such that Eq. (2.220) holds, again $\forall\ k\ge 0$ and $\forall$ pair of disjoint cubes $\Lambda_{L_k}(u)$ and $\Lambda_{L_k}(v)$. Hence, the inequality $p>(3d+s)\alpha$ required in Theorem 2.7.4 can always be achieved, as was explained in the beginning of this subsection; cf. Eq. (2.199) and the discussion following the formulation of Theorem 2.7.4.

This means that, $\forall\, s > 0$, bounded Borel function $\zeta : \mathbb{R} \to \mathbb{R}$ and any finite subset $\mathbb{K} \subset \mathbb{Z}^d$, Eq. (2.203) holds:

- In case (I)—for any bounded interval $I \subset \mathbb{R}$, provided that $|g|$ is large enough;
- In case (II)—for any given $g \neq 0$, with $I = [0, E^*]$ depending[13] upon $|g|$.

2.7.3 Local Statistics of Eigenvalues

In this Section we return to the discussion of the local statistics of eigenvalues of random operators, briefly commented upon in Sect. 2.2.4.

Minami [133] proved the following Theorem 2.7.11 that is considered now as classical in the spectral theory of random operators. In this theorem, δ_c stands for a Dirac measure supported by $c \in \mathbb{R}$.

Theorem 2.7.11 (Cf. [133]). *Consider a sequence of cubes* $\Lambda_L = \Lambda_L(0)$ *for integers* $L \geq 1$. *Let* $E_j = E_j\big(H_{\Lambda_L}(\omega)\big)$ *be the eigenvalues of the operator* $H_{\Lambda_L}(\omega)$. *Given* $E \in \mathbb{R}$, *define the random point process* $\xi(L; E)$ *on* $\mathbb{R}$ *by*

$$\xi(L; E)(\mathrm{d}a) = \sum_{j=1}^{|\Lambda_L|} \delta_{X_j(L,E)}(\mathrm{d}a), \quad a \in \mathbb{R}, \quad \text{where } X_j(L, E) = |\Lambda_L|(E_j - E). \tag{2.221}$$

Suppose that for some $E_0 \in \mathbb{R}$ *the density of states* $\mathcal{N}(E)$ *exists at* $E = E_0$ *and is positive. Suppose also that the fractional moment of the Green's function decays exponentially fast in the following sense.* $\exists\, s_0 \in (0, 1)$, $C \in (0, \infty)$, $m > 0$ *and* $r > 0$ *such that* $\forall\, L$,

$$\mathbb{E}\big[\, |G_{\Lambda_L}(x, y; \zeta)|^{s_0} \,\big] \leq C\mathrm{e}^{-m|x-y|}$$

$\forall\, x, y \in \Lambda_L$ *and* $\zeta \in \mathbb{C}$ *with* $\mathrm{Im}\zeta > 0$, $|\zeta - E_0| < r$. *Under these conditions, the point process* $\xi(\Lambda_L, E_0)$ *defined by* (2.221) *converges weakly, as* $L \to \infty$, *to the Poisson point process* ξ *with intensity measure* $\mathcal{N}(E_0)\mathrm{d}a$.

Later, Nakano [139, 140] obtained a more detailed information on the statistics of eigenvalues, assuming again the results of the localization theory, viz., the results of the MSA. Following [98], Nakano used the notion of the localization centers associated with a normalized ℓ^2-eigenfunction $\psi_j = \psi_j\big(H(\omega)\big) \in \ell^2(\mathbb{Z}^d)$ (cf. (2.207)):

$$\mathbb{X}(\psi_j) := \left\{ x^* \in \mathbb{Z}^d \;:\; |\psi_j(x^*)| = \max_{y \in \mathbb{Z}^d} |\psi_j(y)| \right\}.$$

[13]Clearly, with $g = 0$ there is no disorder in the Hamiltonian $H_0 + 0 \cdot V(x; \omega)$, and for $|g| \ll 1$, the impact of the random potential $gV(\cdot; \omega)$ is very weak.

Although it is natural to expect that $\mathbb{X}(\psi_j)$ contains a single point—the localization centre of ψ_j—possible resonances may give rise to multimodal eigenfunctions having multiple global maxima; a detailed technical discussion of the probability of such events is beyond the scope of this book. In any case, the notion of the localization centers leads naturally to the study of the statistical properties of pairs (E_j, x_j^*), where $E_j = E_j\big(H(\omega)\big)$ is the eigenvalue associated with ψ_j and $x_j^* \in \mathbb{X}(\psi_j)$.

Nakano performed precisely this kind of energy-space statistical analysis. Below we give a summary of results from [139, 140].

- **Sparseness of Localization Centers**

  In paper [140], a basic condition was that the distribution function F has a bounded probability density ρ. To shorten the technical aspects of the presentation, it was also assumed that the bounds (2.32) hold for some $p > 2d$ in a certain interval $I \subset \mathbb{R}$ (implying exponential spectral localization for the LSO $H(\omega)$ in I). The Minami estimate (2.56) was also re-established.

  Theorem 2.7.12 (Cf. [139]). *Let $E \in I$. Set $\eta_k = |\Lambda_{L_k}(0)|^{-1} k^{-2}$ and*

  $$J_k = \left(E - \frac{\eta_k}{2}, E + \frac{\eta_k}{2}\right), \quad k = 1, 2, \ldots.$$

  Then with probability one, there exists $k_0 = k_0(\omega)$ such that, if $k \geq k_0$, there is no eigenfunction of operator $H(\omega)$ with a center of localization in $\Lambda_{L_k}(0)$ and eigenvalue in J_k.

- **Repulsive Nature of the Localization Centers**

  Theorem 2.7.13 (Cf. [139]). *As before, set $\eta_k = |\Lambda_{L_k}(0)|^{-1}k^{-2}$. $\forall\, x \in \mathbb{Z}^d$, $\exists$ $k_1 = k_1(\omega, x)$ such that for $k \geq k_1$ and any interval J_k of the above form there is at most one eigenfunction of $H(\omega)$ with a center of localization in Λ_{L_k} and with the eigenvalue in J_k.*

- **Energy–Space Poisson Distribution**

  Here it is convenient to assume that two properties are satisfied (see Assumptions N1 and N2 below).

  Assumption N1 (*Initial Length Scale Estimate*)
  Let $I \subset \mathbb{R}$ be an open interval where the initial length scale MSA estimate holds: we can find $m > 0$ and $p > 6d$ such that for sufficiently large L_0 we have

  $$\mathbb{P}\{\forall\, E \in I \text{ cube } \Lambda_{L_0}(u) \text{ is } (E, m)\text{–NS}\} \leq L_0^{-p}. \tag{2.222}$$

  Assumption N2 (*Wegner's Estimate*)
  There is a constant $C \in (0, \infty)$ such that for any bounded interval $J \subset \mathbb{R}$ and any cube $\Lambda_L(u)$

$$\mathbb{E}\left[\operatorname{tr} P_J(H_{\Lambda_L(u)})\right] \le C|\Lambda_L(u)|\,|J|, \tag{2.223}$$

where $P_J(H_{\Lambda_L(u)})$ is the spectral projection of operator $H_{\Lambda_L(u)}$ on J.

Note that the LHS in (2.223) gives the expectation of the number of eigenvalues of operator $H_{\Lambda_L(u)}$ in the interval J. It has to be said that our Assumption A (Hölder-continuity of the marginal distribution function F of the on-site random potential $V(\cdot\,;\omega)$) is insufficient for an optimal Wegner estimate to hold (with a volume factor $|\Lambda_L(u)|$. Here one needs a stronger assumption of uniform Lipshitz-continuity of function F. (It would suffice to assume that F has a bounded probability density ρ.)

Theorem 2.7.14 below addresses the energy–space version, $\xi^{\mathrm{E-S}}$, of point process ξ from Eq. (2.221).

Theorem 2.7.14 (Cf. [140]). *Let Assumptions* N1 *and* N2 *hold true. Given a cube $\Lambda = \Lambda_L(0)$, consider the eigenfunctions ψ_j^{per} and eigenvalues $E_j^{\mathrm{per}}(\Lambda)$ of the LSO $H_\Lambda^{\mathrm{per}}(\omega)$ given by*

$$H_\Lambda^{\mathrm{per}}(\omega) = H_\Lambda^{\mathrm{per},0} + V_\Lambda(\omega).$$

Here $H_\Lambda^{\mathrm{per},0}$ is the kinetic energy operator with the periodic boundary conditions. Suppose that the Minami estimate holds in the following form: $\forall$ cube $\Lambda = \Lambda_L(0)$ and bounded interval $I \subset \mathbb{R}$ of length $|I|$,

$$\sum_{l\ge 2} l(l-1)\mathbb{P}\left\{\operatorname{card}\{j : E_j^{\mathrm{per}}(\Lambda) \in I\} = l\right\} \le C|\Lambda|^2|I|^2 \tag{2.224}$$

where $C \in (0,\infty)$ is a constant. Given $E \in \mathbb{R}$, define a random point process $\xi^{\mathrm{E-S}}(L,E)$ on $\mathbb{R}\times\mathbb{B}$ where $\mathbb{B} = [0,1]^d \subset \mathbb{R}^d$:

$$\begin{aligned}\xi^{\mathrm{E-S}}(L,E)(\mathrm{d}a\times\mathrm{d}y) &= \sum_{j=1}^{|\Lambda_L|}\delta_{X_j,S_j}\mathrm{d}a\times\mathrm{d}y,\;\; a\in\mathbb{R},\; y\in\mathbb{K},\\ \text{with } X_j &= |\Lambda_L|(E_j^{\mathrm{per}}(\Lambda_L)-E),\; S_j = L^{-1}x_j^*\end{aligned} \tag{2.225}$$

where x_j^* is a localization center for ψ_j^{per} (defined in a manner as above). Suppose that, for some $E^0 \in \mathbb{R}$, $\exists$ the value of the density of states $\mathcal{N}(E^0) \in (0,\infty)$. Then process $\xi^{\mathrm{E-S}}(L,E^0)$ converges weakly as $L\to\infty$ to $\xi^{\mathrm{E-S}}(E^0)$, where $\xi^{\mathrm{E-S}}(E^0)$ is the Poisson point process on $\mathbb{R}\times\mathbb{K}$ with the intensity measure $\mathcal{N}(E^0)\,\mathrm{d}a\times\mathrm{d}y$.

Remark 2.7.3. The bound (2.224) is essential; without it the result is only that the limiting point process is infinitely divisible.

2.8 The FMM as an Alternative to the MSA

As was stressed earlier, this book concentrates upon the multi-scale analysis of random Anderson-type Hamiltonians. However, we believe that it is useful to say a few words of an alternative approach to the localization phenomena in disordered quantum systems, the fractional moment method (FMM), proposed originally by Aizenman and Molchanov [7] and later generalized in a large number of works.

Speaking informally, the main difference between the MSA and the FMM is that the bounds in probability (on "unwanted" events, such as (E, m)-singularity of a given cube) are replaced in the FMM by bounds on the expectations of (*fractional* powers of) the Green's functions and of EF correlators. In addition, the FMM is—at least, in the context of single-particle models—a *single-scale* method. (Some elements of scale induction appear in the multi-particle variant of the FMM, developed recently by Aizenman and Warzel [9].)

The most striking distinctive technical feature of the FMM, compared to the MSA, is that, if and when it applies, it provides exponential upper bounds on the EF correlators, while the MSA gives rise to a slower rated of decay (at best, sub-exponential).

The first stage of the FMM consists in proving bounds on the Green's functions of the form

$$\mathbb{E}\left[\,\left|G_\Lambda(x, y; E)\right|^s\,\right] \le \mathrm{e}^{-m|x-y|} \tag{2.226}$$

in arbitrarily large, but finite subsets $\Lambda \subset \mathbb{Z}^d$, with the exponent $s \in (0, 1)$ (this explains the name of the method). Additional arguments are required to derive from (2.226) the exponential strong dynamical localization on the entire lattice:

$$\mathbb{E}\left[\sup_{t\in\mathbb{R}} \left|\langle \delta_y\,,\ \mathrm{e}^{-\mathrm{i}tH(\omega)}\,\delta_x\rangle\right|\right] \le \mathrm{Const}\,\mathrm{e}^{-a|x-y|}$$

with $a = a(m) > 0$.

The role of *fractional* moments, $\mathbb{E}\left[\,|G_\Lambda(x, y; E)|^s\,\right]$ with $0 < s < 1$, can be easily understood in the simplest case where $\Lambda = \{x\}$, so that the operator H_Λ is merely a multiplication by $h + V(x;\omega)$, where h is the diagonal element of the lattice Laplacian. Respectively, the only matrix element $G(x, x; E)$ has the form $(h + V(x;\omega) - E)^{-1}$. Assuming for simplicity that the marginal probability distribution of the random field $F(\cdot;\omega)$ admits a bounded, compactly supported probability density $p_V(\cdot)$, it is readily seen that the expectation

$$\mathbb{E}\left[\,\left|G_\Lambda(x, x; E)\right|^s\,\right] = \int_{\mathbb{R}} \frac{p_V(\lambda)}{\left|(h - E) + \lambda\right|^s}\,\mathrm{d}\lambda \tag{2.227}$$

should converge for $s \in (0, 1)$, but diverges for $s \ge 1$. It is clear that the above condition on the marginal probability distribution of the random potential can be substantially relaxed; still, the requirement of the convergence of the integral in the RHS of (2.227) is more restrictive than in the framework of the MSA (which operates with bounds *in probability*). So, this feature of the FMM is at the same time one of its strong points (bounds in expectation) and a weak point (stronger assumptions on the random potential).

The single-scale nature of the FMM is often appreciated by physicists (as well as by mathematicians) for the simpler logical scheme it provides. In addition, this also makes possible applications of the FMM to the localization analysis on graphs other than integer lattices, most notably Bethe lattices (i.e., regular trees), where the volume of a cube of radius L, as well as the cardinality of the sphere of radius L, grows as $\exp(cL)$, $c > 0$. In the course of the MSA induction, we have seen that the factors of the form $|\partial^- \Lambda_{L_k}(u)|$ can be easily absorbed in the exponentially small quantities e^{-mL_k}, provided that $|\partial^- \Lambda_{L_k}(u)|$ grows *slower than exponentially* as $L_k \to \infty$. The FMM bounds, operating with a fixed (and appropriately chosen) scale L_0, can afford an exponential growth of balls (in a tree or more general graph), provided that the initial decay exponent at the scale L_0 is large enough.

There exist today various approaches to the derivation of the eigenfunction correlator decay bounds,

$$\mathbb{E}\Big[\, |\langle \delta_x | \phi(H(\omega)) | \delta_y \rangle| \,\Big] \leq \text{Const } e^{-a|x-y|} \tag{2.228}$$

from the fixed-energy bounds (2.226). In our opinion, the technique by Elgart et al. [84] is one of the simplest and most comprehensible. Furthermore, the analytic techniques of the FMM become particularly simple in the case of a strong disorder ($|g| \gg 1$, cf. [7, 13]). In fact, for $|g|$ large enough, FMM provides a single-site localization condition which, combined with the spectral reduction from [84] and the (finite-volume version of the) Germinet–Klein argument [99], results in a very short and elementary proof of the exponential strong dynamical localization. The first stage of this proof (fixed-energy moment analysis) is presented in [13] (cf. Corollary and Eq. (1.12) in [13]); a detailed elementary proof can be found, e.g., in a paper by Hundertmark [110].

An application of the FMM to multi-particle disordered quantum systems has required a substantial revision of this alternative approach to the phenomenon of Anderson localization. In our opinion, a detailed presentation of the multi-particle FMM, developed by Aizenman and Warzel [9, 10], would require a separate book. In Part II, we will focus on the multi-particle adaptation of the MSA techniques presented in this chapter.

Part II
Multi-particle Localization

Chapter 3
Multi-particle Eigenvalue Concentration Bounds

In this chapter we begin our analysis of localization in multi-particle Anderson tight-binding models with interaction. We already mentioned that the principal difficulty encountered when working with multi-particle systems is the structure of the external random potential term (3.3) in the Hamiltonian combined with the presence of interaction between particles (see Eq. (3.8) below). To tackle this obstacle, we develop a multi-particle version of the MSA (in short, the MPMSA) by scrutinizing and—when necessary—modifying the subsequent steps of the single-particle MSA scheme.

3.1 Basic Notation and Assumptions: The Statement of Localization Results

3.1.1 The Multi-particle Anderson Hamiltonian

In Chaps. 3 and 4 we analyze the spectrum of the N-particle Anderson Hamiltonian in the Hilbert space $\mathcal{H}^{(N)} = \ell^2\left(\left(\mathbb{Z}^d\right)^N\right)$, $N = 2, 3, \ldots$; our ultimate goal is to establish spectral and dynamical localization. Working in the Hilbert space $\mathcal{H}^{(N)}$, we deal with N distinguishable quantum particles on the lattice $\mathbb{Z}^d$. Obviously, localization in $\mathcal{H}^{(N)}$ implies localization in both the bosonic and fermionic subspaces of $\mathcal{H}^{(N)}$ (formed by symmetric and antisymmetric functions, respectively).

In the course of our argument we will have to focus on subsystems of the N-particle system, containing n particles where $n = 1, \ldots, N$. For that reason it is often convenient to refer to an n-particle model, with the phase space $\mathcal{H}^{(n)} = \ell^2((\mathbb{Z}^d)^n)$.

The inner product and the norm in $\mathcal{H}^{(n)}$ are denoted, as before, by $\langle\,\cdot\,,\,\cdot\,\rangle$ and $\|\,\cdot\,\|$. (Other types of norms will be specified at a local level.)
The following objects and notation will be used from this point on:

V. Chulaevsky and Y. Suhov, *Multi-scale Analysis for Random Quantum Systems with Interaction*, Progress in Mathematical Physics 65, DOI 10.1007/978-1-4614-8226-0_3,

- An n-particle configuration represented by a vector $\mathbf{x} = (x_1, \ldots, x_n) \in \left(\mathbb{Z}^d\right)^n$ with components x_j specifying the positions of n particles in the lattice $\mathbb{Z}^d$, $n \geq 1$.

  The max-norm $|\mathbf{x}|$ of a vector $\mathbf{x} = (x_1, \ldots, x_n) \in \left(\mathbb{Z}^d\right)^n$ is defined as $\max\left[|x_j|,\ 1 \leq j \leq n\right]$. It induces in a usual way the distance between points $\mathbf{x} = (x_1, \ldots, x_n)$ and $\mathbf{y} = (y_1, \ldots, y_n)$,

$$\mathrm{dist}(\mathbf{x}, \mathbf{y}) = |\mathbf{x} - \mathbf{y}| = \max\left[\,|x_j - y_j|,\ 1 \leq j \leq n\right],$$

  as well as the distance between subsets of $\left(\mathbb{Z}^d\right)^n$ (in particular, between a point and a subset of $\left(\mathbb{Z}^d\right)^n$) and the diameter of a set in $\left(\mathbb{Z}^d\right)^n$.

  As in Chap. 2, it will be convenient to use the norm $|\mathbf{x}|_1 := |x_1| + \cdots + |x_n|$ in order to describe the natural graph structure of the lattice $\left(\mathbb{Z}^d\right)^n$: points $\mathbf{x}$ and $\mathbf{y}$ are nearest neighbors in $\left(\mathbb{Z}^d\right)^n$ if $|\mathbf{x} - \mathbf{y}|_1 = 1$. The norm $|\cdot|_1$ induces canonically the distance $\mathrm{dist}_1(\cdot, \cdot)$.

  Symbol $\delta_{\mathbf{x}}$ is used for the lattice delta-function at site $\mathbf{x} \in \left(\mathbb{Z}^d\right)^n$: $\delta_{\mathbf{x}}(\mathbf{y}) = \mathbf{1}_{\mathbf{x}}(\mathbf{y}), \mathbf{y} \in \left(\mathbb{Z}^d\right)^n$; the family $\left\{\delta_{\mathbf{x}},\ \mathbf{x} \in \left(\mathbb{Z}^d\right)^n\right\}$ gives an orthonormal basis in $\mathcal{H}^{(n)}$.

  An s-particle sub-configuration in an n-particle configuration $\mathbf{x} \in \left(\mathbb{Z}^d\right)^n$ is defined as a pair $(\mathcal{J}, \mathbf{x}_{\mathcal{J}})$ where $\varnothing \neq \mathcal{J} = \{j_1, \ldots, j_s\} \subset \{1, \ldots, n\}$ and $\mathbf{x}_{\mathcal{J}} = (x_{j_1}, \ldots, x_{j_s}) \in \left(\mathbb{Z}^d\right)^s$, with $s = |\mathcal{J}|$. For the sake of brevity, we will write $\mathbf{x}_{\mathcal{J}}$ instead of $(\mathcal{J}, \mathbf{x}_{\mathcal{J}})$.
- The projection onto the coordinate subspace of the jth particle, defined as the mapping

$$\Pi_j : \mathbf{x} = (x_1, \ldots, x_n) \mapsto x_j, \quad 1 \leq j \leq n.$$

  The support $\Pi\mathbf{x} \subset \mathbb{Z}^d$ of a configuration $\mathbf{x}$ and the full projection $\Pi\mathbf{\Lambda}$ of a subset $\mathbf{\Lambda} \subset \left(\mathbb{Z}^d\right)^n$: these are defined by

$$\Pi\mathbf{x} = \bigcup_{j=1}^{n} \{\Pi_j \mathbf{x}\} \subset \mathbb{Z}^d, \quad \Pi\mathbf{\Lambda} = \bigcup_{j=1}^{n} \Pi_j \mathbf{\Lambda} \subset \mathbb{Z}^d.$$

  Next, a partial projection, $\Pi_{\mathcal{J}}\mathbf{\Lambda}$, determined by a given nonempty subset $\mathcal{J} \subseteq \{1, \ldots, n\}$, is defined by

$$\Pi_{\mathcal{J}}\mathbf{\Lambda} = \bigcup_{j \in \mathcal{J}} \Pi_j \mathbf{\Lambda} \subset \mathbb{Z}^d.$$

- Given a non-empty set $\mathbf{\Lambda} \subsetneq \left(\mathbb{Z}^d\right)^n$, we introduce, as in Chap. 2, its inner boundary $\partial^- \mathbf{\Lambda}$ and the outer boundary $\partial^+ \mathbf{\Lambda}$:

$$\begin{aligned} \partial^- \mathbf{\Lambda} &= \{\mathbf{y} \in \mathbf{\Lambda} : \ \mathrm{dist}_1(\mathbf{y}, \mathbb{Z}^d \setminus \mathbf{\Lambda}) = 1\}, \\ \partial^+ \mathbf{\Lambda} &= \{\mathbf{y} \in (\mathbb{Z}^d)^n \setminus \mathbf{\Lambda} : \ \mathrm{dist}_1(\mathbf{y}, \mathbf{\Lambda}) = 1\} \end{aligned} \tag{3.1}$$

and also the set (usually called the edge boundary of $\mathbf{\Lambda}$)

$$\partial \mathbf{\Lambda} = \left\{(\mathbf{x}, \mathbf{x}') \in \partial^- \mathbf{\Lambda} \times \partial^+ \mathbf{\Lambda} : \ |\mathbf{x} - \mathbf{x}'|_1 = 1\right\} \subset \left(\mathbb{Z}^d\right)^n \times \left(\mathbb{Z}^d\right)^n.$$

- The kinetic energy operator $\mathbf{H}^0 = \mathbf{H}^{0,(n)}$ of an n-particle quantum system on $\mathbb{Z}^d$ given by $\mathbf{H}^0 = -\mathbf{\Delta}$ where $\mathbf{\Delta}$ stands for the lattice Laplacian (on $\left(\mathbb{Z}^d\right)^n$): given $\phi \in \mathcal{H}^{(n)}$,

$$(\mathbf{H}^0 \phi)(\mathbf{x}) = 2dN\phi(\mathbf{x}) - \sum_{\mathbf{y} \in \left(\mathbb{Z}^d\right)^n : \, |\mathbf{x}-\mathbf{y}|_1 = 1} \phi(\mathbf{y}), \ \ \mathbf{x} \in \left(\mathbb{Z}^d\right)^n. \tag{3.2}$$

- The external random potential energy operator $\mathbf{V}(\omega) = \mathbf{V}^{(n)}(\omega)$ specified as the operator of multiplication by the random function

$$\mathbf{x} = (x_1, \ldots, x_n) \in \left(\mathbb{Z}^d\right)^n \mapsto \sum_{j=1}^{n} V(x_j; \omega). \tag{3.3}$$

Here and below—unless stressed otherwise—$(x, \omega) \in \mathbb{Z}^d \times \Omega \mapsto V(x; \omega) \in \mathbb{R}$ is an IID random field on $\mathbb{Z}^d$ satisfying Assumption A from Chap. 2; cf. (2.3). (In the course of presentation, we will mention, for local purposes, some weaker and stronger conditions upon $V(x; \omega)$.)

The probability distribution generated by $V(\cdot\,; \omega)$ is denoted, as earlier, by $\mathbb{P}$, and the expectation relative to $\mathbb{P}$ by $\mathbb{E}$. The probability space is, as before, the triple $(\Omega, \mathfrak{B}, \mathbb{P})$.

- The interaction potential energy operator $\mathbf{U} = \mathbf{U}^{(n)}$ (briefly, the interaction) which is the operator of multiplication by a real-valued function

$$\mathbf{x} \in \left(\mathbb{Z}^d\right)^n \mapsto \mathbf{U}^{(n)}(\mathbf{x}). \tag{3.4}$$

We suppose that function $\mathbf{U}^{(n)}$ is bounded on $\left(\mathbb{Z}^d\right)^n$, although a hard-core component can also be allowed, at the cost of more involved technical procedures. Further, the localization will be established in the main body of Chap. 4 under an assumption that the interaction has a finite range (radius) $r_0 \in \mathbb{N}$. In the simplest (and physically relevant) case where $\mathbf{U}^{(n)}$ is generated by a two-body interaction potential $\Phi^{(2)}$, with

$$\mathbf{U}^{(n)}(\mathbf{x}) = \sum_{1 \le i < j \le n} \Phi^{(2)}(x_i - x_j), \ \ \mathbf{x} = (x_1, \ldots, x_n),$$

this means that $\Phi^{(2)}(x) = 0$ when $|x| > r_0$, $x \in \mathbb{Z}^d$; usually, r_0 is specified as the minimal value with this property. In a more general situation, $\mathbf{U}^{(n)}(\mathbf{x})$ is a sum of k-body potentials

$$\mathbf{U}^{(n)}(\mathbf{x}) = \sum_{k=2}^{n} \sum_{1 \le i_1 < \ldots < i_k \le n} \Phi^{(k)}(x_{i_1}, \ldots, x_{i_k}), \quad \mathbf{x} = (x_1, \ldots, x_n) \in \left(\mathbb{Z}^d\right)^n$$

where $\Phi^{(k)}$ is a function $\left(\mathbb{Z}^d\right)^k \to \mathbb{R}$. Here, the notion of the range of the interaction can be defined as follows. Let a partition of a configuration $\mathbf{x} \in \left(\mathbb{Z}^d\right)^n$ be given, into complementary sub-configurations $\mathbf{x}_{\mathcal{J}} = (x_j,\ j \in \mathcal{J})$ and $\mathbf{x}_{\mathcal{J}^c} = (x_j,\ j \in \{1, \ldots, n\} \setminus \mathcal{J})$, where $\varnothing \ne \mathcal{J} \subsetneq \{1, \ldots, n\}$, with $|\mathcal{J}| = n' \ge 1$, $|\mathcal{J}^c| = n'' \ge 1$ and $n' + n'' = n$. The energy of interaction between $\mathbf{x}_{\mathcal{J}}$ and $\mathbf{x}_{\mathcal{J}^c}$ is determined by

$$\mathbf{U}^{(n',n'')}(\mathbf{x}_{\mathcal{J}} \,|\, \mathbf{x}_{\mathcal{J}^c}) := \mathbf{U}^{(n)}(\mathbf{x}) - \mathbf{U}^{(n')}(\mathbf{x}_{\mathcal{J}}) - \mathbf{U}^{(n'')}(\mathbf{x}_{\mathcal{J}^c}). \tag{3.5}$$

In this definition we adopt the agreement that for $n = 1$, $\mathbf{U}^{(1)} \equiv 0$ (which simply means that the background single-particle potential $\mathbf{U}^{(1)} : \mathbb{Z}^d \to \mathbb{R}$ is incorporated into the random term $V(\cdot\,;\omega)$).

Next, define the distance between two sub-configurations

$$\operatorname{dist}(\mathbf{x}_{\mathcal{J}}, \mathbf{x}_{\mathcal{J}^c}) := \min\left[|x_i - x_j| : i \in \mathcal{J}, j \in \mathcal{J}^c\right]. \tag{3.6}$$

Given $N = 2, 3, \ldots$, we say that the interaction $\mathbf{U}$ described by a sequence of functions $\mathbf{U}^{(n)} : \mathbf{x} \in \left(\mathbb{Z}^d\right)^n \mapsto \mathbb{R}$, $n = 2, \ldots, N$, has range $r_0 \in \mathbb{N}$ if for all $n \in \{2, \ldots, N\}$, $\mathbf{x} \in \left(\mathbb{Z}^d\right)^n$, non-empty subconfigurations $\mathcal{J} \subset \{1, \ldots, n\}$ with $|\mathcal{J}| = n'$ and $\mathcal{J}^c = \{1, \ldots, n\} \setminus \mathcal{J}$ with $|\mathcal{J}^c| = n''$,

$$\operatorname{dist}(\mathbf{x}_{\mathcal{J}}, \mathbf{x}_{\mathcal{J}^c}) > r_0 \text{ implies that } \mathbf{U}^{(n',n'')}(\mathbf{x}_{\mathcal{J}} \,|\, \mathbf{x}_{\mathcal{J}^c}) = 0. \tag{3.7}$$

In terms of functions $\Phi^{(k)}$, this condition means that $\Phi^{(k)}(\mathbf{x}) = 0$ whenever the configuration $\mathbf{x} \in \left(\mathbb{Z}^d\right)^k$ contains a subconfiguration $\mathbf{x}_{\mathcal{J}}$, with $\mathcal{J} \subset \{1, \ldots, k\}$, $1 \le |\mathcal{J}| < k$, such that $\operatorname{dist}(\mathbf{x}_{\mathcal{J}}, \mathbf{x}_{\mathcal{J}^c}) > r_0$.

The properties of translation invariance or symmetry of the interaction are not required for our techniques and will not be assumed, although such properties are natural in many of physical models.

- The n-particle tight-binding Anderson Hamiltonian $\mathbf{H}(\omega) = \mathbf{H}^{(n)}(\omega)$ in $\mathcal{H}^{(n)}$ which is defined as the sum

$$\mathbf{H}(\omega) = \mathbf{H}^0 + \mathbf{U} + g\mathbf{V}(\omega). \tag{3.8}$$

Here, as in the single-particle case, the coupling constant $g \in \mathbb{R}$ serves as a parameter measuring the amplitude of the random potential. The (random) eigenvalues of $\mathbf{H}(\omega)$ (when they exist) are denoted by $E_k\big(\mathbf{H}(\omega)\big)$; the corresponding normalized eigenfunctions are $\mathbf{x} \mapsto \boldsymbol{\psi}_k\big(\mathbf{x}; \mathbf{H}(\omega)\big)$.

- The finite-volume version $\mathbf{H}_{\mathbf{\Lambda}}(\omega) = \mathbf{H}^{(n)}_{\mathbf{\Lambda}}(\omega)$ of the operator $\mathbf{H}^{(n)}(\omega)$, acting in the space $\ell^2(\mathbf{\Lambda}) \simeq \mathbb{C}^{\mathbf{\Lambda}}$:

$$\mathbf{H}_{\mathbf{\Lambda}}(\omega) = \mathbf{H}^0_{\mathbf{\Lambda}} + \mathbf{U}_{\mathbf{\Lambda}} + g\mathbf{V}_{\mathbf{\Lambda}}(\omega). \tag{3.9}$$

Here $\mathbf{\Lambda}$ is a finite subset in $\big(\mathbb{Z}^d\big)^n$; typically $\mathbf{\Lambda} = \mathbf{\Lambda}^{(n)}_L(\mathbf{u})$ where $\mathbf{\Lambda}^{(n)}_L(\mathbf{u})$ is an n-particle cube, centered at point $\mathbf{u} = (u_1, \ldots, u_n) \in \big(\mathbb{Z}^d\big)^n$ and having radius $L \in \mathbb{N}$ and cardinality $|\mathbf{\Lambda}_L(\mathbf{u})| = (2L+1)^{nd}$:

$$\mathbf{\Lambda}_L(\mathbf{u}) = \bigtimes_{j=1}^{n} \Lambda_L(u_j).$$

Further, $\mathbf{H}^0_{\mathbf{\Lambda}_L} = -\mathbf{\Delta}_{\mathbf{\Lambda},\mathrm{Dir}}$ where $\mathbf{\Delta}_{\mathbf{\Lambda},\mathrm{Dir}}$ stands for the discrete Laplacian in $\mathbf{\Lambda}$ with Dirichlet's boundary conditions on $\partial^+\mathbf{\Lambda}$ (the Dirichlet Laplacian in $\mathbf{\Lambda}$, for short). To compare with Eq. (3.2): for $\phi \in \ell^2(\mathbf{\Lambda})$,

$$(-\mathbf{\Delta}_{\mathbf{\Lambda},\mathrm{Dir}}\phi)(\mathbf{x}) = 2dn\phi(\mathbf{x}) - \sum_{\mathbf{y}\in\mathbf{\Lambda}:\, |\mathbf{x}-\mathbf{y}|_1=1} \phi(\mathbf{y}), \ \ \mathbf{x} \in \mathbf{\Lambda}.$$

Next, $\mathbf{V}_{\mathbf{\Lambda}}(\omega)$ and $\mathbf{U}_{\mathbf{\Lambda}}$ in the RHS of Eq. (3.9) stand for the restrictions of the multiplication operators $\mathbf{V}(\omega)$ and $\mathbf{U}$ to $\mathbf{\Lambda}$.

The lattice delta-functions $\delta_{\mathbf{x}}$, $\mathbf{x} \in \mathbf{\Lambda}$, form a natural reference basis in $\ell^2(\mathbf{\Lambda})$; in this basis the operators $\mathbf{H}^0_{\mathbf{\Lambda}}$, $\mathbf{U}_{\mathbf{\Lambda}}$, $\mathbf{V}_{\mathbf{\Lambda}}(\omega)$ and $\mathbf{H}_{\mathbf{\Lambda}}(\omega)$ have Hermitian matrices of size $|\mathbf{\Lambda}|$. The eigenvalues of $\mathbf{H}_{\mathbf{\Lambda}}(\omega)$ are denoted by $E_j\big(\mathbf{H}_{\mathbf{\Lambda}}(\omega)\big)$, $j = 1, \ldots, |\mathbf{\Lambda}|$, and, as in the single-particle case, are supposed to be listed in an increasing order:

$$E_1\big(\mathbf{H}_{\mathbf{\Lambda}}(\omega)\big) \le E_2\big(\mathbf{H}_{\mathbf{\Lambda}}(\omega)\big) \le \ldots \le E_{|\mathbf{\Lambda}|}\big(\mathbf{H}_{\mathbf{\Lambda}}(\omega)\big).$$

The spectrum of $\mathbf{H}_{\mathbf{\Lambda}}(\omega)$, i.e., the collection of the eigenvalues $E_j\big(\mathbf{H}_{\mathbf{\Lambda}_L}(\omega)\big)$ (counted with their multiplicities), is denoted by $\sigma\big(\mathbf{H}_{\mathbf{\Lambda}}(\omega)\big)$. Finally, $\boldsymbol{\psi}_j\big(x; \mathbf{H}_{\mathbf{\Lambda}}(\omega)\big)$, $x \in \mathbf{\Lambda}_L(\mathbf{u})$, stands for the eigenfunction of $\mathbf{H}_{\mathbf{\Lambda}}(\omega)$ with the eigenvalue $E_j\big(\mathbf{H}_{\mathbf{\Lambda}}(\omega)\big)$.

In the basis $\{\delta_{\mathbf{x}}, \mathbf{x} \in \mathbf{\Lambda}\}$, the operators $\mathbf{U}_{\mathbf{\Lambda}}$ and $\mathbf{V}_{\mathbf{\Lambda}_L(\mathbf{u})}(\omega)$ are represented by diagonal matrices.

- As in the case of a single-particle system, an important part will be played by the Green's functions $\mathbf{G}^{(n)}_{\mathbf{\Lambda}_L(\mathbf{u})}(\mathbf{x}, \mathbf{y}; E)$ representing the matrix elements of the resolvent operator $\mathbf{G}_{\mathbf{\Lambda}_L(\mathbf{u})}(E, \omega) = \Big(\mathbf{H}^{(n)}_{\mathbf{\Lambda}_L(\mathbf{u})} - E\Big)^{-1}$ in the delta-basis:

$$\begin{gathered}\mathbf{G}^{(n)}_{\mathbf{\Lambda}_L(\mathbf{u})}(\mathbf{x}, \mathbf{y}; E) = \big\langle \delta_{\mathbf{x}}, \mathbf{G}_{\mathbf{\Lambda}_L(\mathbf{u})}(E, \omega)\delta_{\mathbf{y}} \big\rangle\,, \\ E \in \mathbb{R} \setminus \sigma\left(\mathbf{H}^{(n)}_{\mathbf{\Lambda}_L(\mathbf{u})}(\omega)\right), \ \mathbf{x}, \mathbf{y} \in \mathbf{\Lambda}_L(\mathbf{u}).\end{gathered}$$

In particular, similarly to the single-particle case, the Green's functions for the pair of n-particle cubes $\mathbf{\Lambda} = \mathbf{\Lambda}_L(\mathbf{u})$ and $\tilde{\mathbf{\Lambda}} = \mathbf{\Lambda}_l(\mathbf{v})$ such that $\tilde{\mathbf{\Lambda}} \cup \partial^+ \tilde{\mathbf{\Lambda}} \subset \mathbf{\Lambda}$ satisfy the geometric resolvent equation: $\forall$ configuration $\mathbf{y} \in \Lambda \setminus \tilde{\Lambda}$,

$$\mathbf{G}_{\mathbf{\Lambda}}^{(n)}(\mathbf{v}, \mathbf{y}; E, \omega) = \sum_{(\mathbf{x},\mathbf{x}')\in\partial\tilde{\mathbf{\Lambda}}} \mathbf{G}_{\tilde{\mathbf{\Lambda}}}^{(n)}(\mathbf{v}, \mathbf{x}; E, \omega)\, \mathbf{G}_{\mathbf{\Lambda}}^{(n)}(\mathbf{x}', \mathbf{y}; E, \omega). \tag{3.10}$$

As before, it implies the geometric resolvent inequality (GRI):

$$\begin{aligned} \left|\mathbf{G}_{\mathbf{\Lambda}}^{(n)}(\mathbf{v}, \mathbf{y}; E, \omega)\right| &\le \sum_{(\mathbf{x},\mathbf{x}')\in\partial\tilde{\mathbf{\Lambda}}} \left|\mathbf{G}_{\tilde{\mathbf{\Lambda}}}^{(n)}(\mathbf{v}, \mathbf{x}; E, \omega)\right| \left|\mathbf{G}_{\mathbf{\Lambda}}^{(n)}(\mathbf{x}', \mathbf{y}; E, \omega)\right| \\ &\le \left|\partial\tilde{\mathbf{\Lambda}}\right| \left(\max_{(\mathbf{x},\mathbf{x}')\in\partial\tilde{\mathbf{\Lambda}}} \left|\mathbf{G}_{\tilde{\mathbf{\Lambda}}}^{(n)}(\mathbf{v}, \mathbf{x}; E, \omega)\right| \left|\mathbf{G}_{\mathbf{\Lambda}}^{(n)}(\mathbf{x}', \mathbf{y}; E, \omega)\right| \right) \\ &\le \left|\partial\tilde{\mathbf{\Lambda}}\right| \left\|\mathbf{G}_{\tilde{\mathbf{\Lambda}}}^{(n)}(E, \omega)\right\| \left(\max_{\mathbf{x}'\in\partial^+\tilde{\mathbf{\Lambda}}} \left|\mathbf{G}_{\mathbf{\Lambda}}^{(n)}(\mathbf{x}', \mathbf{y}; E, \omega)\right|. \right) \end{aligned} \tag{3.11}$$

The identity (3.10) is derived in the same fashion as in the single-particle case. The shortest way to convince yourself that Eq. (3.10) holds true is as follows. Note that in the formula (3.9) for the Hamiltonian $\mathbf{H}_{\mathbf{\Lambda}}(\omega)$, (i) the kinetic energy part $\mathbf{H}_{\mathbf{\Lambda}}^0$ is the same as the kinetic energy part $H_{\mathbf{\Lambda}}^0$ for a (Nd)-dimensional single-particle system in the cube $\mathbf{\Lambda}$ considered as a subset of $\mathbb{Z}^{Nd}$ and (ii) the potential energy part $\mathbf{U}_{\mathbf{\Lambda}} + \mathbf{V}_{\mathbf{\Lambda}}$ is a local operator, i.e., diagonal in the delta-basis in $\ell^2(\mathbf{\Lambda})$. These facts are the only properties used in the derivation of Eq. (3.10).

In particular, the geometric resolvent equation for eigenfunctions $\boldsymbol{\psi} = \boldsymbol{\psi}(\,\cdot\,, \mathbf{H}(\omega))$ of $\mathbf{H}(\omega)$ is satisfied: if $\mathbf{H}(\omega)\boldsymbol{\psi} = E\boldsymbol{\psi}$ then, for any $\mathbf{\Lambda} \subset \left(\mathbb{Z}^d\right)^N$,

$$\mathbf{1}_{\mathbf{\Lambda}}\boldsymbol{\psi} = \mathbf{1}_{\mathbf{\Lambda}}\mathbf{G}_{\mathbf{\Lambda}}(E, \omega) T_{\mathbf{\Lambda}}\boldsymbol{\psi},$$

or

$$\boldsymbol{\psi}(\mathbf{x}) = \sum_{(\mathbf{y},\mathbf{y}')\in\partial\mathbf{\Lambda}} \mathbf{G}_{\mathbf{\Lambda}}(\mathbf{x}, \mathbf{y}; E, \omega)\boldsymbol{\psi}(\mathbf{y}'), \quad \mathbf{x} \in \Lambda. \tag{3.12}$$

Equation (3.12) implies the GRIs for the eigenfunctions:

$$\begin{aligned} |\boldsymbol{\psi}(\mathbf{x})| &\le \left(\max_{\mathbf{y}\in\partial^-\mathbf{\Lambda}} |\mathbf{G}_{\mathbf{\Lambda}}(\mathbf{x}, \mathbf{y}; E, \omega)| \right) \sum_{\mathbf{y}'\in\partial^+\mathbf{\Lambda}} |\boldsymbol{\psi}(\mathbf{y}')|, \\ |\boldsymbol{\psi}(\mathbf{x})| &\le \left(\sum_{\mathbf{y}\in\partial^-\mathbf{\Lambda}} |\mathbf{G}_{\mathbf{\Lambda}}(\mathbf{x}, \mathbf{y}; E, \omega)| \right) \max_{\mathbf{y}'\in\partial^+\mathbf{\Lambda}} |\boldsymbol{\psi}(\mathbf{y}')|. \end{aligned} \tag{3.13}$$

These observations open the door to applications, to the multi-particle Green's functions, of the machinery developed in Sect. 2.6 (particularly, Theorem 2.6.1 and 2.6.3 from Sect. 2.6.2).

In what follows, the argument ω will be often omitted, particularly when this would not lead to a confusion.

Families of random operators $\{\mathbf{H}^{(n)}(\omega), \omega \in \Omega\}$ and $\{\mathbf{H}^{(n)}_{\mathbf{\Lambda}_L(\mathbf{u})}(\omega), \omega \in \Omega\}$, with the probability measure $\mathbb{P}$ on the sigma-algebra $\mathfrak{B}$ of subsets of Ω, will be sometimes referred to as ensembles of (n-particle) LSOs, on $\left(\mathbb{Z}^d\right)^n$ and $\mathbf{\Lambda}_L(\mathbf{u})$, respectively.

As was already said before, throughout most of Chaps. 3 and 4 we employ Assumption A upon the IID random potential $V(\,\cdot\,;\omega)$; additional conditions may be also used locally.

As to the interaction $\mathbf{U}$, the condition under which we will prove the multi-particle spectral localization is summarized in

Assumption U. *For a given $N = 2, 3, \ldots$, the interaction $\mathbf{U}^{(N)}$ has a finite range $r_0 \in \mathbb{N}$; cf. Eqs.* (3.5)–(3.7).

We will continue using symbols like H_Λ and $G_\Lambda(x, y; E, \omega)$ for objects related to the single-particle theory.

3.1.2 Multi-particle Localization Results: MPMSA in a Nutshell

In this subsection we state the result on the N-particle localization which will be proved in Chaps. 3 and 4.

Singularity and Separability of Cubes

First, we need the definition that extends the notion of a singular cube to a multi-particle situation; cf. Definition 2.1.1.

Definition 3.1.1. *Given a positive integer $N \geq 2$, values $E \in \mathbb{R}$, $m > 0$, $n \in \{1, \ldots, N\}$ and a sample of potential $V(\,\cdot\,;\omega)$, an n-particle cube $\mathbf{\Lambda}^{(n)}_L(\mathbf{u})$ is called (E, m)-non-singular if*

$$|\mathbf{\Lambda}^{(n)}_L(\mathbf{u})| \max_{\mathbf{x}\in\mathbf{\Lambda}^{(n)}_{L^{1/\alpha}}(\mathbf{u})} \max_{\mathbf{y}\in\partial^-\mathbf{\Lambda}^{(n)}_L(\mathbf{u})} \left|\mathbf{G}_{\mathbf{\Lambda}^{(n)}_L(\mathbf{u})}(\mathbf{x}, \mathbf{y}; E, \omega)\right| \leq \mathrm{e}^{-\gamma(m,L,n)L} \tag{3.14}$$

where

$$\gamma(m, L, n) = \gamma_N(m, L, n) = m\left(1 + L^{-1/8}\right)^{N-n+1}. \tag{3.15}$$

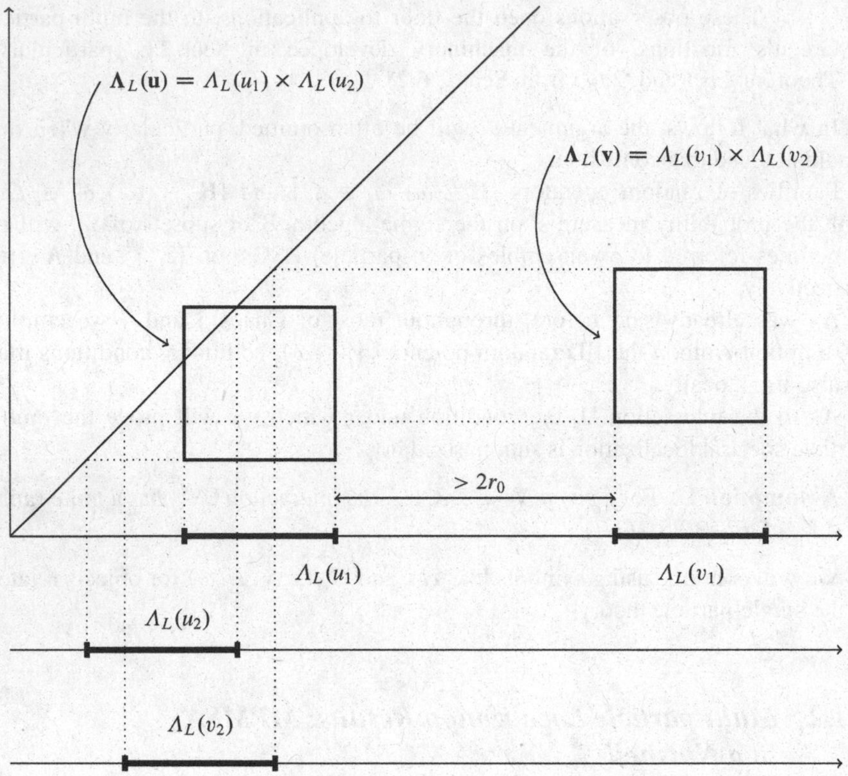

Fig. 3.1 An example of separable two-particle cubes $\mathbf{\Lambda}_L(\mathbf{u}) = \Lambda_L(u_1) \times \Lambda_L(u_2)$ and $\mathbf{\Lambda}_L(\mathbf{v}) = \Lambda_L(v_1) \times \Lambda_L(v_2)$. Here $d = 1$, so these cubes are squares in $\mathbb{Z}^1 \times \mathbb{Z}^1$. In this example, Eq. (3.17) holds true with $J = \{1\}$

Otherwise we call the cube $\mathbf{\Lambda}_L^{(n)}(\mathbf{u})$ (E, m)*-singular. As in Chap. 2, we use abbreviations NS (non-singular) and S (singular).*

The next definition takes into account the specific character of multi-particle Anderson models.

Definition 3.1.2. *(See Fig. 3.1.)*

(A) *A pair of n-particle cubes* $\mathbf{\Lambda}_L^{(n)}(\mathbf{u})$ *and* $\mathbf{\Lambda}_L^{(n)}(\mathbf{v})$ *is called separable if* $|\mathbf{u} - \mathbf{v}| > 11nL$ *and* $\exists$ *a non-empty subset* $\mathcal{J} \subseteq \{1, \dots, n\}$ *with* $\mathcal{J}^c = \{1, \dots, n\} \setminus \mathcal{J}$ *such that either*

$$\Pi_{\mathcal{J}} \mathbf{\Lambda}_{L+r_0}^{(n)}(\mathbf{u}) \cap \left(\Pi_{\mathcal{J}^c} \mathbf{\Lambda}_{L+r_0}^{(n)}(\mathbf{u}) \cup \Pi \mathbf{\Lambda}_{L+r_0}^{(n)}(\mathbf{v}) \right) = \varnothing, \tag{3.16}$$

or

$$\Pi_{\mathcal{J}}\mathbf{\Lambda}_{L+r_0}^{(n)}(\mathbf{v}) \cap \left(\Pi_{\mathcal{J}^c}\mathbf{\Lambda}_{L}^{(n)}(\mathbf{v}) \cup \Pi\mathbf{\Lambda}_{L+r_0}^{(n)}(\mathbf{u})\right) = \varnothing \tag{3.17}$$

where r_0 is the radius of interaction. Equivalently, we say that configurations $\mathbf{u}$ *and* $\mathbf{v}$ *are $(L+r_0)$-separable.*

(B) *We say that* $\mathbf{u}$ *is $(L+r_0)$-separable from* $\mathbf{v}$ *when Eq.* (3.16) *holds true, and* $\mathbf{v}$ *is $(L+r_0)$-separable from* $\mathbf{u}$ *if* (3.17) *is satisfied. (Note that these possibilities do not exclude each other.) When the subset $\mathcal{J}$ should be explicitly mentioned, we use the term $(L+r_0,\mathcal{J})$-separable.*

(C) *In the particular case where $\mathcal{J}=\{1,\dots,n\}$, so that $\Pi_{\mathcal{J}}\equiv\Pi$ and*

$$\Pi\mathbf{\Lambda}_{L+r_0}^{(n)}(\mathbf{u}) \cap \Pi\mathbf{\Lambda}_{L+r_0}^{(n)}(\mathbf{v}) = \varnothing, \tag{3.18}$$

the cubes $\mathbf{\Lambda}_L^{(n)}(\mathbf{u})$ and $\mathbf{\Lambda}_L^{(n)}(\mathbf{v})$ will be called completely separable.

(D) *Given integers $n\geq 1$ and $L\geq 1$, two n-particle cubes $\mathbf{\Lambda}_L^{(n)}(\mathbf{u})$, $\mathbf{\Lambda}_L^{(n)}(\mathbf{v})$ are called* distant *if $|\mathbf{u}-\mathbf{v}|>11nL$.*

The meaning of the condition $\mathrm{dist}(\mathbf{u},\mathbf{v})>11nL$ will be explained in Sect. 4.2. (This condition is somewhat stronger than what is required for the EVC bounds but is technically convenient in our implementation of the MPMSA scheme.)

In the case $n=2$ (two particles), Eqs. (3.16) or (3.17) (or both of them) follows when the cubes are distant from each other (this fact makes the exposition for a two-particle model considerably less technical than for $N\geq 3$).

As with the single-particle MSA, the MPMSA employs the length-scale scheme (cf. Eq. (2.16)) where the initial scale L_0 has to be large enough and $L_{k+1}=\lfloor L_k^\alpha\rfloor$, with parameter $1<\alpha<2$. A convenient choice is again $\alpha=3/2$, although the symbol α will be used throughout relevant calculations in Chap. 4.

In analogy with (2.30), we introduce the quantities

$$p_k(n)=p_k(n,N,p_0,\theta)=4^{N-n}p_0(1+\theta)^k,\quad 1\leq n\leq N,\ k\geq 0, \tag{3.19}$$

depending on parameters p_0 and θ; the assumptions on p_0 and θ in the forthcoming Theorems 3.1.1 and 3.1.2 will be as follows (cf. Eq. (4.1)):

$$p_0>6Nd \ \text{ and } \ \theta:=\frac{1}{6}-\frac{Nd}{p_0}\in\left(0,\frac{1}{6}\right). \tag{3.20}$$

MPMSA Bounds and the MPMSA Strategy in a Nutshell

We will start with the fixed-energy analysis of Green's functions. According to the scheme presented in Sect. 2.4, we will aim to prove the following result, at any scale L_k, $k\geq 0$:

($\mathbf{SS}.N, E, m, p_k, L_k$) $\forall\ n = 1, \ldots, N$ *and any n-particle cube* $\mathbf{\Lambda}^{(n)}_{L_k}(\mathbf{u})$, *the following bound holds true:*

$$\mathbb{P}\left\{ \mathbf{\Lambda}^{(n)}_{L_k}(\mathbf{u}) \text{ is } (E,m)\text{–S} \right\} \le L_k^{-p_k(n)}. \tag{3.21}$$

The derivation of the multi-particle dynamical localization from the bounds ($\mathbf{SS}.N, E, m, p_k, L_k$) will then be obtained by an adaptation of the method of Sect. 2.5 to multi-particle operators.

To establish directly exponential localization of eigenfunctions, we will make use of the variable-energy MPMSA. Following the general strategy of the variable-energy MSA, we aim to prove for all $n = 1, \ldots, N$ and integer $k \ge 0$ the following property of the multi-particle systems:

($\mathbf{DS}.N, n, m, p_k, L_k$) *For any pair of separable n-particle cubes* $\mathbf{\Lambda}^{(n)}_{L_k}(\mathbf{u})$, $\mathbf{\Lambda}^{(n)}_{L_k}(\mathbf{v})$ *the following bound holds true:*

$$\mathbb{P}\left\{ \exists\, E \in \mathbb{R}:\ \mathbf{\Lambda}^{(n)}_{L_k}(\mathbf{u}) \text{ and } \mathbf{\Lambda}^{(n)}_{L_k}(\mathbf{v}) \text{ are } (E,m)\text{–S} \right\} \le L_k^{-2p_k(n)}. \tag{3.22}$$

Property ($\mathbf{DS}.N, n, m, p_k, L_k$) will be referred to as the bound for the probability of double singularity (in a multi-particle system). Alternatively, we will also call it the MPMSA bound(s). The main outcome of Chaps. 3 and 4 are Theorems 3.1.1 and 3.1.2 below.

In contrast with Chap. 2, where the variable-energy MSA could be restricted to a given, bounded interval $I \subset \mathbb{R}$, the multi-particle induction presented in this book cannot operate in a fixed interval I, unless $I = \mathbb{R}$; in the latter case, Anderson localization is ultimately established in the entire real line. Informally, this can be explained as follows. In the course of the induction on the number of particles, an n-particle system is decomposed into sub-systems, say, with $n' \ge 1$ and $n'' \ge 1$ particles, so that $n = n' + n''$. In the analysis of the Green functions of the n-particle system at an energy E in a bounded interval I, one has to assess their counterparts for the subsystems with n' and n'' particles. In absence of interaction between the two subsystems (which is a particularly simple, but unavoidable situation), the total energy is $E = E' + E''$, where E' and E'' are the energies of the respective subsystems. Now it is clear that the possible valucs E' (as well as E'') are not necessarily restricted to the same interval I as the total energy E. As a result, one has to operate with a sequence of energy bands $I^{(n)}$, $n = 1, \ldots, N$.

In our book, we focus mainly on complete localization in strongly disordered systems, which allows us to consider the interval $I = \mathbb{R}$, which justifies the above form of the property (**DS.**N, n, m, p_k, L_k). The analysis of band-edge multi-particle localization requires some adaptations; such a work was done by Ekanga in the framework of his PhD project; cf. [82, 83].

The general strategy of our approach can be presented "in a nutshell" as follows. For the sake of clarity, we assume that $d = 1$, $N = 2$ (as in Fig. 3.1), and discuss only the fixed-energy (two-particle) MSA scheme.
In order to prove the bounds (**SS.**N, E, m, p_k, L_k) for all $k \geq 0$, it would suffice (as we have seen in Sect. 2.4) to assess the probability of the following events relative to a cube $\mathbf{\Lambda}_{L_k}(\mathbf{u})$ and depending upon an arbitrary energy value $E \in I$:

- "$\mathbf{\Lambda}_{L_k}(\mathbf{u})$ is E-resonant";
- "$\mathbf{\Lambda}_{L_k}(\mathbf{u})$ contains ≥ 2 disjoint (E, m)-singular cubes $\mathbf{\Lambda}_{L_{k-1}}(\mathbf{x})$, $\mathbf{\Lambda}_{L_{k-1}}(\mathbf{y})$."

Due to the presence of correlations in the random potential $(x_1, x_2) \mapsto V(x_1; \omega) + V(x_2; \omega)$, we will have to modify the latter step of this strategy (developed originally for single-particle operators) as follows:

- Instead of *disjoint* cubes $\mathbf{\Lambda}_{L_{k-1}}(\mathbf{x})$, $\mathbf{\Lambda}_{L_{k-1}}(\mathbf{y})$, we work with cubes at distance[1] $\geq CL_{k-1}$ from each other, with an appropriate $C > 0$;
- Cubes similar to $\mathbf{\Lambda}_L(\mathbf{u})$ on Fig. 3.1, with centers close to the "diagonal" $\mathbb{D} := \{\mathbf{x} = (x, x), x \in \mathbb{Z}\}$ and far apart from each other, say, $\mathbf{\Lambda}_{L_{k-1}}(0, 0)$ and $\mathbf{\Lambda}_{L_{k-1}}(3L_{k-1}, 3L_{k-1})$, give rise to independent sub-samples of the random potential energy $V(x_1; \omega) + V(x_2; \omega)$; therefore, the respective random Hamiltonians $\mathbf{H}_{\mathbf{\Lambda}_{L_{k-1}}(\mathbf{x})}$, $\mathbf{H}_{\mathbf{\Lambda}_{L_{k-1}}(\mathbf{y})}$ are also independent and can be treated essentially in the same way as in the single-particle case;
- Pairs of cubes like $\mathbf{\Lambda}_L(\mathbf{v})$ on Fig. 3.1, with centers far enough from the diagonal $\mathbb{D}$, present a difficulty, since the respective random Hamiltonians $\mathbf{H}_{\mathbf{\Lambda}_{L_{k-1}}(\mathbf{x})}$, $\mathbf{H}_{\mathbf{\Lambda}_{L_{k-1}}(\mathbf{y})}$ may be correlated. However, we will show that the presence of even one "off-diagonal" (E, m)-S cube has a small probability, sufficient for the fixed-energy scaling procedure.

It may seem surprising that the complication encountered in a *straightforward* adaptation of the single-particle MSA comes from (the pairs of) the cubes like $\mathbf{\Lambda}_L(\mathbf{v})$ on Fig. 3.1 (In Sect. 4.2 we call such cubes partially interactive; cf. Definition 4.2.1). Indeed, physically speaking, they correspond to decomposable subsystems. Unfortunately, this is the price to pay for a frontal attack and (almost) ignoring the fact that single-particle systems in the absence of interaction are already *localized at any scale*. This difficulty has been encountered (in slightly different forms) both in [70, 71] (where the MPMSA was used) and in [9, 10] (based on the multi-particle FMM).

It is to be stressed that the above informal description of the scaling procedure suits only the two-particle systems: for $N \geq 3$, the condition $\operatorname{dist}\left(\mathbf{\Lambda}_{L_{k-1}}(\mathbf{x}), \mathbf{\Lambda}_{L_{k-1}}(\mathbf{y})\right) \geq CL_{k-1}$ becomes insufficient and has to be replaced

[1] This could also be done in the single-particle models, e.g., in the case where the random potential $V(x; \omega)$ is not IID (cf., e.g., [60]).

by the stronger *separability* condition. Some crucial probabilistic estimates also become more complex (and final localization results somewhat weaker and physically less natural) than in the case where $N = 2$.

The variable-energy MPMSA scheme is more involved, but makes use of a similar strategy.

Main Results on Multi-particle Localization

Theorem 3.1.1. *Fix an integer $N \geq 2$ and suppose Assumptions A and 3.1.1 are fulfilled. Take any $m \geq 1$ and $p_0 > 6dN$.*

There exist constants $g^* \in (0, \infty)$, $L_0 \in \mathbb{N}^*$ and $\theta \in (0, 1/6)$ depending upon F, d, m, p_0 and $\mathbf{U}$, such that for any g with $|g| \geq g^*$, $n \in [[1, N]]$ and $k \in \mathbb{N}$ property ($\mathbf{DS}.N, n, m, p_k, L_k$) is satisfied.

Consequently, if $|g| \geq g^*$ then, with probability one, the N-particle Hamiltonian $\mathbf{H}^{(N)}(\omega)$ has pure point spectrum. Moreover, for all eigenfunctions $\boldsymbol{\psi}_j(\cdot;\omega)$ of $\mathbf{H}^{(N)}(\omega)$ we have that

$$|\boldsymbol{\psi}_j(\mathbf{x};\omega)| \leq C_j(\omega)\, \mathrm{e}^{-m|\mathbf{x}|}, \quad \mathbf{x} \in \left(\mathbb{Z}^d\right)^N, \tag{3.23}$$

with random constants $C_j(\omega) \in (0, \infty)$.

Theorem 3.1.1 establishes the complete spectral localization for the Hamiltonian $\mathbf{H}^{(N)}(\omega)$ with a large amplitude g of the external field. Dynamical localization for large g is considered in Theorem 3.1.2. We denote by $\mathscr{B}_1(\mathbb{R})$ the set of all Borel function $\zeta : \mathbb{R} \to \mathbb{C}$ with $\|\zeta\|_\infty \leq 1$.

Theorem 3.1.2. *Let N be an integer ≥ 2 and let Assumptions A and 3.1.1 be fulfilled. Then $\exists\, g^* \in (0, \infty)$ such that if $|g| \geq g^*$ then for all $s > 0$ and any finite subset $\mathbb{K} \subset \left(\mathbb{Z}^d\right)^N$,*

$$\mathbb{E}\left[\sup_{\zeta \in \mathscr{B}_1(\mathbb{R})} \left\| M^s\, \zeta\left(\mathbf{H}^{(N)}(\omega)\right) \mathbf{1}_{\mathbb{K}} \right\| \right] < \infty. \tag{3.24}$$

Here, as in the single-particle case, M stands for the multiplication operator by the norm: $M\phi(\mathbf{x}) = |\mathbf{x}|\phi(\mathbf{x})$, $\mathbf{x} \in \left(\mathbb{Z}^d\right)^N$.

In particular, taking ζ of the form $\zeta_t : \lambda \mapsto \mathrm{e}^{-it\lambda}$, $t \in \mathbb{R}$, one has

$$\mathbb{E}\left[\sup_{t \in \mathbb{R}} \left\| M^s\, \mathrm{e}^{-it\mathbf{H}^{(N)}(\omega)}\, \mathbf{1}_{\mathbb{K}} \right\| \right] < \infty. \tag{3.25}$$

Remark 3.1.1. The reader can see that Theorems 3.1.1 and 3.1.2 address only the issue of multi-particle localization at large amplitudes. Accordingly, as in the single-particle case, we use the range $m \geq 1$ in Theorem 3.1.1; cf. Remark 2.1.9 in

Sect. 2.1.6. Localization at the edge of the spectrum in multi-particle systems by means of the MPMSA is briefly discussed in Sects. 4.7.1 and 4.7.5. (It requires an adaptation of the initial scale MPMSA bound.)

Remark 3.1.2. The value θ in Theorems 3.1.1 and 3.1.2 can be specified as $1/6 - dN/p_0$; see Eq. (3.20).

The proof of Theorems 3.1.1 and 3.1.2 will be our principal goal in Chaps. 3 and 4. The core of the argument, as in the single-particle case, will be the proof of inequalities (3.21) and (3.22). Again, the latter is required for the proof of the spectral and dynamical localization, but the former is simpler to prove directly.

Remark 3.1.3. Theorem 3.1.2 states the strong dynamical localization in its usual form. However, it follows from a slightly different (and stronger) statement where the eigenfunction correlators at large distances are assessed, viz.

$$\mathbb{E}\left[\sup_{t\in\mathbb{R}} \left|\langle \delta_{\mathbf{x}}, \mathrm{e}^{-it\mathbf{H}(\omega)}\,\delta_{\mathbf{y}}\rangle\right|\right] \le C(\mathbf{y})\,\mathrm{e}^{-a\ln^{1+c}|\mathbf{x}-\mathbf{y}|}. \tag{3.26}$$

Indeed, notice first that it suffices to prove (3.25) in the simplest case where $|\mathbb{K}| = 1$, i.e. $\mathbb{K} = \{\mathbf{y}\}$, for some $\mathbf{y} \in (\mathbb{Z}^d)^N$. Fix such a point $\mathbf{y}$. Since M^s in (3.25) is a multiplication operator, we have

$$\left\| M^s\,\mathrm{e}^{-it\mathbf{H}^{(N)}(\omega)}\,\mathbf{1}_{\mathbb{K}}\right\|^2 = \sum_{\mathbf{x}\in(\mathbb{Z}^d)^N} |\mathbf{x}|^{2s} \left|\left\langle \mathbf{1}_{\mathbf{x}}, \mathrm{e}^{-it\mathbf{H}^{(N)}(\omega)}\,\mathbf{1}_{\mathbf{y}}\right\rangle\right|^2 \tag{3.27}$$

and for all large values of $|\mathbf{x}|$, i.e., for all but a finite number of sites $\mathbf{x}$, one has, for example,

$$\frac{1}{2}|\mathbf{x}-\mathbf{y}| \le |\mathbf{x}| \le 2|\mathbf{x}-\mathbf{y}|,$$

so replacing $|\mathbf{x}|$ by the norm-distance $|\mathbf{x}-\mathbf{y}|$ does not affect the convergence of the above series.

Moreover, the factors $|\mathbf{x}|^{2s}$ (or, respectively, $|\mathbf{x}-\mathbf{y}|^{2s}$) are non-random, thus the upper bound on the eigenfunction correlators (3.26) implies (3.25) for any $s > 0$.

In Chap. 4, we will prove the upper bound on the eigenfunction correlators.

3.1.3 *EVC Bounds in the Multi-particle MSA*

Like the single-particle case, an important ingredient of the MPMSA-based proof of localization in multi-particle systems is the eigenvalue concentration (EVC) bound. More precisely, as we have seen in Chap. 2, in the one-particle case the Wegner-type bound for a single cube $\Lambda_L(x)$, $x \in \mathbb{Z}^d$ (with an arbitrary L) was essential for the

MSA procedure. In fact, two kinds of EVC bounds have been used in the course of the scaling analysis: for a single cube and for a pair of disjoint cubes $\Lambda_L(x)$, $\Lambda_L(y)$; cf. Lemma 2.2.2 and Corollary 2.2.4. The bound for the pair of cubes was derived with the help of the IID property of the on-site random external potential $V(\cdot\,;\omega)$.

However, the situation with multi-particle systems is more complicated, because of a highly correlated nature of the multi-particle random potential field (3.3). In fact, random field (3.3) features no decay of correlations at arbitrarily large distances.

For an illustration, consider the case $n = 2$; here, for the two-particle configurations of a particular form $\mathbf{x}' = (x, x')$ and $\mathbf{x}'' = (x, x'')$, the sums

$$V(x;\omega) + V(x';\omega) \ \text{ and } \ V(x;\omega) + V(x'';\omega), \quad x, x', x'' \in \mathbb{Z}^d,$$

are coupled via the random variable $V(x;\omega)$, regardless of the distance between $\mathbf{x}'$ and $\mathbf{x}''$. The same is true for the configurations $\mathbf{x}' = (x', x)$ and $\mathbf{x}'' = (x'', x)$.

In more general terms, the n-particle lattice $\left(\mathbb{Z}^d\right)^n$, $n \geq 2$, cannot be treated as "homogeneous," unlike $\mathbb{Z}^d$ in the single-particle localization theory. In fact, possible positions of n-particle cubes are not equivalent from the point of view of the interaction. For instance, with $n = 2$ and $d = 1$, the configuration $\mathbf{u} = (0, R)$ where $R \in \mathbb{Z}$ and $R \geq r_0$ represents a classical state with two particles at distance R from each other. With $\mathbf{U}^{(2)}(\mathbf{u}) = 0$, these two particles are non-interacting. On the other hand, the configuration $\mathbf{x}_0 = (0, 0)$ corresponds to a pair of particles occupying the same lattice site in $\mathbb{Z}^1$, and the interaction between them may be non-zero.

Similarly, consider (still with $d = 1$) the Hilbert space $\ell^2\big(\mathbf{\Lambda}_L^{(2)}(\mathbf{u})\big)$ and assume the two-particle cube $\mathbf{\Lambda} = \mathbf{\Lambda}_L^{(2)}(\mathbf{u})$ is given by $\mathbf{\Lambda} = \Lambda' \times \Lambda''$ with

$$\Lambda' = \{-L, -L+1, \ldots, L\} \ \text{ and } \ \Lambda'' = \{L+R, L+R+1, \ldots, 3L+R\}$$

where $R \geq r_0$. It describes a system of two particles in distant cubes Λ' and Λ''. Correspondingly, the two-particle Hamiltonian $\mathbf{H}_{\mathbf{\Lambda}}^{(2)}(\omega)$ describes a decoupled system; algebraically, it reads as follows:

$$\mathbf{H}_{\mathbf{\Lambda}}^{(2)}(\omega) = H_{\Lambda'}(\omega) \otimes \mathbf{I}_{\Lambda''} + \mathbf{I}_{\Lambda'} \otimes H_{\Lambda''}(\omega)$$

where $\mathbf{I}_{\Lambda'}$ and $\mathbf{I}_{\Lambda''}$ stand for the identity operators in $\ell^2(\Lambda')$ and $\ell^2(\Lambda'')$, respectively. Accordingly, the eigenvectors of $\mathbf{H}_{\mathbf{\Lambda}}^{(2)} = \mathbf{H}_{\mathbf{\Lambda}}^{(2)}(\omega)$ are given as the tensor products of the eigenvectors of single-particle Hamiltonians $H_{\Lambda'} = H_{\Lambda'}(\omega)$ and $H_{\Lambda''} = H_{\Lambda''}(\omega)$ and can be conveniently labeled by pairs (k_1, k_2) where $k_1, k_2 = 1, \ldots, 2L+1$:

$$\boldsymbol{\psi}_{k_1,k_2}(\mathbf{x}, \mathbf{H}_{\mathbf{\Lambda}}^{(2)}) = \boldsymbol{\psi}_{k_1}(x_1, H_{\Lambda'})\, \boldsymbol{\psi}_{k_2}(x_2, H_{\Lambda''}), \quad \mathbf{x} = (x_1, x_2) \in \mathbf{\Lambda}.$$

The eigenvalues $E_{k_1,k_2}\big(\mathbf{H}_{\mathbf{\Lambda}}^{(2)}(\omega)\big)$ are the sums $E_{k_1}\big(H_{\Lambda'}(\omega)\big) + E_{k_2}\big(H_{\Lambda''}(\omega)\big)$ of eigenvalues of $H_{\Lambda'}(\omega)$ and $H_{\Lambda''}(\omega)$, respectively.

Such a representation is impossible for a "nearly diagonal" square when single-particle projections Λ' and Λ'' are sufficiently close to (or coincide with) each other (e.g., $\mathbf{\Lambda} = \Lambda' \times \Lambda'$).

We come, therefore, to the necessity of a geometrical classification of multi-particle cubes. The reader will see in Chaps. 3 and 4 that the geometrical arguments taking into account the position of multi-particle cubes are crucial to us in the course of our implementation of the multi-particle MSA (MPMSA). This has been manifested in Definition 3.1.2 (separable cubes). The aforementioned geometrical arguments make the whole construction rather complicated, particularly for the number of particles $N \geq 3$. For this reason, we spend a considerable time in adapting the EVC bounds to the multi-particle context.

The structure of Chap. 3 from this point on is as follows.

In Sect. 3.2 we deal with a special case where the (IID) variables $V(x;\omega)$, $x \in \mathbb{Z}^d$, have a probability density analytic in a strip around the real line. This class of probability distributions is rather limited: the two most popular examples are a Cauchy distribution and a Gaussian distributions. More generally, this covers a family of the so-called α-stable probability laws for which the characteristic function (i.e., the inverse Fourier transform of the probability density) has the form $\sim \mathrm{e}^{-|t|^\alpha}$, with $\alpha \in [1, 2]$.

The case of a Cauchy distribution ($\alpha = 1$) is quite intriguing: in the single-particle localization theory, Cauchy distributions often give rise to "exactly solvable" models (a phenomenon observed in a number of problems in probability theory and its applications). Remarkably, the "solvability" for Anderson-type Hamiltonians with a Cauchy-type single-site marginal probability distribution of the external random potential occurs for two "opposite" types of the random field $V(x;\omega)$, $x = (x_1, \ldots, x_d) \in \mathbb{Z}^d$:

(a) When $V(x;\omega)$ is an IID field with Cauchy marginal probability distribution;
(b) For a quasi-periodic potential of the form[2]

$$V(x,\omega) = g \tan\left(\pi\omega + \pi \sum_{1j=1}^{d} \alpha_j, x_j\right), \quad \omega \in \mathbb{T}^1 := \mathbb{R}^1/\mathbb{Z}^1,$$

with an incommensurate frequency vector $\alpha = (\alpha_1, \ldots, \alpha_d) \in \mathbb{T}^d := \mathbb{R}^d/\mathbb{Z}^d$ (the so-called "Maryland model" mentioned in Sect. 1.3); see [91, 93, 142, 152].

The single-particle density of states in the case of the IID random potential with a Cauchy marginal probability distribution can be calculated explicitly (albeit in a somewhat "soft" fashion), with the help of rigorous path integration techniques; cf. [53]. Note also that the proof of Anderson localization for the Maryland model

---

[2]It is straightforward (and well-known) that if a random variable ξ is uniformly distributed in $[0, \pi]$, then $\eta := \tan\xi$ has Cauchy distribution.

is non-perturbative, i.e., the localization occurs for *any non-zero* amplitude $|g|$. However, the multi-particle case with an interaction presents a challenge from the perspective of "complete integrability," and it is interesting to see whether a Cauchy distribution could lead to specific features of localization.

Historically, the two-volume EVC bounds for multi-particle Hamiltonians have been established first for IID random potentials satisfying the aforementioned condition of analyticity; cf. [69]. In accordance with the above observation, the derivation of the two-volume EVC bound makes use of the mutual positions of the cubes in question. It seems difficult to obtain two-volume EVC bounds for all pairs of disjoint multi-particle cubes; in this regard we mention recent papers [56, 57] where some partial results have been obtained in this direction. The approach proposed in [56,57] promises a substantial simplification of the multi-particle MSA scheme and leads to more accurate estimates of decay of the eigenfunctions (albeit under stricter assumptions on F). In this book we follow a compromise, by proving the EVC bounds only for separable pairs of cubes (see Definition 3.1.2) which turns out to be sufficient for the purposes of the MPMSA. However, in Sect. 3.5 we briefly discuss the idea of an EVC bound applicable to any pair of distant cubes, without further elaboration on how it changes the MPMSA scheme.

In Sect. 3.4, we give a multi-particle adaptation of Stollmann's lemma on diagonally monotone families. This in turn will be used to prove a more general two-volume Wegner-type bound under general assumptions upon the marginal probability distribution of the potential $V(\cdot;\omega)$. Again, these bounds are established here only for separable pairs of cubes.

3.2 Molchanov's Formula: Carmona's Argument and Its Generalization

We begin this section by discussing a discrete-space analog of the Feynman path integral proposed by Molchanov in the 1980s. It appears that the first published version of what is called today Molchanov's path integral formula is contained in the monograph [53]. Earlier, Carmona applied Molchanov's formula to the analysis of the density of states in a single-particle model for Gaussian and other analytic random potentials; for details see [53]. Here, we extend Molchanov's formula to n-particle Hamiltonians and use it to prove both the one-volume and two-volume EVC bounds needed in the MPMSA; see Lemma 3.2.3, Theorem 3.2.4 and the comments after this theorem. See also [69].

A particularity of this class of potentials is that one can easily prove the analyticity of the eigenvalue concentration functions, both in the one-volume and two-volume version, with *an optimal dependence* upon the size of the volume.

An optimal *one-volume* EVC bound was proved by Kirsch [114], under the assumption that the marginal probability distribution of the IID random potential admits a bounded, compactly supported probability density.

Unlike the two-volume EVC bound for analytic potentials, its more general version from Sect. 3.4, does not have optimal volume dependence. On the other hand, our proof, based on an adaptation of Stollmann's lemma (cf. Lemma 2.2.1 in Sect. 2.2.2) provides *some* EVC bounds for any (common) marginal distribution of IID random variables $\{V(x;\omega), x \in \mathbb{Z}^d\}$ generating the random potential $V(\omega)$. Naturally, for the obtained EVC bounds to be useful in the framework of the (MP)MSA, the common marginal distribution function F of the random variables $V(x;\omega)$ has to be continuous.[3] Actually, the existing MSA techniques require F to be at least log-Hölder continuous. In our book, we use a stronger Assumption A (Hölder-continuity of F) only to obtain dynamical localization bounds decaying faster than any power-law.

Now we will describe an adaptation of Carmona's method, initially developed for single-particle Anderson-type Hamiltonians, to multi-particle systems.

Recall, a Markov process (X_t) with continuous time $t \geq 0$ on a finite set $\mathbb{A}$ is uniquely determined by the transition rates $q_{a,b} \geq 0$ associated with pairs of distinct points $a, b \in \mathbb{A}$. For our purposes, it suffices to take $\mathbb{A} = \mathbf{\Lambda}_L(\mathbf{u})$ and set the rates of transitions $\mathbf{x} \rightsquigarrow \mathbf{y}$, $\mathbf{x}, \mathbf{y} \in \mathbf{\Lambda}_L(\mathbf{u})$, to be

$$q_{\mathbf{x},\mathbf{y}} = \begin{cases} 1, & \text{if } |\mathbf{y} - \mathbf{x}|_1 = 1 \\ 0, & \text{otherwise,} \end{cases}$$

representing the off-diagonal matrix entries of operator $\mathbf{\Delta}_{\mathbf{\Lambda}_L(\mathbf{u}),\mathrm{Dir}}$ in the Dirac delta-basis $\{\delta_{\mathbf{x}} : \mathbf{x} \in \mathbf{\Lambda}_L(\mathbf{u})\}$ in $\ell^2(\mathbf{\Lambda}_L(\mathbf{u}))$. A trajectory of the process (X_t) is a piecewise constant map $[0, +\infty) \to \mathbf{\Lambda}_L(\mathbf{u})$; it is convenient to refer to X_t as a configuration (occupied) at time t. The points of discontinuity of this map are referred to as jumps of the process (X_t). The probability distribution generated by the process (X_t) is denoted by $\mathbf{P} = \mathbf{P}_{\mathbf{\Lambda}_L(\mathbf{u})}$ and its expectation by $\mathbf{E} = \mathbf{E}_{\mathbf{\Lambda}_L(\mathbf{u})}$. The meaning of the transition rates is explained by the asymptotic formula

$$\mathbf{P}\{X_{t+\epsilon} = \mathbf{y} \mid X_t = \mathbf{x}\} = \epsilon q_{\mathbf{x},\mathbf{y}} + O(\epsilon^2),\ 0 < \epsilon \ll 1.$$

The total exit rate $A(\mathbf{x}) = A_{\mathbf{\Lambda}_L(\mathbf{u})}(\mathbf{x})$ from configuration $\mathbf{x} \in \mathbf{\Lambda}_L(\mathbf{u})$ is defined as $\mathrm{card}\{\mathbf{y} \in \mathbf{\Lambda}_L(\mathbf{u}) : |\mathbf{x} - \mathbf{y}|_1 = 1\}$, i.e., as the number of nearest neighbors of $\mathbf{x}$ in $\mathbf{\Lambda}$. It takes the value in the interval $[dn, 2dn]$, depending upon the coordination number of the point $\mathbf{x} \in \mathbf{\Lambda}_L(\mathbf{u})$.

For details, see Ref. [159], Chap. 2.

We will use Molchanov's path integral formula for the unitary propagator $\exp\left[-it\mathbf{H}^{(n)}_{\mathbf{\Lambda}_L(\mathbf{u})}(\omega)\right]$ generated by the LSO $\mathbf{H}^{(n)}_{\mathbf{\Lambda}_L(\mathbf{u})}(\omega)$. Applied to $\mathbf{H}^{(n)}_{\mathbf{\Lambda}_L(\mathbf{u})}(\omega)$, this formula reads as follows: for any $t > 0$ and any function $f : \mathbf{\Lambda}_L(\mathbf{u}) \to \mathbb{C}$

[3] Recall that Anderson localization on the lattice with Bernoulli IID random potential remains a challenging open problem even for single-particle systems.

$$\left(\exp\left[-\mathrm{i}t\mathbf{H}^{(n)}_{\mathbf{\Lambda}_L(\mathbf{u})}(\omega)\right]f\right)(\mathbf{x})$$
$$= \mathbf{E}\left[f(X_t)(-\mathrm{i})^{J(t)}\exp\left(\int_0^t [A(X_s)-\mathrm{i}W(X_s;\omega)]\,\mathrm{d}s\right)\Big|\, X_0=\mathbf{x}\right]. \tag{3.28}$$

Here, $J(t)$ is the total number of jumps made by the Markov process trajectory $(X_s,\ 0\le s\le t)$, and for all $\mathbf{x}=(x_1,\dots,x_n)\in\mathbf{\Lambda}_L(\mathbf{u})$,

$$W(\mathbf{x};\omega) := \mathbf{U}(\mathbf{x}) + g\sum_{j=1}^{n} V(x_j;\omega). \tag{3.29}$$

For $t<0$, in the right-hand side of (3.28) the value t is to be replaced by $-t$ (i.e., the Markov process runs from time 0 to $-t$) and i by $-$i.

Now, taking expectation over the random potential and using the functional calculus for Hermitian operators, we obtain the representation: for any $\mathbf{x}\in\mathbf{\Lambda}_L(\mathbf{u})$,

$$\mathbb{E}\left[\left\langle \delta_{\mathbf{x}}, \exp\left[\mathrm{i}t\mathbf{H}^{(n)}_{\mathbf{\Lambda}_L(\mathbf{u})}(\omega)\right]\delta_{\mathbf{x}}\right\rangle\right] = \int_{\mathbb{R}} \mathrm{e}^{\mathrm{i}t\lambda}\,\mathrm{d}K_{\mathbf{x},\mathbf{\Lambda}_L(\mathbf{u})}(\lambda), \tag{3.30}$$

where $K_{\mathbf{x},\mathbf{\Lambda}_L(\mathbf{u})}$ is the averaged spectral measure of $\mathbf{H}^{(n)}_{\mathbf{\Lambda}_L(\mathbf{u})}(\omega)$ for the element $\delta_{\mathbf{x}}$. More generally, for any bounded continuous function φ and linear combination $\phi = \sum\limits_{\mathbf{x}\in\mathbf{\Lambda}_L(\mathbf{u})}\langle\phi,\delta_{\mathbf{x}}\rangle\delta_{\mathbf{x}}$, with $||\phi||=1$,

$$\mathbb{E}\left[\left\langle \phi, \varphi(\mathbf{H}^{(n)}_{\mathbf{\Lambda}_L(\mathbf{u})}(\omega))\phi\right\rangle\right] = \int_{\mathbb{R}} \varphi(\lambda)\,\mathrm{d}K_{\phi,\mathbf{\Lambda}_L(\mathbf{u})}(\lambda).$$

We see from Eq. (3.30) that regularity properties of the measure $K_{\mathbf{x},\mathbf{\Lambda}_L(\mathbf{u})}$ are encrypted in the decay properties of the expected value in the left-hand side, and vice versa. In particular, suppose that the function

$$\hat{k}_{\mathbf{x},\mathbf{\Lambda}_L(\mathbf{u})}(t) = \mathbb{E}\left[\left\langle \delta_{\mathbf{x}}, \exp\left[\mathrm{i}t\mathbf{H}^{(n)}_{\mathbf{\Lambda}_L(\mathbf{u})}(\omega)\right]\delta_{\mathbf{x}}\right\rangle\right],\quad t\in\mathbb{R}, \tag{3.31}$$

(which gives the inverse Fourier transform of $K_{\mathbf{x},\mathbf{\Lambda}_L(\mathbf{u})}$) exhibits an exponential decay:

$$\left|\hat{k}_{\mathbf{x},\mathbf{\Lambda}_L(\mathbf{u})}(t)\right| \le C\mathrm{e}^{-a|t|},\quad t\in\mathbb{R},$$

where $a, C>0$. Then the measure $K_{\mathbf{x},\mathbf{\Lambda}_L(\mathbf{u})}$ has a density $k_{\mathbf{x},\mathbf{\Lambda}_L(\mathbf{u})}(\lambda)$ which is analytic in a strip

$$\{\lambda\in\mathbb{C}:\ |\mathrm{Im}\,\lambda|<a\}.$$

Carmona's argument is based on the following observation. The averaged spectral measure $K_{\mathbf{x},\boldsymbol{\Lambda}_L(\mathbf{u})}$ is actually a probability measure on $\mathbb{R}$ (and $\hat{k}_{\mathbf{x},\boldsymbol{\Lambda}_L(\mathbf{u})}(t)$ is its characteristic function, in the probabilistic terminology). The probability density $k_{\mathbf{x},\boldsymbol{\Lambda}_L(\mathbf{u})}(\lambda)$, when it exists, is recovered from $\hat{k}_{\mathbf{x},\boldsymbol{\Lambda}_L(\mathbf{u})}(t)$ via

$$k_{\mathbf{x},\boldsymbol{\Lambda}_L(\mathbf{u})}(\lambda) = \frac{1}{2\pi}\int_{\mathbb{R}} \mathrm{e}^{-\mathrm{i}\lambda t}\hat{k}_{\mathbf{x},\boldsymbol{\Lambda}_L(\mathbf{u})}(t)\mathrm{d}t. \tag{3.32}$$

Lemma 3.2.1. *Suppose that for some $a, b > 0$ and all $t \in \mathbb{R}$ the characteristic function of the marginal probability distribution of the IID random potential $V(x;\omega)$ satisfies*

$$\left|\mathbb{E}\left[\mathrm{e}^{\mathrm{i}tV(0,\omega)}\right]\right| \le b\,\mathrm{e}^{-a|t|}, \quad t \in \mathbb{R}, \tag{3.33}$$

Then, for $|g| > 2d/a$, the quantity $\hat{k}_{\mathbf{x},\boldsymbol{\Lambda}_L(\mathbf{u})}$ defined in (3.31) obeys

$$\left|\hat{k}_{\mathbf{x},\boldsymbol{\Lambda}_L(\mathbf{u})}(t)\right| \le b^n \mathrm{e}^{-B|t|}, \quad t \in \mathbb{R}, \tag{3.34}$$

with $B = (a|g| - 2d)n$, uniformly in $L > 1$, $\mathbf{u} \in \left(\mathbb{Z}^d\right)^n$ and $\mathbf{x} \in \boldsymbol{\Lambda}_L(\mathbf{u})$. Consequently, the probability density $k_{\mathbf{x},\boldsymbol{\Lambda}_L(\mathbf{u})}(\lambda)$ is analytic in a strip around the real line and has a bounded derivative on any finite interval.

Proof. To estimate the absolute value $\left|\hat{k}_{\mathbf{x},\boldsymbol{\Lambda}_L(\mathbf{u})}(t)\right|$, we use Eq. (3.28). For brevity, we only consider $t \ge 0$; the case of $t \le 0$ is similar. We have:

$$\begin{aligned}&\left\langle \delta_{\mathbf{x}}, \exp\left(-\mathrm{i}t\mathbf{H}^{(n)}_{\boldsymbol{\Lambda}_L(\mathbf{u})}(\omega)\right)\delta_{\mathbf{x}}\right\rangle \\ &\quad = \mathbf{E}\left[\mathbf{1}_{X_t}(\mathbf{x})\,(-\mathrm{i})^{J(t)}\exp\left(\int_0^t [A(X_s) - \mathrm{i}W(X_s;\omega)]\,\mathrm{d}s\right)\middle|\, X_0 = \mathbf{x}\right]\end{aligned}$$

and

$$\begin{aligned}\hat{k}_{\mathbf{x},\boldsymbol{\Lambda}_L(\mathbf{u})}(t) &= \mathbf{E}\left[\mathbf{1}_{X_t}(\mathbf{x})\,(-\mathrm{i})^{J(t)}\exp\left(\int_0^t (A(X_s) + \mathrm{i}\mathbf{U}(X_s))\mathrm{d}s\right)\right. \\ &\quad \left.\times\,\mathbb{E}\left[\exp\left(-\mathrm{i}g\int_0^t \sum_{x\in\Pi\boldsymbol{\Lambda}_L(\mathbf{u})} V(x;\,\cdot\,)Q(X_s,x)\,\mathrm{d}s\right)\right]\middle|\, X_0 = \mathbf{x}\right].\end{aligned} \tag{3.35}$$

Here $Q(X_s, x)$, $x \in \Pi\boldsymbol{\Lambda}_L(\mathbf{u})$, stands for the number of particles in configuration X_s with position x. The change of order of integration is justified by the boundedness of the integrand. We can further write:

$$\int_0^t \sum_{x\in\Pi\boldsymbol{\Lambda}_L(\mathbf{u})} V(x;\omega)Q(X_s,x)\,\mathrm{d}s = \sum_{x\in\boldsymbol{\Lambda}_L(\mathbf{u})} V(x;\omega)\tau(x)$$

where $\tau(x)$ is the total time spent at site x by all n particles.

Owing to the independence of the variables $V(x; \cdot)$, we obtain that

$$\mathbb{E}\left[\exp\left(-\mathrm{i}g\int_0^t \sum_{x\in\Pi\mathbf{\Lambda}_L(\mathbf{u})} V(x;\cdot)Q(X_s,x)\,\mathrm{d}s\right)\right] = \prod_{x\in\Pi\mathbf{\Lambda}_L(\mathbf{u})} \mathbb{E}\left[\mathrm{e}^{\mathrm{i}gV(x;\cdot)\tau(x)}\right]$$

In view of condition (3.33), the absolute value of last expression is $\leq b^n \mathrm{e}^{-a|g|tn}$.

Returning to (3.35) and using the bound $A(X_s) \leq 2dn$ we obtain

$$\left|\hat{k}_{\mathbf{x},\mathbf{\Lambda}_L(\mathbf{u})}(t)\right| \leq b^n \mathrm{e}^{(2d-a|g|)tn}.$$

Being equivalent to (3.34), the last bound completes the proof of Lemma 3.2.1. □

Theorem 3.2.2. *Under the assumption* (3.33), $\forall\ E \in \mathbb{R}$, $L \geq 1$, $\mathbf{u} \in \left(\mathbb{Z}^d\right)^n$ *and* $r \in (0,1)$, *the probability that* E *lies at distance at most* r *from the spectrum of* $\mathbf{H}^{(n)}_{\mathbf{\Lambda}_L(\mathbf{u})}(\omega)$ *satisfies*

$$\mathbb{P}\left\{\mathrm{dist}\left[E, \sigma\left(\mathbf{H}^{(n)}_{\mathbf{\Lambda}_L(\mathbf{u})}(\omega)\right)\right] < r\right\} \leq \frac{rb^n}{\pi B}(2L+1)^{nd}, \quad r > 0, \tag{3.36}$$

with the same value of $B > 0$ *as in Eq.* (3.34).

Proof. We begin with an elementary inequality (cf. Proposition VIII.4.11 in [53]). Let $P^{(E-r,E+r)}_{\mathbf{\Lambda}_L(\mathbf{u})}$ be the spectral projection on the subspace spanned by the eigenfunctions of $\mathbf{H}^{(n)}_{\mathbf{\Lambda}_L(\mathbf{u})}(\omega)$ with the eigenvalues from $(E-r, E+r)$. Then

$$\mathbb{P}\left\{\mathrm{dist}\left(E, \sigma\left(\mathbf{H}^{(n)}_{\mathbf{\Lambda}_L(\mathbf{u})}(\omega)\right)\right) < r\right\} \leq \mathbb{E}\left[\mathrm{tr}\ P^{(E-r,E+r)}_{\mathbf{\Lambda}_L(\mathbf{u})}\right].$$

Further, in the Dirac's delta-basis:

$$\mathrm{tr}\ P^{(E-r,E+r)}_{\mathbf{\Lambda}_L(\mathbf{u})} = \sum_{\mathbf{x}\in\mathbf{\Lambda}_L(\mathbf{u})} \left\langle \delta_{\mathbf{x}}, P^{(E-r,E+r)}_{\mathbf{\Lambda}_L(\mathbf{u})}\delta_{\mathbf{x}}\right\rangle,$$

and

$$\mathbb{E}\left[\mathrm{tr}\ P^{(E-r,E+r)}_{\mathbf{\Lambda}_L(\mathbf{u})}\right] = \sum_{\mathbf{x}\in\mathbf{\Lambda}_L(\mathbf{u})} \int_{E-r}^{E+r} k_{\mathbf{x},\mathbf{\Lambda}_L(\mathbf{u})}(s)\mathrm{d}s.$$

Now the assertion of Theorem 3.2.2 follows easily from Lemma 3.2.1 combined with (3.31)–(3.32). □

As was mentioned before, the condition (3.33) is rather restrictive. However, it leads to a volume-type factor $(2L+1)^{nd}$ and the linear order in r in the right-hand side of Eq. (3.36). Moreover, similar patterns are exhibited when we consider conditional probabilities. To illustrate the latter property, we consider below a two-particle model where calculations are straightforward. A general n-particle result also holds true, but its statement and proof are more cumbersome.

Lemma 3.2.3. *Let* $\mathbf{\Lambda}_L(\mathbf{u}) = \Lambda_L(u_1) \times \Lambda_L(u_2)$ *be a two-particle cube with* $\Lambda_L(u_1) \cap \Lambda_L(u_2) = \varnothing$ *and assume condition* (3.33)*. Consider the sigma-algebra* $\mathfrak{B}(\Lambda_L(u_2))$ *generated by the random variables* $\{V(x, \cdot),\ x \in \Lambda_L(u_2)\}$*. For every* $\mathbf{x} \in \mathbf{\Lambda}_L(\mathbf{u})$*, define the conditional expected value*

$$\hat{k}_{\mathbf{x},\mathbf{\Lambda}_L(\mathbf{u})}\Big(t\,\big|\mathfrak{B}(\Lambda_L(u_2))\Big) = \mathbb{E}\Big[\Big\langle \delta_{\mathbf{x}}, \exp\Big[it\mathbf{H}^{(2)}_{\mathbf{\Lambda}_L(\mathbf{u})}(\omega)\Big]\delta_{\mathbf{x}}\Big\rangle\Big|\mathfrak{B}(\Lambda_L(u_2))\Big],\quad t \in \mathbb{R}, \tag{3.37}$$

(the conditional inverse Fourier transform). Then, for $|g| > 2d/a$*,*

$$\max\Big\{\Big|\hat{k}_{\mathbf{x},\mathbf{\Lambda}_L(\mathbf{u})}\Big(t\big|\mathfrak{B}(\Lambda_L(u_2))\Big)\Big| : \mathbf{x} \in \mathbf{\Lambda}_L(\mathbf{u})\Big\} \le b\mathrm{e}^{-Ct},\quad t \in \mathbb{R}, \tag{3.38}$$

with $C = a|g| - 2d$*. This bound holds uniformly in* $\mathbf{u} \in (\mathbb{Z}^d)^2$ *and* $L = 1, 2, \ldots$ *such that* $\Lambda_L(u_1) \cap \Lambda_L(u_2) = \varnothing$*. As a consequence, the Fourier transform*

$$k_{\mathbf{x},\mathbf{\Lambda}}\left(\lambda\big|\mathfrak{B}\left(\Lambda_L(u_2)\right)\right) = \frac{1}{2\pi}\int_{\mathbb{R}} \mathrm{e}^{-\mathrm{i}\lambda t}\hat{k}_{\mathbf{x},\mathbf{\Lambda}_L(\mathbf{u})}\Big(t\big|\mathfrak{B}(\Lambda_L(u_2))\Big)\,\mathrm{d}t$$

gives a (conditional) probability density that is analytic in a strip around the real line.

Proof. For the proof, one could simply repeat the argument used for proving Lemma 3.2.1. Note that, conditional on $\mathfrak{B}(\Lambda_L(u_2))$, the second summand in the sum $V(x_1;\omega) + V(x_2;\omega)$ becomes non-random. This results in the coefficient b instead of b^2 and the exponent $a|g| - 2d$ which is as twice as small as the one in Lemma 3.2.1. The rest of the proof of Lemma 3.2.1 remains valid. □

The above considerations lead us to the following

Theorem 3.2.4. *Under condition* (3.33)*,* $\forall\, E \in \mathbb{R}$*,* $\mathbf{u} = (u_1, u_2) \in (\mathbb{Z}^d)^2$ *and* $L = 1, 2, \ldots$ *such that* $\Lambda_L(u_1) \cap \Lambda_L(u_2) = \varnothing$*, the two-particle Hamiltonian* $\mathbf{H}^{(2)}_{\mathbf{\Lambda}^{(2)}_L(\mathbf{u})}(\omega)$ *in* $\mathbf{\Lambda}^{(2)}_L(\mathbf{u}) = \Lambda_L(u_1) \times \Lambda_L(u_2)$*, obeys*

$$\mathbb{P}\left\{\mathrm{dist}\Big[\,E, \sigma\big(\mathbf{H}^{(2)}_{\Lambda}(\omega)\big)\Big] < r\big|\mathfrak{B}(\Lambda_L(x_2))\right\} \le \frac{4br}{\pi C}(2L+1)^{2d},\quad r > 0,$$

where C*, as in Eq.* (3.38)*, equals* $a|g| - 2d$*.*

It is not difficult to see that Theorem 3.2.4, combined with the notion of separable pairs of cubes, leads to an EVC bound sufficient for the purposes of the multi-scale analysis of two-particle Hamiltonians. We give it here merely as an example of application of Molchanov's path integral formula. In a more general situation (for the number of particles $N > 2$ and for non-analytic marginal distributions), we will use a different approach to the EVC bounds based on Stollmann's lemma on diagonally monotone operator families; see Sect. 3.4.

3.3 Separability of Cubes and EVC Bounds

The notion of separability was introduced in Definition 3.1.2, together with initial comments, and was used in the statement of spectral localization in a multi-particle system; see Theorem 3.1.1. In particular, we noted that in the case of two particles, the cubes that are distant apart are separable. The formal assertion is given in Lemma 3.3.1 below. Before we proceed further, we would like to adopt the following agreement. From now on, we assume that the radius L of a cube $\mathbf{\Lambda}_L$ under consideration is $\geq L_0$ and that in turn, the initial scale L_0 satisfies

$$L_0 > 8r_0 \tag{3.39}$$

where r_0 is the radius of interaction. Although condition (3.39) is not involved in the constructions performed in this chapter, we will see in Chap. 4 that it gives rise to less cumbersome notation and calculations (cf., e.g., the proof of Lemma 4.2.3).

Consider the symmetrized max-norm distance in $\mathbb{Z}^d \times \mathbb{Z}^d$:

$$\mathrm{dist}_{\mathrm{symm}}(\mathbf{u}, \mathbf{v}) := \min\{|\mathbf{u} - \mathbf{v}|, |\mathbf{u} - \pi(\mathbf{v})|\}, \tag{3.40}$$

where $\pi(v_1, v_2) = (v_2, v_1)$, $v_1, v_2 \in \mathbb{Z}^d$.

Lemma 3.3.1. *Let* $\mathbf{\Lambda}_L^{(2)}(\mathbf{u}) = \Lambda_L(u_1) \times \Lambda_L(u_2)$ *and* $\mathbf{\Lambda}_L^{(2)}(\mathbf{v}) = \Lambda_L(v_1) \times \Lambda_L(v_2)$ *be a pair of two-particle cubes. Assume that*

$$\mathrm{dist}_{\mathrm{symm}}(\mathbf{u}, \mathbf{v}) > 8(L + r_0). \tag{3.41}$$

Then at least one of the two separability conditions (3.16), (3.17) *for the cubes* $\mathbf{\Lambda}_L^{(2)}(\mathbf{u})$, $\mathbf{\Lambda}_L^{(2)}(\mathbf{v})$ *is fulfilled.*

Proof. It is readily seen that (at least) one of the conditions (3.16), (3.17) for the cubes $\mathbf{\Lambda}_L^{(2)}(\mathbf{u})$, $\mathbf{\Lambda}_L^{(2)}(\mathbf{v})$ is fulfilled, whenever either

(i) The full projections $\Pi\mathbf{\Lambda}_{L+r_0}^{(2)}(\mathbf{u}) = \Lambda_{L+r_0}(u_1) \cup \Lambda_{L+r_0}(u_2)$ and $\Pi\mathbf{\Lambda}_{L+r_0}^{(2)}(\mathbf{v}) = \Lambda_{L+r_0}(v_1) \cup \Lambda_{L+r_0}(v_2)$ are disjoint or

(ii) One of the four single-particle projections $\Lambda_{L+r_0}(u_1)$, $\Lambda_{L+r_0}(u_2)$, $\Lambda_{L+r_0}(v_1)$, $\Lambda_{L+r_0}(v_2)$ is disjoint from the union of the three remaining ones.

At this moment it is convenient to use the standard embedding $\mathbb{Z}^d \hookrightarrow \mathbb{R}^d$ and refer to the notion of connectedness in the Euclidean space $\mathbb{R}^d$. Specifically, in this proof we treat projections $\Lambda_{L+r_0}(u_1)$, $\Lambda_{L+r_0}(u_2)$, $\Lambda_{L+r_0}(v_1)$, $\Lambda_{L+r_0}(v_2)$ as cubes in $\mathbb{R}^d$, and denote by M, $1 \leq M \leq 4$ the number of connected components in their union.

If neither of possibilities (i), (ii) takes place then $M \leq 2$. (Indeed, if $M = 3$ or $M = 4$, then at least one connected component must contain a single projection; this projection is, therefore, disjoint from the three others.) Consider the two remaining situations:

1. $M = 1$
2. $M = 2$, and neither of the two connected components, C_1 and C_2, is a single one-particle projection cube.[4]

It suffices to show that neither of the cases (1) and (2) is compatible with the assumptions of the lemma.

In case (1), the union of the four aforementioned projections cubes, each having diameter $2(L + r_0)$, is a connected set. Thus, this union has diameter less than or equal to $4(2L + 2r_0)$, thus $|\mathbf{u} - \mathbf{v}| \leq 8(L + r_0)$. This is incompatible with the assumed inequality (3.41)

$$|\mathbf{u} - \mathbf{v}| \geq \mathrm{dist}_{\mathrm{symm}}(\mathbf{u}, \mathbf{v}) > 8(L + r_0).$$

In the case (2), there are exactly two connected components, C_1 and C_2, in the above union; without loss of generality, let C_1 be the component containing $\Lambda_{L+r_0}(u_1)$; it must contain another single-particle projection cube.

Since we ruled out (i), C_1 cannot contain $\Lambda_{L+r_0}(u_2)$: otherwise, one would have $C_1 = \Lambda_{L+r_0}(u_1) \cup \Lambda_{L+r_0}(u_2) = \Pi \mathbf{\Lambda}^{(2)}_{L+r_0}(\mathbf{u})$ and, respectively, $C_2 = \Pi \mathbf{\Lambda}^{(2)}_{L+r_0}(\mathbf{v})$, so that

$$\Pi \mathbf{\Lambda}^{(2)}_{L+r_0}(\mathbf{u}) \cap \Pi \mathbf{\Lambda}^{(2)}_{L+r_0}(\mathbf{v}) = C_1 \cap C_2 = \varnothing,$$

yielding the separability of the pair $\mathbf{\Lambda}^{(2)}_{L+r_0}(\mathbf{u})$, $\mathbf{\Lambda}^{(2)}_{L+r_0}(\mathbf{v})$.

There remain two situations to be ruled out:

- If C_1 contains $\Lambda_{L+r_0}(v_1)$, then we conclude that

$$\exists\, x \in \Lambda_{L+r_0}(u_1) \cap \Lambda_{L+r_0}(v_1).$$

[4] In other words, neither component C_i may have the form $\Lambda_L(u_1)$, or $\Lambda_L(u_2)$, or $\Lambda_L(v_1)$, or $\Lambda_L(v_2)$.

Analogously, considering component $\mathcal{C}_2$,

$$\exists\, y \in \Lambda_{L+r_0}(u_2) \cap \Lambda_{L+r_0}(v_2).$$

Therefore,

$$\exists\,(x, y) \in \mathbf{\Lambda}^{(2)}_{L+r_0}(\mathbf{u}) \cap \mathbf{\Lambda}^{(2)}_{L+r_0}(\mathbf{v})$$

which contradicts the assumption (3.41).

- In a similar fashion, if $\mathcal{C}_1$ contains $\Lambda_{L+r_0}(v_2)$ then we conclude that

$$\exists\, x \in \Lambda_{L+r_0}(u_1) \cap \Lambda_{L+r_0}(v_2) \text{ and } \exists\, y \in \Lambda_{L+r_0}(u_2) \cap \Lambda_{L+r_0}(v_1)$$

so that

$$\exists\,(x, y) \in \mathbf{\Lambda}^{(2)}_{L+r_0}(\mathbf{u}) \cap \pi\mathbf{\Lambda}^{(2)}_{L+r_0}(\mathbf{v})$$

which again contradicts (3.41). □

The reader may wonder why the definition of the separability is not limited to one of the conditions (3.16), (3.17), but includes an additional bound $|\mathbf{u} - \mathbf{v}| > 11nL$. Indeed, either of these conditions would suffice for the proof of the two-volume EVC bound. However, in the course of the MPMSA, the inequality $|\mathbf{u} - \mathbf{v}| > 11nL$ proves useful, so including it into the notion of separability makes unnecessary saying "*... and with* $|\mathbf{u} - \mathbf{v}| > 11nL$ *...*" in a number of intermediate statements.

In fact, it can be shown that for two-particle systems this lower bound can be replaced by a weaker one: $\mathrm{dist}_{\mathrm{symm}}(\mathbf{u}, \mathbf{v}) > 8(L + r_0)$; with $L > 8r_0$, the latter follows, e.g., from $\mathrm{dist}_{\mathrm{symm}}(\mathbf{u}, \mathbf{v}) > 9L$.

The situation for the particle number $n > 2$ is technically more involved. (An example showing that a statement similar to Lemma 3.3.1 cannot be valid for $n \geq 3$ without additional conditions on $\mathbf{\Lambda}^{(n)}_L(\mathbf{u})$ and $\mathbf{\Lambda}^{(n)}_L(\mathbf{v})$ is given in Sect. 3.5.) A general assertion sufficient for our purposes is as follows. Here, again, we add the condition $\mathrm{dist}(\mathbf{u}, \mathbf{v}) > 11nL$ in order to adapt the statement of Lemma 3.3.2 to direct references in Chap. 4.

Lemma 3.3.2. *Let $n \geq 2$, $\mathbf{u}, \mathbf{v} \in \left(\mathbb{Z}^d\right)^n$ and assume that* $\mathrm{dist}(\mathbf{u}, \mathbf{v}) > 11nL$. *Then the following statements hold.*

(A) *Given an n-particle cube $\mathbf{\Lambda}^{(n)}_L(\mathbf{u})$, there exists a finite collection of n-particle cubes $\{\mathbf{\Lambda}^{(n)}_{A(n,L)}(\mathbf{u}^{(i)})\}$ of radius $A(n, L) \leq 2n(L + r_0)$, where i runs from 1 to a number $K(\mathbf{u}, n, L)\}$, with $K(\mathbf{u}, n, L) \leq n^n$, such that if*

$$\mathbf{v} \notin \bigcup_{i=1}^{K(\mathbf{u},n,L)} \mathbf{\Lambda}^{(n)}_{A(n,L)}(\mathbf{u}^{(i)})$$

then the pair $\mathbf{\Lambda}_L^{(n)}(\mathbf{u})$, $\mathbf{\Lambda}_L^{(n)}(\mathbf{v})$ *is separable.*

(B) *If* $|\mathbf{v}| > |\mathbf{u}| + 2(L + r_0)$, *the cubes* $\mathbf{\Lambda}_L(\mathbf{u})$ *and* $\mathbf{\Lambda}_L(\mathbf{v})$ *are separable.*

(C) *In particular, any pair of cubes of the form* $\mathbf{\Lambda}_L^{(n)}(\mathbf{0})$, $\mathbf{\Lambda}_L^{(n)}(\mathbf{v})$ *with* $|\mathbf{v}| > 2(L + r_0)$ *is separable.*

Proof. (A) Given a non-empty set $\mathcal{J} \subseteq \{1, \ldots, n\}$, $R > 0$ and an n-particle configuration $\mathbf{v} = (v_1, \ldots, v_n) \in \left(\mathbb{Z}^d\right)^n$, the sub-configuration $\mathbf{v}_{\mathcal{J}}$ is called an R-cluster (in $\mathbf{v}$) if the union $\cup_{j \in \mathcal{J}} \Lambda_R(v_j)$ is connected, and $\mathcal{J}$ is maximal with that property. (Here it is convenient to identify the lattice cubes $\Lambda_L(v_i) \subset \mathbb{Z}^d$ with their counterparts $\mathrm{B}_L(u) = \{x \in \mathbb{R}^d : |x - u| \leq R\}$ and use the notion of connectedness in $\mathbb{R}^d$.) This definition generates a partition of the configuration $\mathbf{v}$ into R-clusters $\mathbf{v}_{\mathcal{J}_1}, \ldots, \mathbf{v}_{\mathcal{J}_s}$, with $s = s(\mathbf{v})$ (the order of listing does not matter), where $1 \leq s \leq n$ and $1 \leq |\mathcal{J}_l| \leq n$ for all $l = 1, \ldots, s$. Also, the diameter

$$\operatorname{diam} \Pi_{\mathcal{J}} \mathbf{\Lambda}_R(\mathbf{v}) = \max \left[|w - w'| : w, w' \in \Pi \mathbf{\Lambda}_R(\mathbf{v}_{\mathcal{J}}) \right]$$

obeys

$$\operatorname{diam} \Pi_{\mathcal{J}} \mathbf{\Lambda}_R(\mathbf{v}) \leq 2nR.$$

We set $R = L + r_0$ and, for a fixed $\mathbf{u} \in \left(\mathbb{Z}^d\right)^n$, look for a cluster $\mathbf{v}_{\mathcal{J}_l}$ with the property that

$$\Pi_{\mathcal{J}_l} \mathbf{\Lambda}_{L+r_0}(\mathbf{v}) \cap \Pi \mathbf{\Lambda}_{L+r_0}(\mathbf{u}) = \varnothing, \tag{3.42}$$

implying that the configuration $\mathbf{v}$ is $(L + r_0, \mathcal{J}_l)$-separable from $\mathbf{u}$.

A configuration $\mathbf{v}$ that does not have such a cluster $\mathbf{v}_{\mathcal{J}_l}$ satisfies

$$\Pi_{\mathcal{J}_l} \mathbf{\Lambda}_{L+r_0}(\mathbf{v}) \cap \Pi \mathbf{\Lambda}_{L+r_0}(\mathbf{u}) \neq \varnothing, \ \forall\, l = 1, \ldots, s(\mathbf{v}). \tag{3.43}$$

In turn, Eq. (3.43) implies that $\forall\, i = 1, \ldots, n$,

$$v_i \in \Pi \mathbf{\Lambda}_{2n(L+r_0)}(\mathbf{u}).$$

Thus, if a configuration $\mathbf{v}$ has no cluster $\mathbf{v}_{\mathcal{J}_l}$ obeying (3.42) then, $\forall\, i = 1, \ldots, n$, site v_i lies in a single-particle cube $\Pi_j \mathbf{\Lambda}_{2n(L+r_0)}(\mathbf{u})$. There are at most n choices of such a cube for a given i which yields that configurations $\mathbf{v}$ that are not $(L + r_0)$-separable from $\mathbf{u}$ are covered by at most n^n particle cubes. This leads to assertion (A).

For assertion (B), note that when $|\mathbf{v}| > |\mathbf{u}| + 2(L + r_0)$, the diagonal cube $\mathbf{\Lambda}_{|\mathbf{u}|+L+r_0}(\mathbf{0})$ containing $\mathbf{\Lambda}_{L+r_0}(\mathbf{u})$ is disjoint from cube $\mathbf{\Lambda}_{L+r_0}(\mathbf{v})$:

$$\mathbf{\Lambda}_{L+r_0}(\mathbf{v}) \bigcap \mathbf{\Lambda}_{|\mathbf{u}|+L+r_0}(\mathbf{0}) = \varnothing.$$

Then there exists at least one value of $i = 1, \dots, n$ such that

$$\Pi_i \mathbf{\Lambda}_{L+r_0}(\mathbf{v}) \bigcap \Pi_i \mathbf{\Lambda}_{|\mathbf{x}|+L+r_0}(\mathbf{0}) = \varnothing.$$

Since the full projection $\Pi \mathbf{\Lambda}_{|\mathbf{x}|+L+r_0}(\mathbf{0})$ coincides with $\Pi_i \mathbf{\Lambda}_{|\mathbf{x}|+L+r_0}(\mathbf{0})$, configuration $\mathbf{v}$ is $(L + r_0, \mathcal{J})$-separable from $\mathbf{u}$, at least for $\mathcal{J} = \{i\}$.

Assertion (C) follows from (B). This completes the proof of Lemma 3.3.2. □

3.4 Multi-particle Stollmann's Bound

A one-volume Stollmann bound for a multi-particle Anderson tight-binding Hamiltonian is contained in the following

Theorem 3.4.1. *Consider an n-particle cube $\mathbf{\Lambda}_L^{(n)}(\mathbf{u})$ and suppose that the marginal probability distribution function F of the random variables $V(x;\omega)$ has the continuity modulus $\varsigma(\cdot)$:*

$$\varsigma(\epsilon) := \sup_{t\in\mathbb{R}} \big(F(t+\epsilon) - F(t)\big).$$

Then, $\forall\ E \in \mathbb{R}$,

$$\mathbb{P}\left\{ \operatorname{dist}\left(E, \sigma\left(\mathbf{H}^{(n)}_{\mathbf{\Lambda}_L^{(n)}(\mathbf{u})}(\omega)\right)\right) < \epsilon \right\} \le n(2L+1)^{(n+1)d} \varsigma(2\epsilon). \tag{3.44}$$

Proof. It suffices to represent the ensemble $\mathbf{H}_{\mathbf{\Lambda}_L(\mathbf{u})}(\omega)$, as a monotone operator family on an appropriately chosen probability space $(\mathbb{R}^{\mathcal{Q}}, \mu^{(\mathcal{Q})})$; with the aim to apply Stollmann's lemma (Lemma 2.2.1). Here we set $\mathcal{Q} = \Pi \mathbf{\Lambda}_L(\mathbf{u})$ and specify $\mu^{(\mathcal{Q})}$ as the product-measure on $\mathbb{R}^{\mathcal{Q}}$ with the marginals given by the probability distribution of the random variables $V(x;\cdot)$, $x \in \mathbb{Z}^d$. That is, a point from $\mathbb{R}^{\mathcal{Q}}$ is identified with a collection of sample values of random variables $V(y;\cdot)$, $y \in \Pi \mathbf{\Lambda}_L(\mathbf{u})$. Given such a collection, $\underline{\upsilon} = \{\upsilon_y,\ y \in \mathcal{Q}\}$, the potential energy at the point $\mathbf{x} = (x_1, \dots, x_n) \in \mathbf{\Lambda}_L(\mathbf{u})$ reads as

$$W(\mathbf{x}; \underline{\upsilon}) = U(\mathbf{x}) + \sum_{y\in\mathcal{Q}} M_y(\mathbf{x})\upsilon_y \ \text{ where } \ M_y(\mathbf{x}) = \sum_{i=1}^{n} \mathbf{1}_y(x_i).. \tag{3.45}$$

This representation makes it clear that the (non-random) operators $\mathbf{W}(\underline{\upsilon})$ acting by multiplication

$$\mathbf{W}: \phi(\mathbf{x}) \mapsto W(\mathbf{x}; \underline{\upsilon})\phi(\mathbf{x}),\ \phi \in \mathcal{H}^{(n)}_{\mathbf{\Lambda}_L(\mathbf{u})},$$

form a diagonally monotone family relative to the parameter $\underline{\upsilon} \in \mathbb{R}^{\mathcal{Q}}$. Consequently, the operators $\mathbf{H}_{\mathbf{\Lambda}_L(\mathbf{u})}(\underline{\upsilon})$ (again non-random), with $\underline{\upsilon} \in \mathbb{R}^{\mathcal{Q}}$, given by

$$\mathbf{H}_{\mathbf{\Lambda}_L(\mathbf{u})}(\underline{\upsilon}) = \mathbf{H}^0_{\mathbf{\Lambda}_L(\mathbf{u})} + \mathbf{W}(\underline{\upsilon})$$

also form a diagonally monotone family in $\mathcal{H}^{(n)}_{\mathbf{\Lambda}_L(\mathbf{u})}$, relative to $\underline{\upsilon} \in \mathbb{R}^{\mathcal{Q}}$. Moreover, $\forall\, t > 0$,

$$\mathbf{H}_{\mathbf{\Lambda}_L(\mathbf{u})}(\underline{\upsilon} + t\underline{1}) - \mathbf{H}_{\mathbf{\Lambda}_L(\mathbf{u})}(\underline{\upsilon}) \geq (nt)\,\mathbf{I}$$

where $\mathbf{I}$ is the identity operator in $\mathcal{H}^{(n)}_{\mathbf{\Lambda}_L(\mathbf{u})}$ and $\underline{1} = (1, \ldots, 1) \in \mathbb{R}^{\mathcal{Q}}$.

Now the assertion of the theorem follows from Stollmann's estimate for diagonally monotone families (Lemma 2.2.1):

$$\begin{aligned}\mathbb{P}\left\{ \operatorname{dist}\Big(E, \sigma\big(\mathbf{H}^{(n)}_{\mathbf{\Lambda}_L(\mathbf{u})}(\omega)\big)\Big) < \epsilon \right\} &\leq |\mathbf{\Lambda}^{(n)}_L(\mathbf{u})|\,|\mathcal{Q}|\varsigma(2\epsilon) \\ &\leq (2L+1)^{nd} n (2L+1)^d \varsigma(2\epsilon).\end{aligned} \tag{3.46}$$

□

An alternative approach[5] has been developed by Kirsch [114], who proved a one-volume Wegner-type bound for multi-particle lattice Schrödinger operators, with optimal volume-dependence,

$$\mathbb{P}\left\{ \operatorname{dist}\Big(\sigma\big(\mathbf{H}_{\mathbf{\Lambda}}(\omega)\big), E\Big) \leq \epsilon \right\} \leq C|\mathbf{\Lambda}|\, \|\rho\|_\infty\, \epsilon,$$

under an additional assumption of the existence and boundedness of the marginal probability density ρ of the random variable $V(x;\omega)$.

A two-volume multi-particle Stollmann bound is presented in Theorem 3.4.2.

Theorem 3.4.2. *Let $\mathbf{\Lambda}^{(n)}_L(\mathbf{u})$, $\mathbf{\Lambda}^{(n)}_L(\mathbf{v})$ be a separable pair of n-particle cubes. If the marginal probability distribution function F of the random potential V has continuity modulus $\varsigma(\cdot)$, then the n-particle LSOs $\mathbf{H}^{(n)}_{\mathbf{\Lambda}^{(n)}_L(\mathbf{u})}$ and $\mathbf{H}^{(n)}_{\mathbf{\Lambda}^{(n)}_L(\mathbf{v})}$ satisfy*

$$\mathbb{P}\left\{ \operatorname{dist}\Big(\sigma\big(\mathbf{H}^{(n)}_{\mathbf{\Lambda}^{(n)}_L(\mathbf{u})}\big), \sigma\big(\mathbf{H}^{(n)}_{\mathbf{\Lambda}^{(n)}_L(\mathbf{v})}\big)\Big) < \epsilon \right\} \leq n(2L+1)^{(2n+1)d} \varsigma(2\epsilon). \tag{3.47}$$

Proof. For notational brevity, set $\mathbf{\Lambda}' = \mathbf{\Lambda}^{(n)}_L(\mathbf{u})$ and $\mathbf{\Lambda}'' = \mathbf{\Lambda}^{(n)}_L(\mathbf{v})$. Owing to separability of the pair $\mathbf{\Lambda}'$, $\mathbf{\Lambda}''$, we may assume that there exists a non-empty subset $\mathcal{J} \subseteq \{1, \ldots, n\}$ such that

$$\Pi_{\mathcal{J}}\mathbf{\Lambda}' \cap \big(\Pi_{\mathcal{J}^c}\mathbf{\Lambda}' \cup \Pi\mathbf{\Lambda}''\big) = \varnothing.$$

[5] See also [120] where multi-particle models in $\mathbb{R}^d$ were considered.

Further, let $\mathfrak{B} = \mathfrak{B}\left(\Pi_{\mathcal{J}^c}\mathbf{\Lambda}' \cup \Pi\mathbf{\Lambda}''\right)$ be the sigma-algebra generated by the random variables $V(y;\cdot)$, $y \in \Pi_{\mathcal{J}^c}\mathbf{\Lambda}' \cup \Pi\mathbf{\Lambda}''$. Note that the operator $\mathbf{H}_{\mathbf{\Lambda}''}(\omega)$ is $\mathfrak{B}$-measurable, since it is measurable with respect to the smaller sigma-algebra generated by $V(y;\cdot)$ with $y \in \Pi\mathbf{\Lambda}''$. Therefore, conditional on $\mathfrak{B}$, every eigenvalue of $\mathbf{H}_{\mathbf{\Lambda}''}$ is non-random, and we can write

$$\begin{aligned}
&\mathbb{P}\left\{\operatorname{dist}\Big(\sigma\big(\mathbf{H}_{\mathbf{\Lambda}_L(\mathbf{u})}\big), \sigma\big(\mathbf{H}_{\mathbf{\Lambda}_L(\mathbf{v})}\big)\Big) \le \epsilon\right\} \\
&\quad = \mathbb{E}\left[\mathbb{P}\left\{\operatorname{dist}\Big(\sigma\big(\mathbf{H}_{\mathbf{\Lambda}_L(\mathbf{u})}\big), \sigma\big(\mathbf{H}_{\mathbf{\Lambda}_L(\mathbf{v})}\big)\Big) \le \epsilon \,\Big|\, \mathfrak{B}\right\}\right] \\
&\quad \le |\mathbf{\Lambda}''| \sup_{E\in\mathbb{R}} \operatorname{ess\,sup} \mathbb{P}\left\{\operatorname{dist}\Big(E, \sigma\big(\mathbf{H}_{\mathbf{\Lambda}_L(\mathbf{u})}\big)\Big) \le \epsilon \,\Big|\, \mathfrak{B}\right\}.
\end{aligned} \tag{3.48}$$

Here the conditional probability $\mathbb{P}\{\ldots \mid \mathfrak{B}\}$ is, naturally, a $\mathfrak{B}$-measurable, a.s. bounded random variable, and $\operatorname{ess\,sup} \mathbb{P}\{\ldots \mid \mathfrak{B}\}$ stands for its essential supremum; recall that for an a.s. bounded random variable $\xi(\omega)$ one sets

$$\operatorname{ess\,sup} \xi := \sup\{a \in \mathbb{R} : \forall \epsilon > 0\ \mathbb{P}\{\xi(\omega) \ge a - \epsilon\} > 0\}.$$

The probability in the RHS can be assessed with the help of Stollmann's lemma (Lemma 2.2.1) when applied to the conditional probability $\mathbb{P}\{\cdot \mid \mathfrak{B}\left(\Pi_{\mathcal{J}^c}\mathbf{\Lambda}' \cup \Pi\mathbf{\Lambda}''\right)\}$, in the ensemble of operators $\mathbf{H}_{\mathbf{\Lambda}'}(\omega)$. Here we set $\mathcal{Q} = \Pi_{\mathcal{J}}\mathbf{\Lambda}' \subset \mathbb{Z}^d$ and, as before, define $\mu^{(\mathcal{Q})}$ as the product-measure on $\mathbb{R}^{\mathcal{Q}}$ with the same marginals as the random potential $V(x;\cdot)$. If $\underline{v}' = \{v_y,\ y \in \Pi_{\mathcal{J}}\mathbf{\Lambda}'\}$ is a sample value of the external potential over $\Pi_{\mathcal{J}}\mathbf{\Lambda}'$ and $\underline{v}'' = \{v_y,\ y \in \Pi\mathbf{\Lambda}' \setminus \Pi_{\mathcal{J}}\mathbf{\Lambda}'\}$ is that over the complement $\Pi\mathbf{\Lambda}' \setminus \Pi_{\mathcal{J}}\mathbf{\Lambda}'$ then we have the representation

$$\mathbf{H}_{\mathbf{\Lambda}'}(\underline{v}', \underline{v}'') = \mathbf{H}^0_{\mathbf{\Lambda}'} + \mathbf{W}(\underline{v}', \underline{v}'')$$

where $\mathbf{W}(\underline{v}', \underline{v}'')$ is the multiplication operator, by the function

$$\mathbf{x} \mapsto W(\mathbf{x}; \underline{v}', \underline{v}'') = U(\mathbf{x}) + \sum_{y''\in\Pi\mathbf{\Lambda}''\setminus\Pi_{\mathcal{J}}\mathbf{\Lambda}'} M_{y''}(\mathbf{x})v''_{y''} + \sum_{y'\in\Pi_{\mathcal{J}}\mathbf{\Lambda}'} M_y(\mathbf{x})v'_{y'}. \tag{3.49}$$

As before, this yields that the operators $\mathbf{H}_{\mathbf{\Lambda}'}(\underline{v}', \underline{v}'')$ form a diagonally monotone family relative to $\underline{v}'$, with

$$\mathbf{H}_{\mathbf{\Lambda}'}(\underline{v}' + t\underline{1}', \underline{v}'') - \mathbf{H}_{\mathbf{\Lambda}'}(\underline{v}', \underline{v}'') \ge t\,\mathbf{I}$$

$\forall\, t > 0$. Applying Lemma 2.2.1 in combination with Eq. (3.48) completes the proof. □

The following corollary will be helpful in the proof of Lemma 4.4.3.

Corollary 3.1. *Let $\mathbf{\Lambda}_L^{(n)}(\mathbf{u})$, $\mathbf{\Lambda}_L^{(n)}(\mathbf{v})$ be a separable pair of n-particle cubes. Assume that the probability distribution function F satisfies Assumption A. Then, for any given $\beta \in (0, 1)$ and all L large enough,*

$$\mathbb{P}\left\{\operatorname{dist}\left(\sigma(\mathbf{H}^{(n)}_{\mathbf{\Lambda}_L(\mathbf{u})}), \sigma(\mathbf{H}^{(n)}_{\mathbf{\Lambda}_L(\mathbf{v})})\right) < \mathrm{e}^{-L^\beta}\right\} \le \mathrm{e}^{-L^{\beta/2}}. \tag{3.50}$$

Consequently, for any given $Q > 0$ and all L large enough,

$$\mathbb{P}\left\{\operatorname{dist}\left(\sigma(\mathbf{H}^{(n)}_{\mathbf{\Lambda}_L(\mathbf{u})}), \sigma(\mathbf{H}^{(n)}_{\mathbf{\Lambda}_L(\mathbf{v})})\right) < \mathrm{e}^{-L^\beta}\right\} \le L^{-Q}. \tag{3.51}$$

Moreover, for any $Q > 0$ there exists $L^* \in \mathbb{N}$ large enough such that for all $L_0 \ge L^*$, $\theta \in (0, 1]$, $k \in \mathbb{N}$ and any pair of separable cubes $\mathbf{\Lambda}_{L_k}^{(n)}(\mathbf{x})$ and $\mathbf{\Lambda}_{L_k}^{(n)}(\mathbf{y})$

$$\mathbb{P}\left\{\exists\, E \in I : \mathbf{\Lambda}_{L_k}^{(n)}(\mathbf{x}) \text{ and } \mathbf{\Lambda}_{L_k}^{(n)}(\mathbf{y}) \text{ are } E\text{-PR}\right\} < \frac{1}{4} L_k^{-Q(1+\theta)^k} \tag{3.52}$$

where, again, $\{L_k, k \in \mathbb{N}\}$ follow the recursion $L_{k+1} = \lfloor L_k^\alpha \rfloor$.

For the proof of the assertion (3.52), it suffices to notice, as in Sect. 2.2.3 (cf. Corollary 2.2.4), that for any $\beta' > 0$, $\theta \in (0, 1)$ and L_0 large enough, for any $k \in \mathbb{N}$ one has

$$L_k^{-Q(1+\theta)^k} > \mathrm{e}^{-Q 2^k \ln L_k} > \mathrm{e}^{-L_k^{\beta'}}.$$

It is readily seen that the proof of Corollary 3.1, including the choice of all intermediate constants, does not depend in any way upon the form of the operators $\mathbf{H}^0_{\mathbf{\Lambda}'}$, $\mathbf{H}^0_{\mathbf{\Lambda}'}$ and the interaction $\mathbf{U}$. In particular, one can replace these operators by zero operators and still obtain exactly the same bound for the (multiplication) operators

$$\begin{aligned}
\tilde{\mathbf{H}}_{\mathbf{\Lambda}_L(\mathbf{u})} &= 0 \cdot \mathbf{H}^0_{\mathbf{\Lambda}_L^{(n)}(\mathbf{u})} + 0 \cdot \mathbf{U}_{\mathbf{\Lambda}_L^{(n)}(\mathbf{u})} + \mathbf{V}_{\mathbf{\Lambda}_L^{(n)}(\mathbf{u})} \equiv \mathbf{V}_{\mathbf{\Lambda}_L^{(n)}(\mathbf{u})} \\
\tilde{\mathbf{H}}_{\mathbf{\Lambda}_L(\mathbf{v})} &= 0 \cdot \mathbf{H}^0_{\mathbf{\Lambda}_L^{(n)}(\mathbf{v})} + 0 \cdot \mathbf{U}_{\mathbf{\Lambda}_L^{(n)}(\mathbf{v})} + \mathbf{V}_{\mathbf{\Lambda}_L^{(n)}(\mathbf{v})} \equiv \mathbf{V}_{\mathbf{\Lambda}_L^{(n)}(\mathbf{v})}.
\end{aligned}$$

In other words, the following result holds true:

Corollary 3.2. *Let $\mathbf{\Lambda}_L^{(n)}(\mathbf{u})$, $\mathbf{\Lambda}_L^{(n)}(\mathbf{v})$ be a separable pair of n-particle cubes. If the Assumption A is satisfied, then*

$$\begin{aligned}
&\mathbb{P}\left\{\operatorname{dist}\left(\sigma(\mathbf{V}_{\mathbf{\Lambda}_L(\mathbf{u})}(\omega)), \sigma(\mathbf{V}_{\mathbf{\Lambda}_L(\mathbf{v})}(\omega))\right) \le s\right\} \\
&\equiv \mathbb{P}\left\{\min_{\substack{\mathbf{x}\in\mathbf{\Lambda}_L(\mathbf{u}) \\ \mathbf{y}\in\mathbf{\Lambda}_L(\mathbf{v})}} |\mathbf{V}(\mathbf{x};\omega) - \mathbf{V}(\mathbf{y};\omega)| \le s\right\} \le 3^{2dn} L^{2dn} \zeta(2s).
\end{aligned} \tag{3.53}$$

3.5 Extended Wegner-Type Bounds for Distant Pairs of Cubes

According to Lemma 3.3.1, any pair of sufficiently distant two-particle cubes is separable in the sense of Definition 3.1.2. However, starting with the number of particles $N = 3$, the situation becomes more complicated: cf. Lemma 3.3.2. In fact, a direct analog of Lemma 3.3.1 is no longer valid as a simple example below shows. We already noted that this complication makes the whole MPMSA argument for $N \geq 3$ (which we complete in the next chapter) considerably more cumbersome than in the case $N = 2$; cf. [70]. We stress again that the reason for this is that the EVC bound (3.47) in Theorem 3.4.2 has been established under the condition of separability of the pair of cubes, and the latter property is not guaranteed when these cubes are simply at large distance from each other. In this section, we develop an alternative approach that allows to circumvent this obstacle; see [56]. However, we do not pursue this line further in the present text, in part because it needs additional assumptions on F, the probability distribution function of $V(x;\omega)$.

The above-mentioned example is as follows. Let $d = 1$, $N = 3$, $L = 0$ (cubes of radius 0 are single points). Given a positive integer R, consider the following 3-particle configurations in $\mathbb{Z}$ (i.e., sites in $\mathbb{Z}^3$):

$$\mathbf{x} = (0, 0, R), \quad \mathbf{y} = (0, R, R).$$

In physical terms, $\mathbf{y}$ is obtained from $\mathbf{x}$ by moving one particle from the origin 0 to a (distant) point R in $\mathbb{Z}$. Comparing the partial projections of the cubes $\mathbf{\Lambda}' = \mathbf{\Lambda}_0(\mathbf{x})$ and $\mathbf{\Lambda}'' = \mathbf{\Lambda}_0(\mathbf{y})$ (of radius $L = 0$), we observe that

$$\{\Pi_1\mathbf{\Lambda}', \Pi_2\mathbf{\Lambda}', \Pi_3\mathbf{\Lambda}'\} = \{\{0\}, \{R\}\} = \{\Pi_1\mathbf{\Lambda}'', \Pi_2\mathbf{\Lambda}'', \Pi_3\mathbf{\Lambda}''\},$$

no matter how large is the distance $\text{dist}(\mathbf{\Lambda}', \mathbf{\Lambda}'') = |\mathbf{x} - \mathbf{y}| = R$. Consequently, the samples of the random potential $V(\cdot\,;\omega)$ over the support sets $\Pi\mathbf{\Lambda}'$ and $\Pi\mathbf{\Lambda}''$ cannot be separated in any way. This makes operators $\mathbf{H}_{\mathbf{\Lambda}'}(\omega)$ and $\mathbf{H}_{\mathbf{\Lambda}''}(\omega)$ strongly correlated at any distance R. Note that $\mathbf{H}_{\mathbf{\Lambda}'}(\omega)$ and $\mathbf{H}_{\mathbf{\Lambda}''}(\omega)$ are one-dimensional operators (the kinetic part is absent).

However, one can notice that the operators $\mathbf{H}_{\mathbf{\Lambda}'}(\omega)$ and $\mathbf{H}_{\mathbf{\Lambda}''}(\omega)$ are not equally sensitive to the change of the potential $V(x;\omega)$ at the relevant points $x = 0$ and $x = R$. Indeed, the eigenvalues of $\mathbf{H}_{\mathbf{\Lambda}'}(\omega)$ and $\mathbf{H}_{\mathbf{\Lambda}''}(\omega)$ are given by

$$E(\mathbf{H}_{\mathbf{\Lambda}'}) = 2V(0;\omega) + V(R;\omega), \quad E(\mathbf{H}_{\mathbf{\Lambda}''}) = V(0;\omega) + 2V(R;\omega).$$

Consequently,

$$E(\mathbf{H}_{\mathbf{\Lambda}'}) - E(\mathbf{H}_{\mathbf{\Lambda}''}) = V(R;\omega) - V(0;\omega).$$

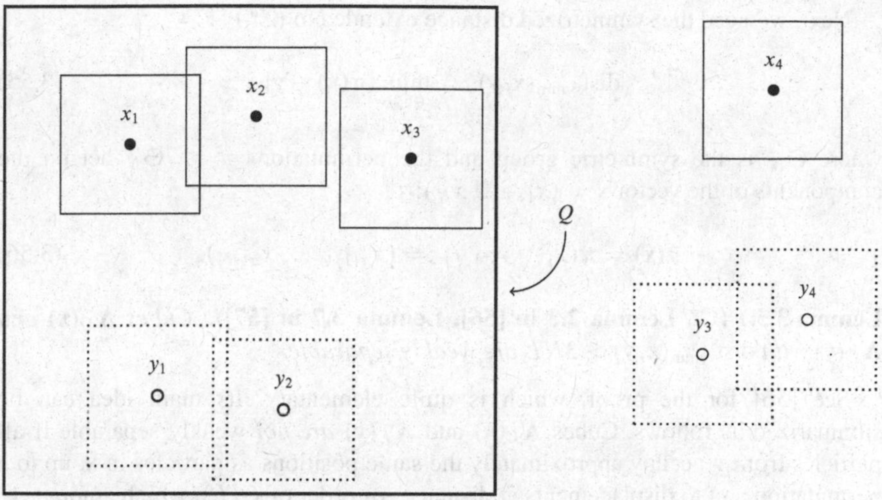

Fig. 3.2 An example for the Definition 3.5.1, with $d = 2$, $N = 4$. Points x_i of configuration $\mathbf{x}$ are depicted by *black dots*, and points y_j of configuration $\mathbf{y}$ by circles. Here $\mathcal{J}_1 = \{1,2,3\}$, $\mathcal{J}_2 = \{1,2\}$. The points x_i, y_j are centers of squares of radius L. 4-particle cubes $\mathbf{\Lambda}_L(\mathbf{x}), \mathbf{\Lambda}_L(\mathbf{y})$ are weakly separable.

As the variables $V(R;\omega)$ and $V(0;\omega)$ have a Hölder-continuous probability distribution function F (see Assumption A), so does the difference $V(R;\omega) - V(0;\omega)$. Hence, $E(\mathbf{H}_{\mathbf{\Lambda}'})$ is close to $E(\mathbf{H}_{\mathbf{\Lambda}''})$ with a small probability. The bottom line is that (in this particular example) the spectra of two operators in distant (albeit not separable) cubes are close with small probability, despite the fact that they are strongly correlated.

Definition 3.5.1. *(See Fig. 3.2) An N-particle cube $\mathbf{\Lambda}_L^{(N)}(\mathbf{x})$ is called weakly separable from $\mathbf{\Lambda}_L^{(N)}(\mathbf{y})$ if there exists a parallelepiped $\mathrm{Q} \subset \mathbb{Z}^d$, of diameter $R \leq 2NL$ and with edges parallel to the coordinate axes, and subsets $\mathcal{J}_1, \mathcal{J}_2 \subset \{1,\ldots,N\}$ such that $|\mathcal{J}_1| > |\mathcal{J}_2|$ (possibly, with $\mathcal{J}_2 = \varnothing$) and*

$$\Pi_{\mathcal{J}_1}\mathbf{\Lambda}_L^{(N)}(\mathbf{x}) \cup \Pi_{\mathcal{J}_2}\mathbf{\Lambda}_L^{(N)}(\mathbf{y}) \subseteq \mathrm{Q},$$

$$\Pi_{\mathcal{J}_2^c}\mathbf{\Lambda}_L^{(N)}(\mathbf{y}) \cap \mathrm{Q} = \varnothing, \tag{3.54}$$

$$\Pi_{\mathcal{J}_1^c}\mathbf{\Lambda}_L^{(N)}(\mathbf{x}) \cap \mathrm{Q} = \varnothing.$$

A pair of cubes $(\mathbf{\Lambda}_L^{(N)}(\mathbf{x})$ and $\mathbf{\Lambda}_L^{(N)}(\mathbf{y}))$ is called weakly separable if at least one of them is weakly separable from the other.

Next, we need the symmetrized distance extended to $(\mathbb{Z}^d)^N$:

$$\operatorname{dist}_{\mathrm{symm}}(\mathbf{x}, \mathbf{y}) = \min_{\pi \in \mathfrak{S}_N} |\pi(\mathbf{x}) - \mathbf{y}| \tag{3.55}$$

where $\mathfrak{S}_N$ is the symmetric group and the permutations $\pi \in \mathfrak{S}_N$ act on the components of the vector $\mathbf{x} = (x_1, \dots, x_N)$:

$$\pi(\mathbf{x}) = \pi(x_1, \dots, x_N) := (x_{\pi(1)}, \dots, x_{\pi(N)}). \tag{3.56}$$

Lemma 3.5.1 (Cf. Lemma 2.3 in [56], Lemma 3.7 in [57]). *Cubes* $\mathbf{\Lambda}_L(\mathbf{x})$ *and* $\mathbf{\Lambda}_L(\mathbf{y})$ *with* $\operatorname{dist}_{\mathrm{symm}}(\mathbf{x}, \mathbf{y}) > 3NL$ *are weakly separable.*

See [56] for the proof, which is quite elementary. Its main idea can be summarized as follows. Cubes $\mathbf{\Lambda}_L(\mathbf{x})$ and $\mathbf{\Lambda}_L(\mathbf{y})$ *are not* weakly separable if all particles from $\mathbf{y}$ occupy approximately the same positions as particles in $\mathbf{x}$, up to a permutation and to displacements at distances of order of $O(L)$, which amounts to a bound $\operatorname{dist}_{\mathrm{symm}}(\mathbf{x}, \mathbf{y}) = O(L)$.

In the above example, we used the fact that the eigenvalues of operators $\mathbf{H}_{\mathbf{\Lambda}'}(\omega)$ and $\mathbf{H}_{\mathbf{\Lambda}''}(\omega)$ were linear polynomials of the variables $V(0;\omega)$, $V(R;\omega)$. This is no longer true when radius $L \geq 1$; here the underlying cubes contain more than one point. In order to exploit the same algebraic idea, introduce the following notation. Given a lattice subset $\mathrm{Q} \subset \mathbb{Z}^d$, we denote by $\xi_{\mathrm{Q}}(\omega)$ the sample mean of the random field V over the Q,

$$\xi_{\mathrm{Q}}(\omega) = |\mathrm{Q}|^{-1} \sum_{x \in \mathrm{Q}} V(x;\omega).$$

Next, define the fluctuations of V relative to the sample mean,

$$\eta_{x,\mathrm{Q}} = V(x;\omega) - \xi_{\mathrm{Q}}(\omega), \quad x \in \mathrm{Q},$$

with

$$V(x;\omega) = \xi_{\mathrm{Q}}(\omega) + \eta_{x,\mathrm{Q}}(\omega), \quad x \in \mathrm{Q}.$$

Denote by $\mathfrak{F}_{\mathrm{Q}}$ the sigma-algebra generated by the random variables $\{\eta_{x,\mathrm{Q}} : x \in \mathrm{Q}\}$, and by $F_{\xi_{\mathrm{Q}}}(\cdot \,|\, \mathfrak{F}_{\mathrm{Q}}) = \left[F_{\xi_{\mathrm{Q}}}(\cdot \,|\, \mathfrak{F}_{\mathrm{Q}})\right](\omega)$ the conditional probability distribution function of ξ_{Q} relative to $\mathfrak{F}_{\mathrm{Q}}$:

$$F_{\xi_{\mathrm{Q}}}(t \,|\, \mathfrak{F}_{\mathrm{Q}}) = \mathbb{P}\left\{ \xi_{\mathrm{Q}} \leq t \,|\, \mathfrak{F}_{\mathrm{Q}} \right\}.$$

From this point on, it is convenient to assume that $\mathrm{Q} \subset \mathbb{Z}^d$ is a lattice parallelepiped, with edges parallel to the coordinate axes. Given $s > 0$, $R \geq 1$, introduce a $\mathfrak{F}_{\mathrm{Q}}$-measurable random variable

$$\nu_R(s;\omega) := \sup\left[\left| F_{\xi_Q}(t+s \mid \mathfrak{F}_Q) - F_{\xi_Q}(t \mid \mathfrak{F}_Q) \right|, t \in \mathbb{R},\ \mathrm{diam}(Q) \le R \right]. \quad (3.57)$$

We will now assume that the random field V, which is not necessarily IID, fulfills the following condition:

Assumption C. *There exist constants* $C, C', A, A', b, b' \in (0, +\infty)$ *such that* $\forall$ $s \in [0,1]$,

$$\mathbb{P}\left\{ \nu_R(s;\omega) \ge C R^A s^b \right\} \le C' R^{A'} s^{b'}. \quad (3.58)$$

It follows from the elementary results of the probability theory that the Assumption C is fulfilled in the particular case of an IID Gaussian random field $V\colon \mathbb{Z}^d \times \Omega \to \mathbb{R}$. Moreover, the bound (3.58) becomes here deterministic (cf. (3.59)).

Indeed, it is well-known (cf., e.g., [159]) that in this particular case the sample average ξ_Q is independent from $\mathfrak{F}_Q$; on the other hand, ξ_Q is itself a Gaussian variable. For the sake of example, let the marginal probability distribution of the field V have zero mean and unit variance; then $\mathbb{E}\left[\, \xi_Q \,\right] = 0$ and $\mathbb{E}\left[\, (\xi_Q)^2 \,\right] = |Q|^{-1}$, yielding

$$F_{\xi_Q}(t+s) - F_{\xi_Q}(t) = \frac{|Q|^{1/2}}{\sqrt{2\pi}} \int_t^{t+s} e^{-y^2/2}\, dy \le \frac{|Q|^{1/2} s}{\sqrt{2\pi}}.$$

Since ξ_Q is $\mathfrak{F}_Q$-independent (hence, $F_{\xi_Q}(\cdot|\mathfrak{F}_Q)$ is nonrandom) and $|Q| \le R^d$, we obtain

$$\mathbb{P}\left\{ \nu_R(s,\omega) \ge (2\pi)^{-1/2} R^{d/2} s \right\} = 0. \quad (3.59)$$

The situation is more complicated already for the IID random field V with a uniform probability distribution on an interval $[a,b]$, $b > a$. It is not difficult to see (cf. [56,63]) that the conditional probability distribution of ξ_Q given $\mathfrak{F}_Q$ may even be singular in this case—for some conditions. However, such conditions have a small probability. More generally, it can be easily shown that Assumption C holds true for IID potentials with uniform marginal probability distribution on $[0,1]$ or, more generally, with marginal probability distribution admitting a piecewise constant, compactly supported probability density. By standard approximation arguments, it can also be extended to IID potentials with smooth, compactly supported probability *density* (cf. [63]). We believe that the requirement of smoothness (thus existence) of the marginal probability *density* can be relaxed, probably to the Hölder continuity of the marginal probability *distribution function*, but at the time of writing these lines, we are unaware of any such result. In any case, the class of smooth (or piecewise smooth) marginal probability densities is sufficiently rich to cover most of the applications to the physical models of the multi-particle disordered systems.

Curiously, the two most popular models of the disorder considered in the physical works on interacting multi-particle systems (cf. [31,105]) are *precisely* the Gaussian and the uniform law. For the former, Assumption C is obviously fulfilled, and even takes a particularly simple and convenient form (cf. (3.59)), while its proof for the latter can be considered as a simple exercise in a standard course of probability theory (cf. [63]). In our opinion, this already justifies using Assumption C in the presentation of the variable-energy MPMSA in Sect. 4.4.

If assumption C holds true, then, taking $s = \mathrm{e}^{-R^{\beta}}$, with R large and $\beta \in (0, 1)$, one can easily deduce from Eq. (3.58) that

$$\mathbb{P}\left\{ \nu_R(s;\omega) \geq CR^A \mathrm{e}^{-R^{\beta'}} \right\} \leq \mathrm{e}^{-R^{\beta''}}, \tag{3.60}$$

with $\beta', \beta'' \in (0, 1)$.

By means of the above concepts, we are prepared to state the following

Theorem 3.5.2 (Cf. [56]). *Suppose that F satisfies Assumption C and cubes $\mathbf{\Lambda}_L(\mathbf{x})$ and $\mathbf{\Lambda}_L(\mathbf{y})$ are weakly separable. Consider cubes $\mathbf{\Lambda}_{L'}(\mathbf{u}) \subseteq \mathbf{\Lambda}_L(\mathbf{x})$, $\mathbf{\Lambda}_{L''}(\mathbf{v}) \subseteq \mathbf{\Lambda}_L(\mathbf{y})$. Then for any $s > 0$ the following bound holds:*

$$\begin{aligned} \mathbb{P}\left\{ \mathrm{dist}\left(\sigma\left(\mathbf{H}_{\mathbf{\Lambda}_{L'}(\mathbf{u})}\right), \sigma\left(\mathbf{H}_{\mathbf{\Lambda}_{L''}(\mathbf{v})}\right)\right) \leq s \right\} &\leq |\mathbf{\Lambda}_{L'}(\mathbf{u})| \cdot |\mathbf{\Lambda}_{L''}(\mathbf{v})| \, \nu_L(2s) \\ &\leq 3^{2Nd} L^{2Nd} \, \nu_L(2s). \end{aligned} \tag{3.61}$$

Taking into account Lemma 3.5.1, we come immediately to the following generalization of the EVC bound given in Lemma 2.2.3:

Corollary 3.3. *Fix an arbitrary integer $N > 1$. Under Assumption C, the bound (3.61) in Theorem 3.5.2 holds true for any pair of $3NL$-distant cubes $\mathbf{\Lambda}_{L'}^{(N)}(\mathbf{u})$, $\mathbf{\Lambda}_{L''}^{(N)}(\mathbf{v})$ with $L', L'' \leq L$.*

In particular, for any $\beta > 0$ there exist $\tilde{\beta} \in (0, \beta)$ and $L_* < \infty$ such that for all $1 \leq n \leq N$, all $L \geq L_*$, and all $\mathbf{u}, \mathbf{v} \in (\mathbb{Z}^d)^n$ with $|\mathbf{u} - \mathbf{v}| > 3nL$, one has

$$\mathbb{P}\left\{ \mathrm{dist}\left(\sigma\left(\mathbf{H}_{\mathbf{\Lambda}_L^{(n)}(\mathbf{u})}\right), \sigma\left(\mathbf{H}_{\mathbf{\Lambda}_L^{(n)}(\mathbf{v})}\right)\right) \leq \mathrm{e}^{-L^{\beta}} \right\} \leq \mathrm{e}^{-L^{\tilde{\beta}}}. \tag{3.62}$$

The above bound is sufficient for the purposes of the MPMSA.

Chapter 4
Multi-particle MSA Techniques

In this chapter we prove Theorems 3.1.1 and 3.1.2. Accordingly, in Sects 4.2–4.6 we suppose that Assumptions A and 3.1.1 are fulfilled.

Our next goal is to derive property (**SS**.N, E, m, p_k, L_k) (see inequality (3.21)), for all $n = 1, \ldots, N$ and $k \in \mathbb{N}$, in the framework of the fixed-energy MPMSA. This is achieved in Sect. 4.3, where we also derive from the fixed-energy bounds their variable-energy counterparts by adapting Theorems 2.5.1 and 2.5.2 to the multi-particle setting.

In Sect. 4.4, we present the variable-energy MPMSA scheme and prove directly the bound (**DS**.N, n, m, p_k, L_k) (see inequality (4.60)) for all $n = 1, \ldots, N$ and $k \in \mathbb{N}$; the final assertion establishing this fact is Theorem 4.4.10.

In both cases, the core of the argument here is an induction in the number of particles $n \in \{1, \ldots, N\}$ combined with the scale induction in k. In a sense, the induction in n provides a bridge between the single-particle and the multi-particle theories. In the course of the inductive argument, we first establish the bound (4.60) for $k = 0$ (i.e., at the length scale L_0) and all $n = 1, \ldots, N$. At the induction step $k \rightsquigarrow k + 1$, we assume (4.60) at level k (that is, for the length scale L_k) for subsequent $n = 1, \ldots, N$ and then move to $k + 1$ (i.e., to the length scale L_{k+1}).

Once the property (**DS**.N, n, m, p_k, L_k) is established for all $n = 1, \ldots, N$ and $k \in \mathbb{N}$, we derive Theorems 3.1.1 and 3.1.2 in Sects. 4.5 and 4.6, respectively. This derivation will require a number of modifications of the standard proof used in the single-particle theory.

Recall, the length scales L_k follow the recursion $L_{k+1} = \lfloor L_k^\alpha \rfloor$. In the previous chapter we defined a sequence $k \mapsto p_k(n)$ where the value $p_k(n) = p_k(n, N, p_0, \theta)$ is of the form $4^{N-n} p_0 (1 + \theta)^k$ and values $p_0, \theta > 0$. In fact, until Sect. 4.7 we will assume that

$$p_0 > 6Nd \text{ and } \theta := \frac{1}{6} - \frac{Nd}{p_0} \in \left(0, \frac{1}{6}\right); \tag{4.1}$$

V. Chulaevsky and Y. Suhov, *Multi-scale Analysis for Random Quantum Systems with Interaction*, Progress in Mathematical Physics 65, DOI 10.1007/978-1-4614-8226-0_4,

cf. Lemma 4.4.6. Parameter p_0 is crucial and analogous to that used in Chap. 2 in the analysis of single-particle systems. Respectively, the quantity $p_k(n)$ will play a role similar to that of the exponent p_k used in the single-particle scale induction.

The above form of the parameter $p_k(n)$ hints that, in the course of the inductive transition $n \rightsquigarrow n+1$, a stronger probabilistic bound of the form (4.60) is assumed for the n-particle system while for the $(n+1)$-particle system it gives rise to a weaker bound. Eventually, this results in more restrictive bounds upon the threshold g^* in Theorems 3.1.1 and 3.1.2, with g^* rapidly increasing with N. This is not a mere technicality: in fact, mathematical methods known at the time of writing these lines (cf. [9, 71]) do not have the capacity to obtain uniform localization bounds for a large number of particles. In other words, at present it is unclear how (and seems difficult) to prove localization in an arbitrarily large volume $\Lambda_L(0) \subset \mathbb{Z}^d$ in the physical configuration space, for a system of $N = \varrho|\Lambda_L(0)|$ particles, with a given value of the spatial particle density $\varrho > 0$.

As we have already seen in Chap. 2, in the single-particle localization theory, the MSA bounds imply strong dynamical localization. In Sect. 4.6 we adapt this argument to the multi-particle setting. In fact, owing to the MPMSA bounds (4.60) with the exponent $p_k(n, p, \theta) \to \infty$ as $k \to \infty$, we prove multi-particle strong dynamical localization with the decay rate of the eigenfunction correlators faster than polynomial, essentially in the same way as it was done in Sect. 2.7.2.

Our last remark is that the multi-particle version of the FMM approach developed by Aizenman and Warzel [9] also uses an induction on the number of particles. Moreover, in contrast with the single-particle FMM in its various versions, the multi-particle FMM developed in [9] employs elements of scale induction as well, incorporating them into certain technical statements.

4.1 A Remark on Two-Particle Systems

Before we embark on the formal analysis of a general multi-particle Anderson model, we would like to stress a special place occupied by the two-particle theory.

A series of discussions that the authors have had with a number of colleagues across both the physics and maths communities have supplied evidence that for the physical camp the analysis of two-particle systems may often be of a greater importance than even a general case where the number of particles $3 \leq N \ll |\Lambda|$. Some arguments supporting such an opinion can be summarized as follows.

- Anderson-type Hamiltonians are usually applied in the framework of physical models of disordered media where quantum particles constitute a low-density "gas" with rare collisions. This explains why single-particle models give quite an accurate description of transport/localization phenomena both in theory and in applications.

- In a "gas" of particles of low density, it is natural to expect that two-particle processes are more significant than those involving $N \geq 3$ particles. (Certainly, this is an extremely delicate statement which we cannot discuss in this book.)
- In the physical literature it has been recently discussed, in the framework of various models and methods (cf. [31, 32, 105]; here the order of citation is alphabetical), that the Anderson localization phenomenon persists in presence of certain inter-particle interactions. While the existing mathematically rigorous analysis of N-particle models firmly establishes the presence of localization (admittedly, under technical assumptions that are yet too restrictive from the physical point of view), it is far from clear whether the case of $N = 3$ or $N = 10^3$ is conceptually more informative than $N = 2$. The next radical step would be to analyze $N = \varrho|\Lambda|$, with the density of particles $\varrho > 0$ and with localization bounds uniform in $|\Lambda|$.

It should be stressed that in the previous paragraph we merely cite arguments conveyed to us, leaving it to the readers to forge their own opinion on the issue. As far as the rigorous argument is concerned, the main ingredient of the MPMSA required to extend the two-particle approach is the induction in the number of particles, which is the topic of this chapter. Yet, the analysis of two-particle systems (cf. [70]) is technically less involved and more streamlined. It can be considered as a good starting point for learning the multi-particle MSA techniques. For the authors of this book the two-particle tight-binding model has been a significant step in the analysis of localization phenomenon in interacting disordered systems, more important (at least, psychologically) than the adaptation to a general model with $N \geq 3$ particles. A reader familiar with the paper [70] can see that the principal argument from that work can be interpreted as an induction step based upon well-known localization properties (viz., MSA bounds) established for the single-particle Anderson model.

4.2 Some Geometric Notions and Facts

In this section we present a number of auxiliary assertions of a geometric nature characterizing the so-called partially and fully interactive cubes; see Definition 4.2.1 below. Consider the following subset of the lattice $(\mathbb{Z}^d)^n$ which we will call the diagonal:

$$\mathbb{D}_0 = \{\mathbf{x} = (x, \dots, x),\ x \in \mathbb{Z}^d\}.$$

Recall that the diameter of a bounded subset $X \subset \mathbb{Z}^d (\hookrightarrow \mathbb{R}^d)$, induced by the max-norm distance, is defined by

$$\operatorname{diam} X = \max_{x,y \in X} \operatorname{dist}(x, y) = \max_{x,y \in X} |x - y|. \tag{4.2}$$

Lemma 4.2.1. *For any* $\mathbf{x} = (x_1, \dots, x_n) \in (\mathbb{Z}^d)^n$

$$\operatorname{dist}(\mathbf{x}, \mathbb{D}_0) \leq \operatorname{diam} \Pi\mathbf{x} \leq 2 \operatorname{dist}(\mathbf{x}, \mathbb{D}_0). \tag{4.3}$$

Proof. By definition of $\mathbb{D}_0$,

$$\operatorname{dist}(\mathbf{x}, \mathbb{D}_0) = \min_{u\in\mathbb{Z}^d} \max_j |x_j - u|.$$

Since each x_i is a particular choice of $u \in \mathbb{Z}^d$, the last expression is

$$\leq \max_{i,j} |x_j - x_i| = \operatorname{diam} \Pi\mathbf{x},$$

proving the left-hand inequality in (4.3). On other hand, if $\mathbf{u} = (u, \dots, u) \in (\mathbb{Z}^d)^n$ is the closest point to $\mathbf{x}$ in $\mathbb{D}_0$, then, for $1 \leq i, j \leq n$,

$$|x_i - x_j| \leq |x_i - u| + |u - x_j| \leq 2\operatorname{dist}(\mathbf{x}, \mathbb{D}_0)$$

yielding

$$\max_{i,j} |x_j - x_i| \leq 2\operatorname{dist}(\mathbf{x}, \mathbb{D}_0). \qquad \square$$

Definition 4.2.1. *Let r_0 be the range of the interaction* $\mathbf{U}$. *An n-particle cube $\mathbf{\Lambda}_L^{(n)}(\mathbf{x})$ is called fully interactive if* $\operatorname{dist}(\mathbf{x}, \mathbb{D}_0) \leq 2n(L + r_0)$*, and partially interactive otherwise.*

In the sequel we will use abbreviations FI (fully interactive) and PI (partially interactive).

Admittedly, the term *partially interactive* may be seen as misleading; perhaps, a less ambiguous alternative would be to call PI cubes partially non-interacting. However, in our opinion, the shorter term PI fits the system better. One could also call PI cubes decomposable (into two or more subsystems not interacting with each other), and FI cubes—non-decomposable; cf. Lemma 4.2.2 below.

Observe that if a cube $\mathbf{\Lambda}_L(\mathbf{x})$ is PI/FI, then so are all cubes $\mathbf{\Lambda}_L(\pi\mathbf{x})$, $\pi \in \mathfrak{S}_n$, obtained from $\mathbf{\Lambda}_L(\mathbf{x})$ by permuting the positions of individual particles (see (3.56)).

The following simple geometrical statement explains the main *raison d'être* of Definition 4.2.1 and the choice of the lower bound $\operatorname{dist}(\mathbf{x}, \mathbb{D}_0) > 2n(L + r_0)$ for the partially interactive cubes.

Lemma 4.2.2. *If an n-particle cube $\mathbf{\Lambda}_L(\mathbf{x})$ is* PI, *then there exist two complementary non-empty index subsets $\mathcal{J}, \mathcal{J}^c \subset \{1, \dots, n\}$ such that*

$$\Pi_{\mathcal{J}} \mathbf{\Lambda}_{L+r_0}(\mathbf{x}) \cap \Pi_{\mathcal{J}^c} \mathbf{\Lambda}_{L+r_0}(\mathbf{x}) = \varnothing. \tag{4.4}$$

Proof. It is convenient to use the standard embedding $\mathbb{Z}^d \hookrightarrow \mathbb{R}^d$ and the notion of connectedness in $\mathbb{R}^d$. We have

$$\Pi\mathbf{\Lambda}_{L+r_0}(\mathbf{x}) = \bigcup_{i=1}^{n} \Pi_i \mathbf{\Lambda}_{L+r_0}(\mathbf{x}) = \bigcup_{i=1}^{n} \Lambda_{L+r_0}(x_i), \tag{4.5}$$

and if the union of cubes $\Lambda_{L+r_0}(x_i)$ in the RHS, considered as cubes in $\mathbb{R}^d$, is not a connected set, then there exists a decomposition of the required form (4.4). So, it suffices to show that if the cube $\mathbf{\Lambda}_L(\mathbf{u})$ is partially interactive, the union of cubes in (4.5) cannot be a connected set.

Indeed, assume otherwise. Then

$$\operatorname{diam} \Pi \mathbf{\Lambda}_{L+r_0}(\mathbf{x}) \leq \sum_{i=1}^{n} \operatorname{diam} \Pi \Lambda_{L+r_0}(x_i) = n \cdot 2(L + r_0). \tag{4.6}$$

Observe that

$$\operatorname{diam} \Pi \mathbf{x} \leq \operatorname{diam} \Pi \mathbf{\Lambda}_{L+r_0}(\mathbf{x}).$$

By assumption, $\operatorname{dist}(\mathbf{x}, \mathbb{D}_0) > 2n(L + r_0)$, so that we can write, starting with the inequality (4.6) and using Lemma 4.2.1,

$$\begin{aligned} 2n(L + r_0) &\geq \operatorname{diam} \Pi \mathbf{\Lambda}_{L+r_0}(\mathbf{x}) \geq \operatorname{diam} \Pi \mathbf{x} \\ &\geq \operatorname{dist}(\mathbf{x}, \mathbb{D}_0) \\ &> 2n(L + r_0). \end{aligned}$$

The obtained contradiction completes the proof. □

For a given PI cube $\mathbf{\Lambda}_L^{(n)}(\mathbf{u})$, there may be several decompositions

$$\{1, \dots, n\} = \mathcal{J} \cup \mathcal{J}^{\mathrm{c}} \text{ with } 1 \leq |\mathcal{J}|, |\mathcal{J}^{\mathrm{c}}| < n \text{ and } \mathcal{J} \cap \mathcal{J}^{\mathrm{c}} = \varnothing \tag{4.7}$$

corresponding to the representations

$$\mathbf{\Lambda}_L^{(n)}(\mathbf{u}) = \mathbf{\Lambda}_L^{(n')}(\mathbf{u}') \times \mathbf{\Lambda}_L^{(n'')}(\mathbf{u}''), \quad n' = |\mathcal{J}|, \; n'' = |\mathcal{J}^{\mathrm{c}}|, \tag{4.8}$$

satisfying (4.4) and corresponding to subsystems not interacting with each other. Here $\mathbf{u}' = \mathbf{u}_{\mathcal{J}}$, $\mathbf{u}'' = \mathbf{u}_{\mathcal{J}^{\mathrm{c}}}$, and, obviously, $n' + n'' = n$. We will assume that one such decomposition of the form (4.8) is fixed for each PI cube; it will be referred to as the canonical decomposition of the PI cube $\mathbf{\Lambda}_L^{(n)}(\mathbf{u})$.

We have seen in Lemma 4.2.2 how the PI property of a given cube can be turned into an advantage. The following statement shows that pairs of FI cubes also have some specific feature useful for the MPMSA scheme.

Lemma 4.2.3. *Let $n \geq 2$. If the n-particle cubes $\mathbf{\Lambda}_L^{(n)}(\mathbf{x})$, $\mathbf{\Lambda}_L^{(n)}(\mathbf{y})$ are* FI, *and if $|\mathbf{x} - \mathbf{y}| > n(10L + 8r_0)$ then $\Pi \mathbf{\Lambda}_L^{(n)}(\mathbf{x}) \cap \Pi \mathbf{\Lambda}_L^{(n)}(\mathbf{y}) = \varnothing$.*

Proof. If, for some $R > 0$,

$$R < |\mathbf{x} - \mathbf{y}| = \max_{i,j} |x_j - y_j|,$$

then there exists $j_0 \in \{1, \dots, n\}$ such that $|x_{j_0} - y_{j_0}| > R$. Since both cubes are FI, $\text{dist}(\mathbf{x}, \mathbb{D}_0) \le 2n(L + r_0)$, $\text{dist}(\mathbf{y}, \mathbb{D}_0) \le 2n(L + r_0)$, so we can use (4.3) for the centers $\mathbf{x}$, $\mathbf{y}$, namely,

$$\max_{i,j} |x_i - x_j| \le 2\,\text{dist}(\mathbf{x}, \mathbb{D}_0), \quad \max_{i,j} |y_i - y_j| \le 2\,\text{dist}(\mathbf{y}, \mathbb{D}_0)$$

and write:

$$|x_{j_0} - x_i| \le 2 \cdot 2n(L + r_0), \; |y_{j_0} - y_j| \le 2 \cdot 2n(L + r_0).$$

By the triangle inequality, for any $i, j \in \{1, \dots, n\}$ and $R = 10L + 8r_0$, we have

$$\begin{aligned} |x_i - y_j| &\ge |x_{j_0} - y_{j_0}| - |x_{j_0} - x_i| - |y_{j_0} - y_j| \\ &> n(10L + 8r_0) - 8n(L + r_0) = 2nL. \end{aligned}$$

Therefore,

$$\begin{aligned} \text{dist}\big(\Pi \mathbf{\Lambda}_L^{(n)}(\mathbf{x}), \Pi \mathbf{\Lambda}_L^{(n)}(\mathbf{y})\big) &= \min_{1 \le i,j \le n} \text{dist}\big(\Lambda_L(x_i), \Lambda_L(y_j)\big) \\ &\ge \min_{1 \le i,j \le n} |x_i - y_j| - 2L > 2nL - 2L > 0. \end{aligned}$$

This means that $\Pi \mathbf{\Lambda}_L^{(n)}(\mathbf{x}) \cap \Pi \mathbf{\Lambda}_L^{(n)}(\mathbf{y}) = \varnothing$. □

Owing to the condition (3.39), we can replace the expression $10L + 8r_0$ by a simpler one, $11L$, with $L \ge L_0$.

Corollary 4.1. *Suppose that $L_0 > 8r_0$ (cf. (3.39)). If $|\mathbf{x} - \mathbf{y}| \ge 11nL_k$, $k \ge 0$, and n-particle cubes $\mathbf{\Lambda}_{L_k}(\mathbf{x})$, $\mathbf{\Lambda}_{L_k}(\mathbf{y})$ are* FI *then these cubes are separable. Moreover, $\Pi \mathbf{\Lambda}_{L_k}(\mathbf{x}) \cap \Pi \mathbf{\Lambda}_{L_k}(\mathbf{y}) = \varnothing$ and, therefore, the random operators $\mathbf{H}_{\mathbf{\Lambda}_{L_k}(\mathbf{x})}$ and $\mathbf{H}_{\mathbf{\Lambda}_{L_k}(\mathbf{y})}$ are independent (as the values of random potential $V(x;\omega)$ are* IID*).*

The condition $L_0 > 8r_0$ will always be assumed below, along with a number of other technical lower bounds on L_0.

4.3 Fixed-Energy MPMSA

In accordance with Remark 3.1.1, the assumption

$$m \ge 1 \tag{4.9}$$

will be in force until the end of Sect. 4.4. (The argument in Sect. 4.3.1 (see Theorem 4.3.2) implies that this assumption can be treated as formally established.)

Also, we assume that an integer $N > 1$ (the number of particles) is fixed. The value of N can be arbitrarily large, but several key parameters of the MPMSA scheme depend upon N, explicitly or implicitly.

4.3.1 Initial Scale Bounds

In this section we obtain initial scale bounds required for the fixed-energy scaling analysis and its variable-energy variant (considered in Sect. 4.4).

Recall, in (3.19) we have defined the sequence of positive numbers depending on parameters θ and p_0:

$$p_k(n) = p_k(n, N, p_0, \theta) = 4^{N-n} p_0 (1+\theta)^k, \quad 1 \le n \le N, \; k \ge 0. \tag{4.10}$$

In particular, at $k = 0$, we have $p_0(n) = 4^{N-n} p_0$.

The MPMSA combines the scale induction (in L_k, $k = 0, 1, \dots$) with the induction in the number of particles (in $n = 1, 2, \dots, N$). Put differently, our method is perturbative (as the MSA usually is), and one treats

(i) The kinetic energy operator $\mathbf{H}_0 = -\boldsymbol{\Delta}$
(ii) The interaction operator $\mathbf{U}$

as perturbations of the random external potential energy operator $\mathbf{V}(\omega)$. The latter is diagonal in the δ-basis in $\left(\mathbb{Z}^d\right)^n$, thus has "ultimately localized" eigenfunctions. The scale induction takes care of the perturbation (i), while the perturbation (ii) requires—in our method—the induction in the number of particles. This explains the fact that the strongest assumption on the disorder amplitude (guaranteeing the validity of (**SS.**$N, \mathbb{R}, m, p_0, L_0$)) is made for the 1-particle systems ($n = 1$).

Definition 4.3.1. *Given $E \in \mathbb{R}$, an n-particle cube $\mathbf{\Lambda}_{L_{k+1}}(\mathbf{u})$ is called E-non-resonant (E-NR) if the resolvent $\mathbf{G}_{\mathbf{\Lambda}_{L_{k+1}}(\mathbf{u})}(E, \omega)$ is defined and satisfies*

$$\left\| \mathbf{G}_{\mathbf{\Lambda}_{L_{k+1}}(\mathbf{u})}(E, \omega) \right\| \le \mathrm{e}^{L_{k+1}^{\beta}} \tag{4.11}$$

and E-resonant (E-R), otherwise. A cube $\mathbf{\Lambda}_{L_{k+1}}(\mathbf{u})$ is called E-completely non-resonant (E-CNR) if it is E-NR and does not contain any E-R cube $\mathbf{\Lambda}_L(\mathbf{w})$ with $L \ge L_k$ (for which $\left\| \mathbf{G}_{\mathbf{\Lambda}_L(\mathbf{w})}(E, \omega) \right\| > \mathrm{e}^{L^{\beta}}$). Otherwise, $\mathbf{\Lambda}_{L_{k+1}}(\mathbf{u})$ is said to be E-partially resonant (E-PR).

For the purposes of this subsection only, introduce also the following

Definition 4.3.2. *Let be given $E \in \mathbb{R}$, $\eta > 0$, and $\omega \in \Omega$. A cube $\mathbf{\Lambda}_L^{(n)}(\mathbf{x})$ is called (E, η)-R if*

$$\operatorname{dist}\Big(\sigma\big(\mathbf{H}_{\mathbf{\Lambda}_L(\mathbf{x})}(\omega)\big), E\Big) < \eta, \tag{4.12}$$

and (E,η)-NR, *otherwise. A cube* $\mathbf{\Lambda}_L^{(n)}(\mathbf{x})$ *is called* (E,η)-good *if*

$$\min_{\mathbf{x}\in\mathbf{\Lambda}_{L_0}(\mathbf{u})} |g\mathbf{V}(\mathbf{x};\omega) - E| \geq \left\| \mathbf{H}^0_{\mathbf{\Lambda}_{L_0}(\mathbf{u})} + \mathbf{U}_{\mathbf{\Lambda}_{L_0}(\mathbf{u})} \right\| + \eta, \tag{4.13}$$

and (E,η)-bad, *otherwise.*

Lemma 4.3.1. (A) *If a cube* $\mathbf{\Lambda}_L^{(n)}(\mathbf{x})$ *is* (E,η)-good, *then it is also* (E,η)-NR.
(B) *If a cube* $\mathbf{\Lambda}_L^{(n)}(\mathbf{x})$ *is* (E,η)-NR, *with* $\eta > 4Nd$, *then*

$$\max_{\mathbf{y}\in\partial^-\mathbf{\Lambda}_L^{(n)}(\mathbf{x})} |G_{\mathbf{\Lambda}_L(\mathbf{u})}(\mathbf{x},\mathbf{y};E)| \leq \mathrm{e}^{-\tilde{m}L}, \quad \tilde{m} = \frac{1}{2}\ln\left(\frac{\eta}{4Nd}\right), \tag{4.14}$$

and, consequently, cube $\mathbf{\Lambda}_L^{(n)}(\mathbf{x})$ *is* (E,m)-NS *with* $m = 2^{-N}\tilde{m}$.

Proof. (A) This assertion follows directly from the min-max principle, since the eigenvalues of the multiplication operator $\mathbf{V}(\omega)$ are given by the values of the function $\mathbf{y} \mapsto \mathbf{V}(\mathbf{y};\omega)$ on $\mathbf{\Lambda}_L^{(n)}(\mathbf{x})$, and $\mathbf{H}_{\mathbf{\Lambda}_L^{(n)}(\mathbf{x})}(\omega)$ can be considered as perturbation of $\mathbf{V}(\omega)$ by $\mathbf{H}^0_{\mathbf{\Lambda}_{L_0}(\mathbf{u})} + \mathbf{U}_{\mathbf{\Lambda}_{L_0}(\mathbf{u})}$.

(B) Inequality (4.14) follows directly from the Combes–Thomas bound (2.90). For the cube $\mathbf{\Lambda}_L^{(n)}(\mathbf{x})$ to be (E,m)-NS, one needs that (cf. 3.15)

$$\tilde{m} \geq m\left(1 + L^{-1/8}\right)^{N-n}$$

Since $L \geq 1$ and $N - n \leq N$, the bound (4.14) implies indeed the (E,m)-NS property of the cube $\mathbf{\Lambda}_L^{(n)}(\mathbf{x})$ with $m = 2^{-N}\tilde{m}$, although the latter estimate is quite crude for large L. □

Now we formulate a result which is stronger than required for the fixed-energy multi-particle MSA. (Indeed, while it would be sufficient to prove Theorem 4.3.2 for a fixed value $E \in \mathbb{R}$, i.e., for an interval reduced to a single point, $I = [E,E]$, its proof, well-known in the single-particle theory, naturally extends to the case where $0 < |I| < \infty$.) In the framework of the variable-energy MPMSA (considered in Sect. 4.4), this result (or rather Corollary 4.2) provides the initial length scale estimate for Hamiltonians $\mathbf{H}(\omega)$ with *bounded* random potential.

Theorem 4.3.2. *Suppose that the random variables* $\{V(x;\omega), x \in \mathbb{Z}^d\}$ *generating the random potential are IID, and their common marginal distribution is continuous. Fix a positive integer* L_0 *and a bounded interval* $I \subset \mathbb{R}$ (possibly *reduced to a single point). For any* $m > 0$ *and* $p_0 > 0$, *there is a number* $g_0 > 0$ *such that if* $|g| \geq g_0$, *then for all* $n \in [[1,N]]$ *and any n-particle cube* $\mathbf{\Lambda}_{L_0}^{(n)}(\mathbf{u})$ *property* **(SS.**N, I, m, p_0, L_0*) is satisfied:*

$$\mathbb{P}\{\exists E \in I : \ \mathbf{\Lambda}_{L_0}(\mathbf{u}) \text{ is } (E,m)\text{-S}\} \leq L_0^{-2p_0(n)}. \tag{4.15}$$

Proof. Fix a bounded interval $I = [E_1 - \delta, E_1 + \delta]$, possibly with $\delta = 0$. Further, fix $N \geq 2$, a cube $\mathbf{\Lambda}^{(n)}_{L_0}(\mathbf{u})$, $m > 0$ and $p_0 > 0$. Set, as in (4.10), $p_0(n) = 4^{N-n} p_0$. Further, let

$$\eta := 4Nd\, e^{2^N m}, \quad s := \left\| \mathbf{H}^0_{\mathbf{\Lambda}^{(n)}_{L_0}(\mathbf{u})} + \mathbf{U}_{\mathbf{\Lambda}^{(n)}_{L_0}(\mathbf{u})} \right\| + \eta + \delta.$$

(Clearly, $\eta > 4Nd$.) Owing to Lemma 4.3.1, it suffices to prove that, for all $|g| \geq g_0(m)$ and $n \in [[1, N]]$,

$$\mathbb{P}\left\{ \mathbf{\Lambda}^{(n)}_{L_0}(\mathbf{u}) \text{ is } (E_1, \eta + \delta)\text{-bad} \right\} \leq L_0^{-2p_0(n)}. \tag{4.16}$$

Indeed, if $\mathbf{\Lambda}^{(n)}_{L_0}(\mathbf{u})$ is $(E_1, \eta + \delta)$-good, then, obviously, it is also (E, η)-good for all $E \in I$.

The event in the LHS of this bound holds provided that no value of the sum $g \sum_i V(x_i; \omega)$ for $\mathbf{x} \in \mathbf{\Lambda}_{L_0}(\mathbf{u})$ belongs to the interval $I_s := [E_1 - s, E_1 + s]$. In other words, the following event occurs:

$$\mathcal{R}_g = \left\{ \omega : \exists\, \mathbf{x} \in \mathbf{\Lambda}^{(n)}_{L_0}(\mathbf{u}) : \; g\mathbf{V}(\mathbf{x}; \omega) \in I_s \right\} \tag{4.17}$$

Its probability can be assessed as follows:

$$\mathbb{P}\{ \mathcal{R}_g \} \leq |\mathbf{\Lambda}^{(n)}_{L_0}(\mathbf{u})| \max_{\mathbf{x} \in \mathbf{\Lambda}^{(n)}_{L_0}(\mathbf{u})} \mathbb{P}\{ g\mathbf{V}(\mathbf{x}; \omega) \in I_s \}. \tag{4.18}$$

Observe that if the projection $\Pi \mathbf{x} = \{y_1, \dots, y_{n'}\}$, $1 \leq n' \leq n$, then

$$g\mathbf{V}(\mathbf{x}; \omega) = g \sum_{i=1}^{n} V(x_i; \omega) = g \sum_{i=1}^{n'} \mathcal{N}(y_i) V(y_i; \omega),$$

where $\mathcal{N}(y) = \text{card}\{ j : x_j = y \}$, with $\sum_i \mathcal{N}(y_i) = n$, and the random variables $\{V(y_i; \omega), 1 \leq i \leq n'\}$ are IID. The convolution of several probability measures, of which at least one is continuous, is continuous. Therefore, we have a control over the probability in the RHS of Eq. (4.18). More precisely, this probability tends to zero as $|g| \to \infty$, since

$$\{ \omega : g\mathbf{V}(\mathbf{x}; \omega) \in I_s \} = \{ \omega : \mathbf{V}(\mathbf{x}; \omega) \in I_{s,g} \}$$

where the rescaled interval

$$I_{s,g} := \left[\frac{E_1}{g} - \frac{s}{g}, \frac{E_1}{g} + \frac{s}{g} \right]$$

has length $2|g|^{-1}s \to 0$ as $|g| \to \infty$. As a result, for any fixed values $m > 0$ and $p_0 > 0$, the probability in (4.17) is bounded by $L_0^{-2p_0(n)}$, if $|g|$ is large enough. This completes the proof. □

The starting point of the variable-energy MPMSA presented in Sect. 4.4 is an analog of the estimate (4.15) for two distinct multi-particle cubes and all energies $E \in I \subseteq \mathbb{R}$, where I is an interval. For bounded intervals I, the required bound follows easily from Theorem 4.3.2.

Corollary 4.2. *Under the assumptions of Theorem 4.3.2, fix a positive integer L_0 and a bounded interval $I \subset \mathbb{R}$. For any $m > 0$ and $p_0 > 0$, there is a number $g_0 > 0$ such that if $|g| \ge g_0$, then for any $n \in [[1, N]]$ and any pair of n-particle cubes $\mathbf{\Lambda}_{L_0}(\mathbf{u})$ and $\mathbf{\Lambda}_{L_0}(\mathbf{v})$, one has*

$$\mathbb{P}\left\{\exists\, E \in I :\ \mathbf{\Lambda}_{L_0}^{(n)}(\mathbf{u}) \text{ and } \mathbf{\Lambda}_{L_0}^{(n)}(\mathbf{v}) \text{ are } (E,m)\text{-S}\right\} \le L_0^{-2p_0(n)}. \tag{4.19}$$

Proof. Regardless of the relative positions of the cubes $\mathbf{\Lambda}_{L_0}^{(n)}(\mathbf{u})$ and $\mathbf{\Lambda}_{L_0}^{(n)}(\mathbf{v})$, the event in the LHS of (4.19) is smaller than that in the LHS of (4.15). By Theorem 4.3.2, the probability of the latter is bounded by $L_0^{-2p_0(n)}$. □

If the random potential is a.s. bounded, i.e., for some $A < \infty$

$$\mathbb{P}\{\,|V(x;\omega)| \le A\,\} = 1,$$

then the spectrum of $\mathbf{H}(\omega)$, as well as spectra of finite-volume approximants $\mathbf{H}_{\mathbf{\Lambda}_L^{(n)}(\mathbf{u})}(\omega)$, are contained in some bounded interval $I \subset \mathbb{R}$, so it suffices to prove all MPMSA bounds (starting from the initial length scale bound) in such a bounded energy interval. This shows that Corollary 4.2 suffices for the purposes of the variable-energy MPMSA, in the particular case where the random potential is a.s. *bounded*.

Now we turn to a two-volume initial length scale bound in the situation where $I = \mathbb{R}$, which arises naturally when the random potential is a.s. *unbounded*.

Theorem 4.3.3. *Suppose that the random variables $\{V(x;\omega), x \in \mathbb{Z}^d\}$ generating the random potential are IID, and their common marginal distribution is continuous. Fix a positive integer L_0. For any $m > 0$ and $p_0 > 0$, there is a number $g_0 > 0$ such that if $|g| \ge g_0$, then for all $n \in [[1, N]]$ and any pair of separable n-particle cubes $\mathbf{\Lambda}_{L_0}^{(n)}(\mathbf{u})$, $\mathbf{\Lambda}_{L_0}^{(n)}(\mathbf{v})$ property $(\mathbf{DS}.N, n, m, p_0, L_0)$ is satisfied:*

$$\mathbb{P}\left\{\exists\, E \in \mathbb{R} :\ \mathbf{\Lambda}_{L_0}^{(n)}(\mathbf{u}) \text{ and } \mathbf{\Lambda}_{L_0}^{(n)}(\mathbf{v}) \text{ are } (E,m)\text{-S}\right\} \le L_0^{-2p_0(n)}. \tag{4.20}$$

Proof. Fix two separable cubes $\mathbf{\Lambda}_{L_0}^{(n)}(\mathbf{u})$ and $\mathbf{\Lambda}_{L_0}^{(n)}(\mathbf{v})$, $m > 0$, and set

$$\eta := 4Nd\, \mathrm{e}^{2^N m}$$

and

$$s = \eta + \max\left\{ \|\mathbf{H}^0_{\mathbf{\Lambda}^{(n)}_{L_0}(\mathbf{u})} + \mathbf{U}_{\mathbf{\Lambda}^{(n)}_{L_0}(\mathbf{u})}\|, \|\mathbf{H}^0_{\mathbf{\Lambda}^{(n)}_{L_0}(\mathbf{v})} + \mathbf{U}_{\mathbf{\Lambda}^{(n)}_{L_0}(\mathbf{v})}\| \right\}.$$

With our choice of η and s, by Lemma 4.3.1 applied to each of these cubes,

$$\begin{aligned}
&\mathbb{P}\left\{\exists\, E \in \mathbb{R} : \mathbf{\Lambda}^{(n)}_{L_0}(\mathbf{u}) \text{ and } \mathbf{\Lambda}^{(n)}_{L_0}(\mathbf{v}) \text{ are } (E,m)\text{-S}\right\} \\
&\le \mathbb{P}\left\{\exists\, E \in \mathbb{R} : \mathbf{\Lambda}^{(n)}_{L_0}(\mathbf{u}) \text{ and } \mathbf{\Lambda}^{(n)}_{L_0}(\mathbf{v}) \text{ are } (E,\eta)\text{-bad}\right\} \\
&\le \mathbb{P}\left\{ \min_{\mathbf{x}\in\mathbf{\Lambda}^{(n)}_{L_0}(\mathbf{u})} \min_{\mathbf{y}\in\mathbf{\Lambda}^{(n)}_{L_0}(\mathbf{v})} |g\mathbf{V}(\mathbf{x};\omega) - g\mathbf{V}(\mathbf{y};\omega)| \le 2s \right\} \\
&\le \sum_{\mathbf{x}\in\mathbf{\Lambda}^{(n)}_{L_0}(\mathbf{u})} \sum_{\mathbf{y}\in\mathbf{\Lambda}^{(n)}_{L_0}(\mathbf{v})} \mathbb{P}\{\,|g\mathbf{V}(\mathbf{x};\omega) - g\mathbf{V}(\mathbf{y};\omega)| \le 2s\,\} \\
&\le 3^{2nd} L_0^{2nd} \max_{\mathbf{x}\in\mathbf{\Lambda}^{(n)}_{L_0}(\mathbf{u}),\, \mathbf{y}\in\mathbf{\Lambda}^{(n)}_{L_0}(\mathbf{v})} \mathbb{P}\{\,|g\mathbf{V}(\mathbf{x};\omega) - g\mathbf{V}(\mathbf{y};\omega)| \le 2s\,\}.
\end{aligned} \tag{4.21}$$

By hypothesis, $\mathbf{\Lambda}^{(n)}_{L_0}(\mathbf{u})$ and $\mathbf{\Lambda}^{(n)}_{L_0}(\mathbf{v})$ are separable (cf. Definition 3.1.2), so we can assume without loss of generality that for some subset $\varnothing \neq \mathcal{J} \subseteq \{1,\dots,n\}$

$$\Pi_{\mathcal{J}}\mathbf{\Lambda}^{(n)}_{L+r_0}(\mathbf{u}) \cap \left(\Pi_{\mathcal{J}^c}\mathbf{\Lambda}^{(n)}_{L+r_0}(\mathbf{u}) \cup \Pi\mathbf{\Lambda}^{(n)}_{L+r_0}(\mathbf{v})\right) = \varnothing, \tag{4.22}$$

(here $\mathcal{J}^c = \{1,\dots,n\} \setminus \mathcal{J}$). To simplify the argument, we suppose that for some $j \in \{1,\dots,N\}$, setting $\mathcal{J}_{\neq j} = \{1,\dots,N\} \setminus \{j\}$,

$$\Lambda_{L+r_0}(u_j) \cap \left(\Pi_{\mathcal{J}_{\neq j}}\mathbf{\Lambda}^{(n)}_{L_0}(\mathbf{u}) \cup \Pi\mathbf{\Lambda}^{(n)}_{L}(\mathbf{v})\right) = \varnothing, \tag{4.23}$$

hence the random variables $\{V(x;\omega),\, x \in \Lambda_{L_0}(u_j)\}$ are independent from the sigma-algebra $\mathfrak{B} = \mathfrak{B}(\mathbf{u},\mathbf{v},L_0,j)$ generated by the random variables

$$\{V(x;\omega),\ x \in \Pi_{\mathcal{J}_{\neq j}}\mathbf{\Lambda}^{(n)}_{L_0}(\mathbf{u}) \cup \Pi\mathbf{\Lambda}^{(n)}_{L}(\mathbf{v})\}.$$

Return to the last bound in (4.21). For any fixed $\mathbf{x} \in \mathbf{\Lambda}^{(n)}_{L_0}(\mathbf{u})$, $\mathbf{y} \in \mathbf{\Lambda}^{(n)}_{L_0}(\mathbf{v})$,

$$\begin{aligned}
&\mathbb{P}\{\,|g\mathbf{V}(\mathbf{x};\omega) - g\mathbf{V}(\mathbf{y};\omega)| \le 2s\,\} \\
&= \mathbb{P}\left\{\left|V(x_j;\omega) - \left(\mathbf{V}(\mathbf{y};\omega) - \textstyle\sum_{i\neq j} V(x_i;\omega)\right)\right| \le 2g^{-1}s\right\} \\
&= \operatorname{ess\,sup} \mathbb{P}\left\{\left|V(x_j;\omega) - \left(\mathbf{V}(\mathbf{y};\omega) - \textstyle\sum_{i\neq j} V(x_i;\omega)\right)\right| \le 2g^{-1}s \,\middle|\, \mathfrak{B}\right\}
\end{aligned}$$

$$\le \sup_{\lambda\in\mathbb{R}} \operatorname{ess\,sup} \ \mathbb{P}\left\{ \left|V(x_j;\omega) - \lambda\right| \le 2g^{-1}s \mid \mathfrak{B} \right\}$$

$$= \sup_{\lambda\in\mathbb{R}} \mathbb{P}\left\{ \left|V(x_j;\omega) - \lambda\right| \le 2g^{-1}s \right\} \le \varsigma(4g^{-1}s) \underset{g\to+\infty}{\longrightarrow} 0.$$

Thus for $|g|$ large enough, the RHS of (4.21) is bounded by $L_0^{-2p_0(n)}$. □

We wish to stress that the initial scale bounds in Theorems 4.3.2 and 4.3.3 require very weak assumptions on the on-site random potential $V(\cdot\,;\omega)$ (merely continuity of its marginal distribution function F) and the interaction $\mathbf{U}^{(N)}$. The latter has only to be bounded, without any decay properties (let alone the finite range condition). We also mentioned earlier that the boundedness condition can be further relaxed, by allowing a hard core condition. (In that case the rest of the MPMSA scheme would need some technical modifications.) On the other hand, in our argument establishing the induction step of the MPMSA, the condition that the interaction $\mathbf{U}^{(N)}$ has a short range (see Assumption 3.1.1 at the end of Sect. 3.1.1) is essential.

4.3.2 Tunneling

As before, we call two n-particle cubes $\mathbf{\Lambda}_L^{(n)}(\mathbf{u})$ and $\mathbf{\Lambda}_L^{(n)}(\mathbf{v})$ distant if their centers obey $|\mathbf{u}-\mathbf{v}| > 11nL$.

Definition 4.3.3 (**(E,m)-tunneling**). *Fix numbers $E \in \mathbb{R}$ and $m \ge 1$. We say that an n-particle cube $\mathbf{\Lambda}_{L_{k+1}}^{(n)}(\mathbf{u})$ is (E,m)-tunneling ((E,m)-T) if there exist two distant* FI *cubes $\mathbf{\Lambda}_{L_k}^{(n)}(\mathbf{x})$, $\mathbf{\Lambda}_{L_k}^{(n)}(\mathbf{y}) \subset \mathbf{\Lambda}_{L_{k+1}}^{(n)}(\mathbf{u})$ which are (E,m)-S.*

Lemma 4.3.4. *Suppose that for some $E \in \mathbb{R}$, $n \in [1,N]$ and $k \ge 0$, the property (SS.N, E, m, p_k, L_k) holds true, viz.: $\forall\ n = 1,\dots,N$ and any n-particle cube $\mathbf{\Lambda}_{L_k}^{(n)}(\mathbf{x})$, the following bound holds true:*

$$\mathbb{P}\left\{ \mathbf{\Lambda}_{L_k}^{(n)}(\mathbf{x}) \text{ is } (E,m)-\mathrm{S} \right\} \le L_k^{-p_k(n)}. \tag{4.24}$$

Then for any $E \in \mathbb{R}$ and any n-particle cube $\mathbf{\Lambda}_{L_{k+1}}^{(n)}(\mathbf{u})$

$$\mathbb{P}\left\{ \mathbf{\Lambda}_{L_{k+1}}^{(n)}(\mathbf{u}) \text{ is } (E,m)\text{-}T \right\} \le \frac{3^{2nd}}{2} L_{k+1}^{-\frac{2p_k(n)}{\alpha}+2nd}. \tag{4.25}$$

Proof. By Corollary 4.1, for any pair of distant FI cubes $\mathbf{\Lambda}_{L_k}(\mathbf{x})$, $\mathbf{\Lambda}_{L_k}(\mathbf{y})$, their supports are disjoint:

$$\Pi\mathbf{\Lambda}_{L_k}(\mathbf{x}) \cap \Pi\mathbf{\Lambda}_{L_k}(\mathbf{y}) = \varnothing.$$

Therefore, the respective operators $\mathbf{H}_{\mathbf{\Lambda}_{L_k}(\mathbf{x})}$, $\mathbf{H}_{\mathbf{\Lambda}_{L_k}(\mathbf{y})}$ are independent, and so are the events

$$\{\mathbf{\Lambda}_{L_k}(\mathbf{x}) \text{ is } (E,m)-\text{S}\}, \quad \{\mathbf{\Lambda}_{L_k}(\mathbf{y}) \text{ is } (E,m)-\text{S}\}.$$

Now the claim follows from the hypothesis (4.24) and an elementary upper bound on the number of pairs $(\mathbf{x},\mathbf{y})$ in $\mathbf{\Lambda}_{L_{k+1}}(\mathbf{u})$. □

A convenient particularity of the fixed-energy MPMSA is that, as we shall see, one does not have to treat pairs of PI cubes. More precisely, the presence even of a single (E,m)-singular PI cube $\mathbf{\Lambda}^{(n)}_{L_k}(\mathbf{x}) \subset \mathbf{\Lambda}^{(n)}_{L_{k+1}}(\mathbf{x})$ can be ruled out with high probability, sufficient for the proof of the inductive bound $(\mathbf{SS}.N, E, m, p_k, L_k)$ for all $k \geq 0$. For this reason, the above definition of (E,m)-tunneling concerns only pairs of FI cubes.

4.3.3 Localization in Decoupled Systems at a Fixed Energy

The following statement gives a hint of how the PI property of a cube $\mathbf{\Lambda}^{(n)}_{L_{k+1}}(\mathbf{u})$ will be used in the course of the induction step $k \rightsquigarrow k+1$.

Lemma 4.3.5. *Fix an energy $E \in \mathbb{R}$. Consider an n-particle* PI *cube with canonical decomposition $\mathbf{\Lambda}^{(n)}_{L}(\mathbf{u}) = \mathbf{\Lambda}^{(n')}_{L}(\mathbf{u}') \times \mathbf{\Lambda}^{(n'')}_{L}(\mathbf{u}'')$. Assume that*

(a) *$\mathbf{\Lambda}^{(n)}_{L}(\mathbf{u})$ is E-*NR;
(b) *$\forall$ eigenvalue $E''_j \in \sigma(\mathbf{H}_{\mathbf{\Lambda}^{(n'')}_{L_k}(\mathbf{u}'')})$, $\mathbf{\Lambda}^{(n')}_{L_k}(\mathbf{u}')$ is $(E-E''_j, m)$-*NS;
(c) *Also, $\forall\, E'_i \in \sigma(\mathbf{H}_{\mathbf{\Lambda}^{(n')}_{L_k}(\mathbf{u}')})$, $\mathbf{\Lambda}^{(n'')}_{L_k}(\mathbf{u}'')$ is $(E-E'_i, m)$-*NS.

*Then $\mathbf{\Lambda}^{(n)}_{L_k}(\mathbf{u})$ is (E,m)-*NS.

Proof. Since $\mathbf{\Lambda}^{(n)}_{L}(\mathbf{u})$ is PI, $\mathbf{H}_{\mathbf{\Lambda}^{(n)}_{L}(\mathbf{u})}$ admits the decomposition

$$\mathbf{H}_{\mathbf{\Lambda}^{(n)}_{L}(\mathbf{u})} = \mathbf{H}_{\mathbf{\Lambda}^{(n')}_{L_k}(\mathbf{u}')} \otimes \mathbf{1}^{(n'')} + \mathbf{1}^{(n')} \otimes \mathbf{H}_{\mathbf{\Lambda}^{(n')}_{L_k}(\mathbf{u}'')}$$

thus its eigenvalues are the sums $E_{i,j} = E'_i + E''_j$, where $E'_i \in \sigma(\mathbf{H}_{\mathbf{\Lambda}_{L_k}(\mathbf{u}')})$ and, respectively, $E''_j \in \sigma(\mathbf{H}_{\mathbf{\Lambda}_{L_k}(\mathbf{u}'')})$. Eigenvectors of $\mathbf{H}_{\mathbf{u},k}$ can be chosen in the form $\mathbf{\Psi}_{i,j} = \boldsymbol{\phi}_i \otimes \boldsymbol{\psi}_j$ where $\{\boldsymbol{\phi}_a\}$ are eigenvectors of $\mathbf{H}_{\mathbf{\Lambda}^{(n')}_{L_k}(\mathbf{u}')}$ and $\{\boldsymbol{\psi}_b\}$ are eigenvectors of $\mathbf{H}_{\mathbf{\Lambda}^{(n'')}_{L_k}(\mathbf{u}'')}$. For each pair (E'_i, E''_j), the non-resonance assumption $|E - (E'_i + E''_j)| \geq \mathrm{e}^{-L_k^{\beta}}$ reads as $|(E-E'_i) - E''_j| \geq \mathrm{e}^{-L_k^{\beta}}$ and also as $|(E-E''_j) - E'_i| \geq \mathrm{e}^{-L_k^{\beta}}$. Therefore, we can write

$$\mathbf{G}(\mathbf{u},\mathbf{y};E) = \sum_{E'_i}\sum_{E''_j} \frac{\boldsymbol{\phi}_i(\mathbf{u}')\boldsymbol{\phi}_i(\mathbf{y}')\,\boldsymbol{\psi}_j(\mathbf{u}'')\boldsymbol{\psi}_j(\mathbf{y}'')}{(E'_i+E''_j)-E} \tag{4.26}$$

$$= \sum_{E'_i} \mathbf{P}'_a(\mathbf{u}',\mathbf{y}')\,\mathbf{G}_{\boldsymbol{\Lambda}_{L_k}(\mathbf{u}'')}(\mathbf{u}'',\mathbf{y}'';E-E'_i) \tag{4.27}$$

$$= \sum_{E''_j} \mathbf{P}''_b(\mathbf{u}'',\mathbf{y}'')\,\mathbf{G}_{\boldsymbol{\Lambda}_{L_k}(\mathbf{u}')}(\mathbf{u}',\mathbf{y}';E-E''_j), \tag{4.28}$$

where

$$\|\mathbf{G}_{\boldsymbol{\Lambda}_{L_k}(\mathbf{u}')}(E-E''_j)\| \le \mathrm{e}^{L_k^\beta}, \quad \|\mathbf{G}_{\boldsymbol{\Lambda}_{L_k}(\mathbf{u}'')}(E-E'_i)\| \le \mathrm{e}^{L_k^\beta}.$$

Next, by assumption, for every $E''_j \in \sigma(\mathbf{H}_{\boldsymbol{\Lambda}_{L_k}(\mathbf{u}'')})$, the projection cube $\boldsymbol{\Lambda}_{L_k}(\mathbf{u}')$ is $(E-E''_j,m)$-NS, and for every $E'_i \in \sigma(\mathbf{H}_{\boldsymbol{\Lambda}_{L_k}(\mathbf{u}')})$, the projection cube $\boldsymbol{\Lambda}_{L_k}(\mathbf{u}'')$ is $(E-E'_i,m)$-NS.

For any $\mathbf{y} \in \partial^-\boldsymbol{\Lambda}_{L_k}(\mathbf{u})$, either $|\mathbf{u}'-\mathbf{y}'| = L_k$, in which case we infer from (4.28), combined with $(E-E''_j,m)$-NS property of the cube $\boldsymbol{\Lambda}_{L_k}(\mathbf{u}')$, that

$$\left|\mathbf{G}(\mathbf{u},\mathbf{y};E)\right| \le |\boldsymbol{\Lambda}_{L_k}(\mathbf{u}'')|\,\mathrm{e}^{-\gamma(m,L_k,n-1)L_k+L_k^\beta} \tag{4.29}$$

or $|\mathbf{u}''-\mathbf{y}''| = L_k$, and then we have by (4.27)

$$\left|\mathbf{G}(\mathbf{u},\mathbf{y};E)\right| \le |\boldsymbol{\Lambda}_{L_k}(\mathbf{u}')|\,\mathrm{e}^{-\gamma(m,L_k,n-1)L_k+2L_k^\beta}.$$

In either case, the LHS is bounded by

$$\exp\left(-m(1+L_k^{-1/8})^{N-(n-1)+1}L_k + 2L_k^{1/4} + \text{Const}\ln L_k\right) < \tfrac{1}{2}\mathrm{e}^{-\gamma(m,L_k,n)L_k},$$

for L_0 large enough, since $m \ge 1$. □

Theorem 4.3.6. *Suppose that property (SS.n,E,m,p_k,L_k) holds true for some $n \in [2,N]$, any $k \in \mathbb{N}$ and all $E \in \mathbb{R}$. Fix any $k \in \mathbb{N}$. Then, for L_0 large enough and any* PI *cube $\boldsymbol{\Lambda}_{L_k}^{(n)}(\mathbf{x})$, one has*

$$\mathbb{P}\left\{\boldsymbol{\Lambda}_{L_k}^{(n)}(\mathbf{x}) \text{ is } (E,m)\text{-S}\right\} \le \frac{1}{4}L_k^{-2p_k(n)}. \tag{4.30}$$

Proof. Let $\mathcal{S}_k = \{\boldsymbol{\Lambda}_{L_k}^{(n)}(\mathbf{x}) \text{ is } (E,m)\text{-S}\}$. Consider the canonical decomposition $\boldsymbol{\Lambda}_{L_k}^{(n)}(\mathbf{x}) = \boldsymbol{\Lambda}_{L_k}^{(n')}(\mathbf{x}') \times \boldsymbol{\Lambda}_{L_k}^{(n'')}(\mathbf{x}'')$. We start by noting that

$$\mathbb{P}\{\mathcal{S}_k\} < \mathbb{P}\left\{\boldsymbol{\Lambda}_{L_k}^{(n)}(\mathbf{x}) \text{ is } E\text{-R}\right\} + \mathbb{P}\left\{\boldsymbol{\Lambda}_{L_k}^{(n)}(\mathbf{x}) \text{ is } E\text{-NR and } (E,m)\text{-S}\right\}.$$

The first term in the RHS is bounded with the help of Corollary 3.1 (cf. Eq. (3.52)). Specifically, we can pick $Q > 0$ large enough in (3.52) and bound the first term by $\frac{1}{8}L_k^{-2p_k(n)}$. So we focus on the second term.

Apply Lemma 4.3.5: since the option (a) is ruled out, it remains to assess the probability of events listed in options (b) and (c). Consider the former:

$$\begin{aligned}
&\mathbb{P}\left\{ \exists E''_j \in \sigma\left(\mathbf{H}_{\mathbf{\Lambda}^{(n'')}_{L_k}(\mathbf{x}'')}\right) : \mathbf{\Lambda}^{(n')}_{L_k}(\mathbf{x}') \text{ is } (E - E''_j, m)\text{-S} \right\} \\
&= \mathbb{E}\left[\mathbb{P}\left\{ \exists E''_j \in \sigma\left(\mathbf{H}_{\mathbf{\Lambda}^{(n'')}_{L_k}(\mathbf{x}'')}\right) : \mathbf{\Lambda}^{(n')}_{L_k}(\mathbf{x}') \text{ is } (E - E''_j, m)\text{-S} \,|\, \mathfrak{F}_{\mathbf{\Lambda}^{(n'')}_{L_k}(\mathbf{x}'')} \right\} \right] \\
&\le |\mathbf{\Lambda}^{(n'')}_{L_k}(\mathbf{x}'')| \sup_{\lambda'' \in \mathbb{R}} \mathbb{P}\left\{ \mathbf{\Lambda}^{(n')}_{L_k}(\mathbf{x}') \text{ is } (\lambda'', m)\text{-S} \right\} \\
&\le (2L_k + 1)^{nd} L_k^{-p_k(n-1)} \le 3^{nd} L_k^{-4p_k(n)+nd} \\
&\le \frac{3^{nd}}{L_k^{p_k(n)}} L_k^{-2p_k(n)} \le \frac{1}{16} L_k^{-2p_k(n)},
\end{aligned} \tag{4.31}$$

with $L_0 \ge 7$, since $p_k(n) \ge p_0 > 6nd$ (cf. (3.20)).

Similarly, one obtains

$$\mathbb{P}\left\{ \exists E'_i \in \sigma(\mathbf{H}_{\mathbf{\Lambda}^{(n')}_{L_k}(\mathbf{x}')}) : \mathbf{\Lambda}^{(n'')}_{L_k}(\mathbf{x}'') \text{ is } (E - E'_i, m)\text{-S} \right\} \le \frac{1}{16} L_k^{-2p_k(n)}. \tag{4.32}$$

Collecting (4.31), (4.32), and (3.52), the assertion follows. □

4.3.4 Scale Induction at a Fixed Energy

Theorem 4.3.7. *Suppose that*

$$p_0 > \frac{2\alpha N d}{2 - \alpha} \quad \text{and} \quad 0 < \theta < \min\left\{ \frac{2 - \alpha}{\alpha} - \frac{2Nd}{p_0}, \frac{1}{6} \right\}. \tag{4.33}$$

*Then the property (**SS.**n, E, m, p_k, L_k) implies (**SS.**$n, E, m, p_{k+1}, L_{k+1}$) for any $k \ge 0$, provided that L_0 is large enough.*

Proof. By Theorem 2.6.1, applicable to an arbitrary LSO regardless of the structure of its potential energy (hence, also to a multi-particle operator $\mathbf{H}_{\mathbf{\Lambda}^{(n)}_{L_{k+1}}(\mathbf{u})}$), the cube $\mathbf{\Lambda}^{(n)}_{L_{k+1}}(\mathbf{u})$ is (E, m)-NS, provided that it is E-NR and all singular cubes (if any) $\mathbf{\Lambda}^{(n)}_{L_k}(\mathbf{u}) \subset \mathbf{\Lambda}^{(n)}_{L_{k+1}}(\mathbf{u})$ can be covered by one cube of radius $O(L_k)$.

First of all, note that, by Theorem 3.4.1, for any $Q > 0$ and $\theta > 0$,

$$\mathbb{P}\left\{\mathbf{\Lambda}_{L_{k+1}}^{(n)}(\mathbf{u}) \text{ is } E\text{-R}\right\} \le \frac{1}{4} L_{k+1}^{-Q(1+\theta)^k}, \tag{4.34}$$

provided that L_0 is large enough. Taking $Q = 4^{N-n} p(1+\theta)$, we obtain

$$\mathbb{P}\left\{\mathbf{\Lambda}_{L_{k+1}}^{(n)}(\mathbf{u}) \text{ is } E\text{-R}\right\} \le \frac{1}{4} L_{k+1}^{-p_k(n)\cdot(1+\theta)} = \frac{1}{4} L_{k+1}^{-p_{k+1}(n)}. \tag{4.35}$$

Further, by virtue of Theorem 4.3.6, the probability to have at least one (E,m)-S PI cube $\mathbf{\Lambda}_{L_k}^{(n)}(\mathbf{v}) \subset \mathbf{\Lambda}_{L_{k+1}}^{(n)}(\mathbf{u})$ is bounded by

$$|\mathbf{\Lambda}_{L_{k+1}}^{(n)}(\mathbf{u})| \cdot \frac{1}{4} L_{k+1}^{-2p_k(n)} \le \frac{1}{2} L_{k+1}^{-\frac{2p_k(n)}{\alpha}+2nd} \tag{4.36}$$

(counting possible positions of the center $\mathbf{v}$ and applying (4.30) to each $\mathbf{v}$).

Next, by Lemma 4.3.4, the probability to find at least two distant FI (E,m)-S cubes of radius L_k inside $\mathbf{\Lambda}_{L_{k+1}}^{(n)}(\mathbf{u})$ is bounded by

$$\frac{3^{2nd}}{2} L_{k+1}^{-\frac{2p_k(n)}{\alpha}+2nd}. \tag{4.37}$$

Collecting (4.35)–(4.37), the claim follows by a simple calculation:

$$\begin{aligned}
&\mathbb{P}\left\{\mathbf{\Lambda}_{L_{k+1}}^{(n)}(\mathbf{u}) \text{ is } (E,m)\text{-S}\right\} \\
&\quad \le \frac{1}{4} L_{k+1}^{-p_{k+1}(n)} + \frac{1}{2} L_{k+1}^{-\frac{2p_k(n)}{\alpha}+2nd} + \frac{3^{2nd}}{2} L_{k+1}^{-\frac{2p_k(n)}{\alpha}+2nd} \\
&\quad \le \frac{1}{4} L_{k+1}^{-p_{k+1}(n)} + 3^{2nd} L_{k+1}^{-\frac{2}{\alpha}4^{N-n}p(1+\theta)^k+2nd} \\
&\quad \le \frac{1}{4} L_{k+1}^{-p_{k+1}(n)} + 3^{2nd} L_{k+1}^{-\frac{2}{\alpha(1+\theta)}p_{k+1}(n)+2nd} \le L_{k+1}^{-p_{k+1}(n)},
\end{aligned}$$

if L_0 is large enough and (4.33) is fulfilled. Indeed, the last inequality,

$$\frac{1}{4} L_{k+1}^{-p_{k+1}(n)} + 3^{2nd} L_{k+1}^{-\frac{2}{\alpha(1+\theta)}p_{k+1}(n)+2nd} \le L_{k+1}^{-p_{k+1}(n)}$$

can be shown as follows:

$$\begin{aligned}
&\frac{2p_{k+1}(n)}{\alpha(1+\theta)} > p_{k+1}(n) + 2nd \\
&\Leftrightarrow \frac{2 \cdot 4^{N-n} p_0(1+\theta)^{k+1}}{\alpha(1+\theta)} > 4^{N-n} p_0(1+\theta)^{k+1} + 2nd
\end{aligned}$$

$$\Leftrightarrow \frac{2 \cdot 4^{N-n}}{\alpha} > 4^{N-n}(1+\theta) + \frac{2nd}{p_0(1+\theta)^k}$$

$$\Leftrightarrow \frac{2-\alpha}{\alpha} - \frac{2nd}{4^{N-n}\, p_0(1+\theta)^k} > \theta$$

which derives from the hypothesis (4.33). □

4.3.5 Conclusion of the Fixed-Energy MPMSA

In this subsection, we make a stronger assumption on the value of the parameter p_0: given the Hölder exponent $s > 0$ from Assumption A (cf. (2.3)), we require that

$$p_0 \geq \frac{30nd}{s}, \tag{4.38}$$

so that for all $k \geq 0$,

$$\frac{p_k s}{10} \geq 3nd. \tag{4.39}$$

Taking into account the initial scale bound (**SS.**n, E, m, p_0, L_0), we come by induction in k to the main result of the fixed-energy MPMSA. Namely, Theorems 4.3.2 and 4.3.7 imply, by induction in $k = 0, 1, \ldots$

Theorem 4.3.8. *For any $m \geq 1$ there exist $L_0^* \in \mathbb{N}$ and $g^* \in (0, \infty)$ such that for $L_0 = L_0^*$ and $\forall\, g$ with $|g| \geq g^*$, property (**SS.**N, m, p_k, L_k) holds true $\forall\, k \geq 0$.*

4.3.6 From Fixed-Energy MPMSA to Dynamical Localization

Following the same strategy as in Sect. 2.5, we will obtain first a derivation of eigenfunction correlator bounds in a fixed bounded interval, which is sufficient for the proof of multi-particle strong dynamical localization in the case where the random potential is a.s. bounded, i.e., for some $K < \infty$

$$\mathbb{P}\{\, |V(0;\omega)| \leq K \,\} = 1,$$

so the a.s. spectrum $\sigma(\mathbf{H}^{(n)}(\omega))$ as well as the spectra of all finite-volume approximants, $\sigma(\mathbf{H}^{(n)}_{\mathbf{\Lambda}_L^{(n)}(\mathbf{x})}(\omega))$, are contained in the bounded interval

$$J = \Big[-4nd - nK - \|\mathbf{U}\|, 4nd + nK + \|\mathbf{U}\| \Big].$$

Then we adapt this derivation to unbounded potentials satisfying the Assumption A (cf. (2.142) in Chap. 2 and (4.50) below), using an analog of Corollary 2.5.5 (cf. Theorem 4.3.11 below).

From Fixed to Variable Energy

Recall that in Sect. 3.1.2 we have introduced the notion of separability of cubes $\mathbf{\Lambda}_L^{(n)}(\mathbf{u})$, $\mathbf{\Lambda}_L^{(n)}(\mathbf{v})$ (cf. Definition 3.1.2). Assertion (B) of Lemma 3.3.2 gives a sufficient condition for two n-particle cubes $\mathbf{\Lambda}_L^{(n)}(\mathbf{u})$, $\mathbf{\Lambda}_L^{(n)}(\mathbf{v})$ to be separable:

$$|\mathbf{u}| > |\mathbf{v}| + 2(L + r_0) \quad \text{or} \quad |\mathbf{v}| > |\mathbf{u}| + 2(L + r_0).$$

In particular (cf. assertion (C) of Lemma 3.3.2), cubes $\mathbf{\Lambda}_L^{(n)}(\mathbf{u})$, $\mathbf{\Lambda}_L^{(n)}(\mathbf{0})$ are separable, provided that

$$|\mathbf{u}| > 2(L + r_0). \tag{4.40}$$

We will see that for an adaptation of the proofs of dynamical localization, used in Chap. 2 in the single-particle context, to multi-particle Hamiltonians $\mathbf{H}(\omega)$, it suffices to work with pairs of cubes $\mathbf{\Lambda}_L^{(n)}(\mathbf{u})$, $\mathbf{\Lambda}_L^{(n)}(\mathbf{0})$ obeying the above geometrical condition (4.40). For this reason, we adapt the argument from Sect. 2.5 to such pairs of cubes.

Let us recapitulate the general set-up of Sect. 2.5, adapting the notation to multi-particle operators.

As in Sect. 2.5, we work with a bounded interval $I \subset \mathbb{R}$ which we assume to have a unit length, until the end of this section. Given an integer $L \geq 0$ and a cube $\mathbf{\Lambda}_L = \mathbf{\Lambda}_L^{(n)}(\mathbf{x})$, set

$$\mathrm{M}_{\mathbf{x}}(E) = \mathrm{M}_{\mathbf{x},L}(E) = \max_{\mathbf{v} \in \partial^- \mathbf{\Lambda}_L(\mathbf{x})} |G_{\mathbf{\Lambda}_L}(\mathbf{x}, \mathbf{v}; E)|, \tag{4.41}$$

and again introduce subsets of I parameterized by $a > 0$:

$$\mathscr{E}_{\mathbf{x}}(a) = \{E \in I : \mathrm{M}_{\mathbf{x}}(E) \geq a\}.$$

The next assertion, Theorem 4.3.9, is an adaptation of Theorem 2.5.1. Its proof is not related to the structure of the potential energy and applies both to single- and multi-particle LSOs. For this reason, the formal proof of Theorem 4.3.9 is omitted.

Theorem 4.3.9. *Take a positive integer L and a cube $\mathbf{\Lambda}_L = \mathbf{\Lambda}_L^{(n)}(\mathbf{x})$ and consider the Hamiltonian $\mathbf{H}_{\mathbf{\Lambda}_L}$. Let $\{E_j,\ 1 \leq j \leq |\mathbf{\Lambda}_L|\}$ be the eigenvalues of $\mathbf{H}_{\mathbf{\Lambda}_L}$. Fix a bounded interval $I \subset \mathbb{R}$ of length $|I|$. Let numbers a, b, c and $q_L > 0$ be such that*

$$b \le \min\{|\mathbf{\Lambda}_L(u)|^{-1}ac^2, c\}, \tag{4.42}$$

and for all $E \in I$

$$\mathbb{P}\{\mathrm{M}_{\mathbf{x}}(E) \ge a\} \le q_L. \tag{4.43}$$

Then there is an event $\mathcal{B} = \mathcal{B}_{\mathbf{x}} \subset \Omega$ with $\mathbb{P}\{\mathcal{B}_{\mathbf{x}}(b)\} \le |I|\, b^{-1} q_L$ such that for all $\omega \notin \mathcal{B}$, the set

$$\mathscr{E}_{\mathbf{x}}(2a) = \left\{E : \mathrm{M}_{\mathbf{x}}(E) \ge 2a\right\}$$

is contained in a union of intervals $\cup_{j=1}^{K} I_j$ where $I_j := \{E : |E - E_j| \le 2c\}$ and $K \le |\mathbf{\Lambda}_L(\mathbf{u})|$.

Remark 4.3. Similarly to Remark 2.5.1, it can be seen that condition (4.42) is satisfied when $a = a(L_k)$, $b = b(L_k)$, $c = c(L_k)$, and $q_L = q_{L_k}$ have the form

$$a(L_k) = L_k^{-\frac{p_k}{5}}, \ \ b(L_k) = L_k^{-\frac{4p_k}{5}}, \ \ c(L_k) = 3^{\frac{nd}{2}} L_k^{-\frac{p_k}{5}}, \ \ q_{L_k} = L_k^{-p_k}, \tag{4.44}$$

provided that $p_k \ge 5nd$. (Recall that in our scaling scheme $p_k = p_k(n) = 4^{N-n} p_0(1+\theta)^k > p_0 > 6nd > 5nd$.)

Below we use notations similar to those used in Sect. 2.5: given an n-particle cube $\mathbf{\Lambda}_L^{(n)}(\mathbf{u})$, we denote by $\mathrm{M}_{\mathbf{u}}$ the function defined by

$$\mathrm{M}_{\mathbf{u}} : E \mapsto \max_{\mathbf{v} \in \partial^- \mathbf{\Lambda}_L(\mathbf{u})} |\mathbf{G}_{\mathbf{\Lambda}_L(\mathbf{u})}(\mathbf{u}, \mathbf{v}; E)|, \quad E \in I. \tag{4.45}$$

Theorem 4.3.10. *Suppose that the following bound holds true for some integer $L > 0$, positive p, bounded interval $I \subset \mathbb{R}$, any $E \in I$ and any $\mathbf{u} \in (\mathbb{Z}^d)^n$:*

$$\mathbb{P}\{\mathrm{M}_{\mathbf{u}}(E) > a\} \le q_L. \tag{4.46}$$

Let the quantities a, b, c and q_L satisfy (4.42) and (4.43). Then for any pair of separable n-particle cubes $\mathbf{\Lambda}_L^{(n)}(\mathbf{x})$ and $\mathbf{\Lambda}_L^{(n)}(\mathbf{y})$,

$$\mathbb{P}\left\{\exists\, E \in I : \min\left(\mathrm{M}_{\mathbf{x}}(E), \mathrm{M}_{\mathbf{y}}(E)\right) > a\right\} \le C L^{(2n+1)d}\, (8c)^s + \frac{2\,|I|\, q_L}{b}, \tag{4.47}$$

where constant $C = C(n, d, s) \in (0, +\infty)$ and parameter $s \in (0, 1]$ is the Hölder exponent from Assumption A (see Eq. (2.3)).

Proof. The proof essentially repeats that of Theorem 2.5.2. Take the cubes $\mathbf{\Lambda}_L^{(n)}(\mathbf{x})$ and $\mathbf{\Lambda}_L^{(n)}(\mathbf{y})$ as in the statement of the theorem. Let the events $\mathcal{B}_{\mathbf{x}}(b)$, $\mathcal{B}_{\mathbf{y}}(b)$ be defined as in Theorem 4.3.9. Set $\mathcal{B} = \mathcal{B}_{\mathbf{x}} \cup \mathcal{B}_{\mathbf{y}}$, then we have

$$\mathbb{P}\{\mathcal{E}_{\mathbf{x}}(2a) \cap \mathcal{E}_{\mathbf{y}}(2a) \neq \varnothing\} \leq \mathbb{P}\{\mathcal{B}\} + \mathbb{P}\{\{\mathcal{E}_{\mathbf{x}}(2a) \cap \mathcal{E}_{\mathbf{y}}(2a) \neq \varnothing\} \cap \mathcal{B}^{c}\}$$
$$\leq 2b^{-1}q_L\,|I| + \mathbb{P}\{\{\mathcal{E}_{\mathbf{x}}(2a) \cap \mathcal{E}_{\mathbf{y}}(2a) \neq \varnothing\} \cap \mathcal{B}^{c}\}. \quad (4.48)$$

It remains to assess the last probability in the RHS. For $\omega \in \mathcal{B}^{c}$, each of the sets $\mathcal{E}_{\mathbf{x}}(a)$, $\mathcal{E}_{\mathbf{y}}(a)$ is covered by intervals of width $4c(L)$ around the respective eigenvalues $E_i \in \sigma(H_{\mathbf{\Lambda}_L(\mathbf{x})})$ and $E'_j \in \sigma(H_{\mathbf{\Lambda}_L(\mathbf{y})})$. By hypothesis, the cubes $\mathbf{\Lambda}_L(\mathbf{x})$ and $\mathbf{\Lambda}_L(\mathbf{y})$ are separable. Therefore, one can apply the EVC bound given by Theorem 3.4.2 (cf. Eq. (3.47)):

$$\mathbb{P}\{\{\mathcal{E}_{\mathbf{x}}(a) \cap \mathcal{E}_{\mathbf{y}}(a) \neq \varnothing\} \cap \mathcal{B}^{c}\} \leq \mathbb{P}\{\operatorname{dist}(\sigma(\mathbf{H}_{\mathbf{\Lambda}_L(\mathbf{x})}), \sigma(\mathbf{H}_{\mathbf{\Lambda}_L(\mathbf{y})})) \leq 4c\}$$
$$\leq n(2L+1)^{(2n+1)d}\varsigma(2\cdot 4c)$$

(here $\varsigma(\cdot)$ is the continuity modulus of the PDF F_V, cf. (2.3))

$$\leq C(n,d,s)L^{(2n+1)d}\cdot(8c)^{s}, \; s \in (0,1]. \quad (4.49)$$

Collecting (4.3.6) and (4.49), the assertion follows. □

An adaptation of this technique to a large class of unbounded IID random potentials $V(\cdot;\omega)$ can be carried out in the same way as Sect. 2.5, where we introduced the Assumption A1 (cf. (2.142)):
The common marginal distribution function

$$F(a) = \mathbb{P}\{V(x;\omega) \leq a\}, \;\; x \in \mathbb{Z}^d,$$

of the IID random variables $\{V(x;\cdot), x \in \mathbb{Z}^d\}$ satisfies a uniform Hölder continuity condition: $\exists$ constants $A > 0$ and $C \in (0,+\infty)$, such that $\forall\, t \geq 1$,

$$F(-t) + (1 - F(t)) \equiv \mathbb{P}\{|V(0;\omega)| \geq t\} \leq Ct^{-A}. \quad (4.50)$$

For the reader's convenience, we formulate below a direct analog of Theorem 2.5.6 for n-particle systems. We keep the same notations; in particular, we use the same length scale $\{L_k, k \in \mathbb{N}\}$ and the sequence of exponents $\{p_k = p(1+\theta)^k, k \in \mathbb{N}\}$ as before.

Theorem 4.3.11. *Let the random field V satisfy the Assumption A (cf. (2.3)) and Assumption A1 (cf. (2.142)). Suppose that the following bound holds true for some integer $k \geq 0$, some $q_{L_k} \in (0,1]$, any $u \in \mathbb{Z}^d$ and all $E \in \mathbb{R}$:*

$$\mathbb{P}\{\; \Lambda_{L_k}(u) is (E,m)\text{-S}\} \leq q_{L_k}.$$

Given $\mathbf{x}, \mathbf{y} \in (\mathbb{Z}^d)^n$ and $m > 0$, consider the following functions:

$$E \mapsto \mathbf{M}_{\mathbf{x},L_k}(E,\omega) = \max_{\mathbf{v}\in\partial^-\mathbf{\Lambda}_{L_k}(\mathbf{x})} |\mathbf{G}_{\mathbf{\Lambda}_{L_k}(\mathbf{x})}(\mathbf{x},\mathbf{v};E,\omega)|,$$

$$E \mapsto \mathbf{M}_{\mathbf{y},L_k}(E,\omega) = \max_{\mathbf{v}\in\partial^-\mathbf{\Lambda}_{L_k}(\mathbf{x})} |\mathbf{G}_{\mathbf{\Lambda}_{L_k}(\mathbf{y})}(\mathbf{y},\mathbf{v};E,\omega)|.$$

Assume that positive numbers $a(L_k) \geq \mathrm{e}^{-mL_k}$ and $b(L_k) \leq c(L_k)$ are given such that

$$b(L_k) \leq (2L_k+1)^{-2nd}\, a(L_k)\,(c(L_k))^2.$$

Then for any pair of separable cubes $\mathbf{\Lambda}^{(n)}_{L_k}(\mathbf{x})$, $\mathbf{\Lambda}^{(n)}_{L_k}(\mathbf{y})$ the following bound holds true:

$$\begin{aligned} &\mathbb{P}\left\{\exists\, E\in\mathbb{R} : \min\left(\mathbf{M}_{\mathbf{x},L_k}(E),\mathbf{M}_{\mathbf{y},L_k}(E)\right) > a(L_k)\right\} \\ &\leq 3^{2nd}\, C_s\, L_k^{2nd}\,(8c(L_k))^s + \frac{4\left(L_k^{-\frac{1}{10}p_k} + 4nd\,\mathrm{e}^{2m}\right) q_{L_k}}{b(L_k)} + L_k^{-\frac{4}{20}p_k} \end{aligned} \tag{4.51}$$

In particular, if

$$a(L_k) = L_k^{-\frac{p_k}{5}},\; b(L_k) = L_k^{-\frac{4p_k}{5}},\; c(L_k) = 3^{\frac{nd}{2}}\, L_k^{-\frac{p_k}{5}},\; q_{L_k} \leq L_k^{-p_k},$$

then the RHS in (4.51) *is bounded by*

$$\mathrm{Const}\, L_k^{-c(1+\theta)^k}, \quad \theta, c = c(A,s,n,d,p) > 0. \tag{4.52}$$

Here the exponent $s \in (0,1]$ and the constant C_s are those given in (2.3)*, and $A \in (0,1]$ is the exponent figuring in the Assumption A1 (cf.* (2.142)*).*

Eigenfunction Correlators in Finite Cubes

Like Theorem 4.3.9 above, the assertion (and the proof) of Theorem 4.3.12 is an adaptation of its single-particle counterpart, Theorem 2.5.3. We therefore omit the proof of this theorem. Notĕ, however, that the assertion of Theorem 4.3.12 is to be applied to pairs of *separable* cubes $\mathbf{\Lambda}^{(n)}_{L_k}(\mathbf{x})$ and $\mathbf{\Lambda}^{(n)}_{L_k}(\mathbf{y})$, for the proof of *its hypothesis* (cf. (4.53)) is achieved with the help of the multi-particle two-volume EVC bound, proven only for separable pairs of cubes. This suffices for the proof of the multi-particle strong dynamical localization.

Theorem 4.3.12. *Fix an integer $k \geq 0$ and assume that the following bound holds for a given pair of cubes $\mathbf{\Lambda}^{(n)}_{L_k}(\mathbf{x})$ and $\mathbf{\Lambda}^{(n)}_{L_k}(\mathbf{y})$ with $|\mathbf{x}-\mathbf{y}| > 2L_k+1$:*

$$\mathbb{P}\left\{\exists\, E\in\mathbb{R} : \min\left(\mathbf{M}_{\mathbf{x}}(E),\mathbf{M}_{\mathbf{y}}(E)\right) > a(L_k)\right\} \leq h(L_k). \tag{4.53}$$

Then for any finite connected set $\mathbf{\Lambda} \subset (\mathbb{Z}^d)^n$ *such that* $\mathbf{\Lambda} \supset \mathbf{\Lambda}^{(n)}_{L_k}(\mathbf{x}) \cup \mathbf{\Lambda}^{(n)}_{L_k}(\mathbf{y})$ *and any Borel function* $\phi \in \mathscr{B}_1(\mathbb{R})$

$$\mathbb{E}\Big[\Big|\langle \delta_{\mathbf{x}}, \phi(\mathbf{H}_{\mathbf{\Lambda}}(\omega))\, \delta_{\mathbf{y}}\rangle\Big|\Big] \le C\, L_k^{nd}\, a(L_k) + h(L_k). \tag{4.54}$$

Conclusion: Strong Decay of Eigenfunction Correlators

We have seen in Chap. 2 (cf. Sect. 2.5.3) that our methods and results can be easily extended to a class of unbounded random potentials, satisfying Assumption A1, i.e., a power-law decay bound on the tail probabilities of the IID random variables $V(x;\omega)$:

$$\exists\, A \in (0,1],\ C \in (0,\infty)\ \ \forall\, t \ge 1 \quad \mathbb{P}\{\, |V(x;\omega)| \ge t\,\} \le \mathrm{Const}\, t^{-A}.$$

Recall that the condition $A \le 1$ does not mean that a faster decay cannot be *allowed* (it is actually quite welcome), but merely that it *is not used*; this makes some arguments simpler.

For this reason, below we focus only on the case where the IID random variables $V(x;\omega)$ are a.s. bounded: $\mathbb{P}\{\,|V(x;\omega)| \le K\,\} = 1$, for some $K < \infty$. Consequently, there exists a bounded interval I containing the a.s. spectra of the random operators $\mathbf{H}^{(n)}(\omega)$, $1 \le n \le N$, as well as the a.s. spectra of all their finite-volume approximants $\mathbf{H}^{(n)}_{\mathbf{\Lambda}}(\omega)$, $\mathbf{\Lambda} \subset (\mathbb{Z}^d)^n$, $|\mathbf{\Lambda}| < \infty$. Naturally, it suffices to take

$$I = \big[\, -N(4d+K) - \|\mathbf{U}\|,\ N(4d+K) + \|\mathbf{U}\|\,\big].$$

This interval will be fixed below.

Recall that the main argument in Sect. 2.5.4 makes use of the vague convergence of the spectral measure $\mu^{\mathbf{\Lambda}_L}_{\mathbf{x},\mathbf{y}}$ of the operator $\mathbf{H}_{\mathbf{\Lambda}_L(\mathbf{0})}$ defined by integrals of bounded Borel functions ϕ,

$$\int_{\mathbb{R}} \mathrm{d}\mu^{\mathbf{\Lambda}_L}_{\mathbf{x},\mathbf{y}}(\lambda)\, \phi(\lambda) = \langle \delta_{\mathbf{x}}, \phi(\mathbf{H}_{\mathbf{\Lambda}_L(\mathbf{0})})\, \delta_{\mathbf{y}}\rangle \tag{4.55}$$

to the similar spectral measures for the operator $\mathbf{H}(\omega)$ on the entire lattice. This fact applies to arbitrary LSO, regardless of a particular structure of the potential energy (e.g., single- or multi-particle).

Therefore, applying the Fatou lemma on convergent measures, combined with upper bounds on EF correlators in arbitrarily large finite cubes, we obtain in the same way as in Sect. 2.5.4 our main result about strong decay of the multi-particle eigenfunction correlators:

Theorem 4.3.13. *There exists* $g^* \in (0,+\infty)$ *such that for all* $|g| \ge g^*$ *the eigenfunction correlators of the operator* $\mathbf{H}(\omega)$ *admit the following bound: there* *exist constants* $a, c \in (0, +\infty)$ *such that for all* $\mathbf{x}, \mathbf{y} \in (\mathbb{Z}^d)^n$ *and for some* $C(\mathbf{x}) \in (0, +\infty)$

$$\mathbb{E}\left[\sup_{t\in\mathbb{R}} \left|\langle \delta_{\mathbf{x}}, \mathrm{e}^{-it\mathbf{H}(\omega)}\, \delta_{\mathbf{y}}\rangle\right|\right] \le C(\mathbf{x})\, \mathrm{e}^{-a \ln^{1+c} |\mathbf{x}-\mathbf{y}|}. \tag{4.56}$$

Proof. **Step 1.** By Theorem 4.3.8, for all $k \ge 0$,

$$\mathbb{P}\{\, \mathbf{\Lambda}_{L_k}(\mathbf{u}) \text{ is } (E, m)\text{-S} \,\} \le L_k^{-p(1+\theta)^k}.$$

Step 2. As in Remark 4.3 (and Remark 2.5.1 before), set

$$a(L_k) = L_k^{-\frac{1}{5}p_k},\ b(L_k) = L_k^{-\frac{4}{5}p_k},\ c(L_k) = 3^{\frac{nd}{2}} L_k^{-\frac{1}{5}p_k},\ q_{L_k} = L_k^{-p_k}.$$

These quantities fulfill the condition

$$b(L_k) < \left|\mathbf{\Lambda}_{L_k}^{(n)}\right|^{-1} a(L_k) c(L_k)^2.$$

Then a direct application of Theorem 4.3.10 gives, for *separable* cubes $\mathbf{\Lambda}_{L_k}(\mathbf{x})$ and $\mathbf{\Lambda}_{L_k}(\mathbf{y})$

$$\begin{aligned}
\mathbb{P}&\left\{ \exists\, E \in I : \min\left(\mathrm{M}_{\mathbf{x}}, \mathrm{M}_{\mathbf{y}}\right) > L_k^{-\frac{p_k}{5}} \right\} \\
&\le C L_k^{(2n+1)d - \frac{1}{5}sp_k} + 2\,|I|\, L_k^{-p_k + \frac{4}{5}p_k} \\
&\le C_1 \left(L_k^{3nd - \frac{1}{5}sp_k} + L_k^{-\frac{1}{5}p_k} \right)
\end{aligned}$$

with $C_1 = C_1(n, d, s, |I|)$. Since $3nd \le \frac{sp_k}{10}$ (cf. (4.39)), we obtain

$$\begin{aligned}
\mathbb{P}\left\{ \exists\, E \in I : \min\left(\mathrm{M}_{\mathbf{x}}, \mathrm{M}_{\mathbf{y}}\right) > L_k^{-\frac{p_k}{5}} \right\} &\le C_1 \left(L_k^{\frac{sp_k}{10} - \frac{sp_k}{5}} + L_k^{-\frac{1}{5}p_k} \right) \\
&\le C_2 L_k^{-\tilde{p}_k},
\end{aligned}$$

with some $C_2 = C_2(n, d, s, |I|)$ and

$$\tilde{p}_k = \tilde{p}_k(s) = \tilde{p}(1+\theta)^k,\ \ \tilde{p} = \tilde{p}(s) := \frac{sp_0}{10},\ \theta > 0.$$

The separability condition is important here. Recall, however, that by virtue of assertion (B) of Lemma 3.3.2, for any $\mathbf{x}$ and $\mathbf{y}$ such that $|\mathbf{v}| > |\mathbf{u}| + 2(L + r_0)$, the cubes $\mathbf{\Lambda}_{L_k}(\mathbf{x})$ and $\mathbf{\Lambda}_{L_k}(\mathbf{y})$ are separable. As a result, given a point $\mathbf{x}$, we can obtain required bounds on the EF correlators only for $\mathbf{y}$

sufficiently distant from $\mathbf{x}$. (For the remaining, finite set of points $\mathbf{y}$ this can be absorbed in a factor of the form $C(\mathbf{x})$ (and not a uniform constant) in the RHS of Eq. (4.56).)

Step 3. By Theorem 4.3.12, for any bounded continuous function ϕ with $\|\phi\|_\infty \leq 1$, we have that, for $\mathbf{y}$ sufficiently distant from $\mathbf{x}$,

$$\begin{aligned}\mathbb{E}\left[\left|\langle \delta_\mathbf{x}, \phi(\mathbf{H}_{\mathbf{\Lambda}_{L_k}(0)})\,\delta_\mathbf{y}\rangle\right|\right] &\leq C_2 L_k^{-\tilde{p}_k} + O(L_k^{nd}) L_k^{-\frac{p_k}{5}} \\ &\leq C_3 L_k^{-\tilde{p}_k}, \quad C_3 < \infty.\end{aligned} \tag{4.57}$$

In the proof of dynamical localization, we have to work with the functions $t \mapsto \mathrm{e}^{-\mathrm{i}t\mathbf{H}}$. Since the a.s. spectrum of $\mathbf{H}(\omega)$ is contained in the bounded interval I, one can replace $\mathrm{e}^{-\mathrm{i}t\lambda}$ by $\phi_t : \lambda \mapsto \mathrm{e}^{-\mathrm{i}t\lambda}\chi_I(t)$, where χ_I with $\|\chi_I\|_\infty = 1$ is continuous, compactly supported and equals 1 on I.

Step 4. Recall that in Sect. 2.5.4 we stated Lemma 2.5.7, which applies, in fact, to arbitrary LSO on lattices $\mathbb{Z}^D$, for any $D \geq 1$, regardless of the structure of their potential energy, hence also to the multi-particle LSO $\mathbf{H}(\omega)$. As a result, we can argue as in Sect. 2.5.4: by virtue of the Fatou lemma, the same bound as in (4.57) holds true for the operator $\mathbf{H}(\omega)$:

$$\mathbb{E}\left[\left|\langle \delta_\mathbf{x}, \phi(\mathbf{H})\,\delta_\mathbf{y}\rangle\right|\right] \leq C_3 L_k^{-\tilde{p}(1+\theta)^k}.$$

Step 5. Assume first that $R := |\mathbf{x}-\mathbf{y}| > 3L_0$. Then there is $k_\circ \geq 0$ such that $R \in (3L_{k_\circ}, 3L_{k_\circ+1}]$. Let $c = \frac{\ln(1+\theta)}{\ln\alpha} > 0$, then

$$(1+\theta)^{k_\circ}\alpha^{k_\circ} = \exp\left(k_\circ \ln\alpha \cdot \frac{\ln\alpha + \ln(1+\theta)}{\ln\alpha}\right) = \left(\alpha^{k_\circ}\right)^{1+c}.$$

Since $L_{k_\circ} \geq \mathrm{e}^{C'\alpha^{k_\circ}\ln L_0}$ for some $C' > 0$ (cf. Sect. 2.5.5, Step 5), we obtain

$$\begin{aligned}-\ln L_{k_\circ}^{-\tilde{p}(1+\theta)^{k_\circ}} &\geq C'\tilde{p}\,(\alpha(1+\theta))^{k_\circ}\ln L_0 \\ &= \frac{C'\tilde{p}}{\ln^c L_0}\cdot\left(\alpha^{k_\circ}\ln L_0\right)^{1+c}\end{aligned}$$

On the other hand, $|\mathbf{x}-\mathbf{y}| \leq 3L_{k_\circ}^\alpha$, hence

$$\alpha^{k_\circ}\ln L_0 \geq \frac{1}{\alpha}\ln\left(\frac{|\mathbf{x}-\mathbf{y}|}{3}\right).$$

This yields that for some $a > 0$,

$$L_k^{-\tilde{p}(1+\theta)^k} \leq \mathrm{e}^{-a\ln^{1+c}|\mathbf{x}-\mathbf{y}|}.$$

For pairs $\mathbf{x}, \mathbf{y}$ with $|\mathbf{x}-\mathbf{y}| \leq 3L_0$, the required bound can be absorbed in a sufficiently large factor $C(\mathbf{x}) > 0$.

As was already noticed in Chap. 2 (cf. Sect. 2.5.5), our bounds depend upon ϕ only through $\|\phi\|_\infty$, so making the above notations more cumbersome, we could have been writing the estimates not for an individual function ϕ but for the supremum over $\phi \in \mathscr{B}_1$, thus proving the claim. □

4.3.7 *From Eigenfunction Correlators to Strong Dynamical Localization: Proof of Theorem 3.1.2*

As was explained in Chap. 2 (cf. Remark 2.5.5 in Sect. 2.5.5), the uniform bounds on eigenfunction correlators imply the traditional form of strong dynamical localization; the particularity of the multi-particle lattice Schrödinger operators resides in the form of the potential energy, and the presence of a non-trivial interaction makes much more difficult to assess the eigenfunction correlators (EFC). However, once sufficiently strong upper bounds on the EFC are obtained, they lead to the strong dynamical localization for the multi-particle system in question. Therefore, Theorem 4.3.13 implies multi-particle strong dynamical localization of all orders $s > 0$, as stated in Theorem 3.1.2.

4.4 Variable-Energy MPMSA

In this section, we describe a multi-particle analog of the variable-energy MSA presented in Chap. 2. Admittedly, the variable-energy MPMSA argument has a more sophisticated logical structure than its fixed-energy counterpart. But, since the multi-particle localization theory is still at its early stage, we believe that developing a variety of possible approaches to localization could be beneficial.

In order to reduce the analytic complexity of the variable-energy approach to the multi-particle systems, we will use in this section a stronger hypothesis on the random potential $V(\cdot;\omega)$—Assumption C introduced in Sect. 3.5 (cf. (3.58)).

Another reason for using Assumption C is that it leads to the localization estimates more natural from the physical point of view (cf. Remark 4.5.1). Recall that Corollary 3.3, following from Theorem 3.5.2 and Lemma 3.5.1, establishes a two-volume Wegner-type bound for any pair of sufficiently distant n-particle cubes. More precisely:

for any pair of cubes $\mathbf{\Lambda}^{(n)}_{L'}(\mathbf{u})$, $\mathbf{\Lambda}^{(n)}_{L''}(\mathbf{v})$ with $|\mathbf{u}-\mathbf{v}| > 3nL$ and $L', L'' \leq L$

$$\mathbb{P}\left\{ \operatorname{dist}\left(\sigma\left(\mathbf{H}^{(n)}_{\mathbf{\Lambda}^{(n)}_{L'}(\mathbf{u})}\right), \sigma\left(\mathbf{H}^{(n)}_{\mathbf{\Lambda}^{(n)}_{L''}(\mathbf{v})}\right)\right) \leq s \right\} \leq 3^{2Nd} L^{2Nd}\, \nu_L(2s). \tag{4.58}$$

In particular, for any $\beta \in (0,1)$ and some $\tilde{\beta} \in (0,\beta)$, $L_* < \infty$, for all $L \ge L_*$ and any pair of $3nL$-distant n-particle cubes,

$$\begin{gathered}\text{for any pair of cubes } \mathbf{\Lambda}_L^{(n)}(\mathbf{u}),\ \mathbf{\Lambda}_L^{(n)}(\mathbf{v}) \text{ with } |\mathbf{u}-\mathbf{v}| > 3nL \\ \mathbb{P}\left\{ \operatorname{dist}\left(\sigma(\mathbf{H}^{(n)}_{\mathbf{\Lambda}_L^{(n)}(\mathbf{u})}), \sigma(\mathbf{H}^{(n)}_{\mathbf{\Lambda}_L^{(n)}(\mathbf{v})})\right) \le \mathrm{e}^{-L^{\beta}} \right\} \le \mathrm{e}^{-L^{\tilde{\beta}}}. \end{gathered} \tag{4.59}$$

The strength of this bound, compared to the Wegner-bound used in the previous section, is expressed by the fact that it applies to all *distant* pairs, and not only to *separable* pairs of cubes. The price that one has to pay is a stronger assumption on the common marginal probability distribution of the IID random field $V(\cdot\,;\omega)$.

4.4.1 Modified (Stronger) Double Singularity Bound

In order to avoid any confusion with the property $(\mathbf{DS}.N,n,m,p_k,L_k)$ (cf. Eq. (4.60) in Sect. 3.1.2) related to the pairs of *separable* n-particle cubes, we introduce its counterpart $(\widetilde{\mathbf{DS}}.N,n,m,p_0,L_0)$ for the pairs of *distant* cubes. Since the class of distant cubes is much larger than that of separable ones, this requirement is much stronger than $(\mathbf{DS}.N,n,m,p_0,L_0)$.

$(\widetilde{\mathbf{DS}}.N,n,m,p_k,L_k)$ *For any pair of* **distant** *n-particle cubes $\mathbf{\Lambda}_{L_k}^{(n)}(\mathbf{u})$, $\mathbf{\Lambda}_{L_k}^{(n)}(\mathbf{v})$ the following bound holds true:*

$$\mathbb{P}\left\{ \exists\, E \in \mathbb{R}:\ \mathbf{\Lambda}_{L_k}^{(n)}(\mathbf{u}) \text{ and } \mathbf{\Lambda}_{L_k}^{(n)}(\mathbf{v}) \text{ are } (E,m)-\mathrm{S} \right\} \le L_k^{-2p_k(n)}. \tag{4.60}$$

4.4.2 Initial Scale Bounds: Variable-Energy Analysis

The starting point for the variable-energy multi-particle MSA procedure is the property $(\widetilde{\mathbf{DS}}.N,n,m,p_0,L_0)$, validity of which must be established for all $n \in [[1,N]]$, given values of N, L_0, m, p_0. Recall that this was done in Theorem 4.3.3 for arbitrary (not necessarily bounded) random potentials $V(x;\omega)$, satisfying Assumption A.

The proof is simpler for a.s. bounded random potentials: in that case, it follows easily from a fixed-energy initial length scale bound (cf. Corollary 4.2).

It is worth noticing that for the proof of the double singularity bound at the initial scale L_0 under the assumption of strong disorder, the distinction between the conditions $(\mathbf{DS}.N,n,m,p_0,L_0)$ and $(\widetilde{\mathbf{DS}}.N,n,m,p_0,L_0)$ is irrelevant, since both of them are derived from the single singularity bound. Indeed, the presence of two E-singular cubes with $E \in I \subset \mathbb{R}$ and centers $\mathbf{x}$ and $\mathbf{y}$, no matter distant or not, distinct or not, implies the presence of at least one E-singular cube with $E \in I \subset \mathbb{R}$. For large $|g|$ and a continuous probability distribution of the random potential $V(\cdot;\omega)$, the spread of the probability distribution of $gV(\cdot;\omega)$ is also large. This results in a small probability to have at least one eigenvalue E_i in a given bounded interval $I \subset \mathbb{R}$, regardless of the presence of the interaction.

4.4.3 The Inductive Scheme: Variable-Energy Version

In the course of making the inductive step (in both n and k) in the proof of the bound (4.60), we will distinguish between the following three types of pairs of n-particle cubes $\mathbf{\Lambda}_L(\mathbf{u})$ and $\mathbf{\Lambda}_L(\mathbf{v})$.

(I) Both $\mathbf{\Lambda}_L(\mathbf{u})$ and $\mathbf{\Lambda}_L(\mathbf{v})$ are PI (a PI pair).
(II) Both $\mathbf{\Lambda}_L(\mathbf{u})$ and $\mathbf{\Lambda}_L(\mathbf{v})$ are FI (an FI pair).
(III) One of the cubes, $\mathbf{\Lambda}_L(\mathbf{u})$ or $\mathbf{\Lambda}_L(\mathbf{v})$, is FI while the other is PI (a mixed pair).

In case (III) the analysis given in Sect. 4.4.7 requires a combination of techniques used in cases (I) and (II).

Note that, according to Definition 3.1.2, if n-particle cubes $\mathbf{\Lambda}_{L_k}(\mathbf{x})$, $\mathbf{\Lambda}_{L_k}(\mathbf{y})$ are separable, then $|\mathbf{x}-\mathbf{y}| \geq 11nL_k$. Consequently, for any separable pair of FI cubes $\mathbf{\Lambda}_{L_k}(\mathbf{x})$ and $\mathbf{\Lambda}_{L_k}(\mathbf{y})$ the respective operators $\mathbf{H}_{\mathbf{\Lambda}_{L_k}(\mathbf{x})}$ and $\mathbf{H}_{\mathbf{\Lambda}_{L_k}(\mathbf{y})}$ are independent (cf. Corollary 4.1 from Sect. 4.2). While one might think that analyzing FI pairs of cubes could be a difficult task, the fact is that, in the course of the MPMSA, they can be analyzed essentially in the same way as in the conventional, single-particle MSA. Speaking informally, this can be explained as follows: a clustered system of n particles can be considered as a composite quantum object moving in a disordered environment. Since two such objects at a sufficiently large distance are under impact of independent local subsamples of the random potential, one may expect their quantum transport properties to be similar to those of two single quantum particles. Corollary 4.1 will allow us to transform this rather vague argument into a rigorous mathematical assertion.

4.4.4 Localization in Decoupled Systems

In this section we follow the formula for the length scales $L_{k+1} = \lfloor L_k^{\alpha} \rfloor$. The content of the section is adapted to the induction step $k \rightsquigarrow k+1$ in the course of proving Eq. (4.60), i.e., deriving property $(\widetilde{\mathbf{DS}}.N, n, m, p_{k+1}, L_{k+1})$ from $(\widetilde{\mathbf{DS}}.N, \tilde{n}, m, p_k, L_k)$, $\tilde{n} \in [[1, n]]$, for pairs of PI cubes $\mathbf{\Lambda}^{(n)}_{L_{k+1}}(\mathbf{u})$ and $\mathbf{\Lambda}^{(n)}_{L_{k+1}}(\mathbf{v})$.

We would like to emphasize that we work now under the Assumption C and use the eigenvalue concentration bound (4.59) for all sufficiently distant cubes $\mathbf{\Lambda}^{(n)}_{L}(\mathbf{u})$,

$\mathbf{\Lambda}_L^{(n)}(\mathbf{v})$, viz.: $|\mathbf{u}-\mathbf{v}| > 3nL$. The latter condition will replace in this section the (stringer) separability condition used in the course of the fixed-energy analysis; it will appear in a number of formulae and arguments.

Recall that we introduced in Sect. 3.1.2 the notion of distant pairs of n-particle cubes of radius L. For the reader's convenience, and taking into account that this notion now becomes important, and supersedes the notion of separability, we repeat item (D) of Definition 3.17 as a separate

Definition 4.4.1. *A pair of n-particle cubes* $\mathbf{\Lambda}_L^{(n)}(\mathbf{u})$, $\mathbf{\Lambda}_L^{(n)}(\mathbf{v})$ *is called* distant *if* $|\mathbf{u}-\mathbf{v}| > 11nL$.

Observe that the above notion depends upon two integer parameters: the number of particles $1 \le n \le N$ and the radius of the cubes L. Taking into account a considerable number of definitions and abbreviations used in our book, we will sometimes write explicitly "$11nL$-distant cubes" or, where appropriate, "$3nL$-distant cubes".

As we have seen in Sect. 4.3, distant pairs of FI cubes $\mathbf{\Lambda}_L^{(n)}(\mathbf{u})$, $\mathbf{\Lambda}_L^{(n)}(\mathbf{v})$ are automatically completely separable (cf. Definition 3.17, item (C)), i.e., their full projections $\Pi\mathbf{\Lambda}_L^{(n)}(\mathbf{u})$, $\Pi\mathbf{\Lambda}_L^{(n)}(\mathbf{v})$ are disjoint, rendering independent the respective Hamiltonians $\mathbf{H}^{(n)}_{\mathbf{\Lambda}_L^{(n)}(\mathbf{u})}(\omega)$ and $\mathbf{H}^{(n)}_{\mathbf{\Lambda}_L^{(n)}(\mathbf{v})}(\omega)$. This explains why it is more convenient to keep the stronger of the two conditions, $|\mathbf{u}-\mathbf{v}| > 3nL$ and $|\mathbf{u}-\mathbf{v}| > 11nL$, in the Definition 4.4.1.

The notion of tunneling introduced in Sect. 4.3 (cf. Definition 4.3.3 in Sect. 4.3.2) has to be modified; in this section, Definition 4.3.3 will be replaced by the following

Definition 4.4.2 (**(m, I)-partial tunneling**). *Let* $I \subseteq \mathbb{R}$ *be an interval and* $m \ge 1$.

(i) *We say that an n-particle cube* $\mathbf{\Lambda}_{L_{k+1}}^{(n)}(\mathbf{u})$ *is* (m, I)*-tunneling* ((m, I)-T) *if there exists* $E \in I$ *and two distant* (E, m)-S *cubes* $\mathbf{\Lambda}_{L_k}^{(n)}(\mathbf{v}_j) \subset \mathbf{\Lambda}_{L_{k+1}}^{(n)}(\mathbf{u})$, $j = 1, 2$ *(i.e., with* $|\mathbf{u}-\mathbf{v}| > 3nL$*).*

(ii) *An n-particle* PI *cube* $\mathbf{\Lambda}_{L_{k+1}}^{(n)}(\mathbf{u})$, $n \ge 2$, *with the canonical decomposition*

$$\mathbf{\Lambda}_{L_{k+1}}^{(n)}(\mathbf{u}) = \mathbf{\Lambda}_{L_{k+1}}^{(n')}(\mathbf{u}') \times \mathbf{\Lambda}_{L_{k+1}}^{(n'')}(\mathbf{u}'') \tag{4.61}$$

is called (m, I)*-partially tunneling* ((m, I)-PT) *if at least one of the cubes* $\mathbf{\Lambda}_{L_{k+1}}^{(n')}(\mathbf{u}')$, $\mathbf{\Lambda}_{L_{k+1}}^{(n'')}(\mathbf{u}'')$ *is* (m, I)*-tunneling. Otherwise, the cube* $\mathbf{\Lambda}_{L_{k+1}}^{(n)}(\mathbf{u})$ *is called* (m, I)*-non partially tunelling* ((m, I)-NPT).

As a development of an observation made in Sect. 2.1, it is worth mentioning that the notion of tunneling does not refer to a specific value of energy E, but is associated with an entire (uncountable) energy interval I. Consequently, it may seem surprising that having a single (m, I)-tunneling cube is an event of a small probability (operating with countable intersections or unions of events related to individual values of E would not yield it). This can be explained by the fact that the definition involves presence of pairs of (E, m)-singular cubes for some $E \in I$. In other words, it is double singularity that occurs at a small probability and makes the definition working.

In this chapter, we focus mainly on the multi-particle localization for strongly disordered systems ($|g| \gg 1$), and in this case spectral (and dynamical) localization occurs for all energies. In other words, the random operators $\mathbf{H}^{(n)}(\omega)$, $1 \le n \le N$, have p.p. spectrum with probability one. For this reason, it will be convenient to consider $I = \mathbb{R}$.

The following statement gives a hint of how the PI property of a cube $\mathbf{\Lambda}^{(n)}_{L_{k+1}}(\mathbf{u})$ will be used in the course of the induction step $k \rightsquigarrow k+1$. It is quite similar to Lemma 4.3.5.

Lemma 4.4.1. *Fix integers $N \ge 2$ and $n \in \{2, \dots N\}$. For a given integer $k \ge 0$, consider an n-particle* PI *cube $\mathbf{\Lambda}^{(n)}_{L_{k+1}}(\mathbf{u})$ with a canonical decomposition $\mathbf{\Lambda}' \times \mathbf{\Lambda}''$ where $\mathbf{\Lambda}' = \mathbf{\Lambda}^{(n')}_{L_{k+1}}(\mathbf{u}')$, $\mathbf{\Lambda}'' = \mathbf{\Lambda}^{(n'')}_{L_{k+1}}(\mathbf{u}'')$, and*

$$\operatorname{dist}(\Pi\mathbf{\Lambda}', \Pi\mathbf{\Lambda}'') > r_0,$$

Assume that for some $m \ge 1$

(a) $\mathbf{\Lambda}^{(n)}_{L_{k+1}}(\mathbf{u})$ *is $(m, \mathbb{R})$-*NPT,
(b) $\mathbf{\Lambda}^{(n)}_{L_{k+1}}(\mathbf{u})$ *is E-*NR *for some $E \in \mathbb{R}$.*

*Then $\exists\ L_0^* = L_0^*(N, d) \in (0, \infty)$ such that if $L_0 \ge L_0^*$, the cube $\mathbf{\Lambda}^{(n)}_{L_{k+1}}(\mathbf{u})$ is (E, m)-*NS.

Proof. Consider the eigenvalues $E'_a = E'_a(\mathbf{H}')$, $E''_a = E'_b(\mathbf{H}'')$ and normalized eigenvectors $\boldsymbol{\psi}'_a = \boldsymbol{\psi}'_a(\mathbf{H}')$, $\boldsymbol{\psi}''_b = \boldsymbol{\psi}''_b(\mathbf{H}'')$ of the operators $\mathbf{H}' = \mathbf{H}^{(n')}_{\mathbf{\Lambda}'}(\omega)$ and $\mathbf{H}'' = \mathbf{H}^{(n'')}_{\mathbf{\Lambda}''}(\omega)$, respectively:

$$\begin{aligned} &\left(E'_a(\mathbf{H}'), \boldsymbol{\psi}'_a(\mathbf{H}')\right), \quad a = 1, \dots, |\mathbf{\Lambda}'|, \\ &\left(E''_b(\mathbf{H}''), \boldsymbol{\psi}''_b(\mathbf{H}'')\right), \quad b = 1, \dots, |\mathbf{\Lambda}''|. \end{aligned}$$

Write:

$$E - E'_a(\mathbf{H}') - E''_b(\mathbf{H}'') = \left(E - E'_a(\mathbf{H}')\right) - E''_b(\mathbf{H}'').$$

Further, by the hypothesis of the lemma, $\mathbf{\Lambda}^{(n)}_{L_{k+1}}(\mathbf{u})$ is E-NR. Therefore, for all $E_a(\mathbf{H}')$, the n''-particle cube $\mathbf{\Lambda}^{(n'')}_{L_{k+1}}(\mathbf{u}'')$ is $(E - E_a(\mathbf{H}'))$-NR. In addition, by the assumption of $(m, \mathbb{R})$-NPT, for any $E \in \mathbb{R}$ the cube $\mathbf{\Lambda}^{(n'')}_{L_{k+1}}(\mathbf{u}'')$ cannot contain two distant $(E - E_a(\mathbf{H}'), m)$-S cubes of size L_k.

As a result, if

- There exists a cube $\mathbf{\Lambda}_{L_k}^{(n'')}(\mathbf{w}) \subset \mathbf{\Lambda}_{L_{k+1}}^{(n'')}(\mathbf{u}'')$ which is $(E - E_a'(\mathbf{H}'), m)$-S, and in addition
- No distant pair of cubes of radius L_k lying in $\mathbf{\Lambda}_{L_{k+1}}^{(n'')}(\mathbf{u}'')$ is (E, m)-S,

then all $(E - E_a'(\mathbf{H}'), m)$-S cubes of radius L_k lying in $\mathbf{\Lambda}_{L_{k+1}}^{(n'')}(\mathbf{u}'')$ can be covered by a single cube of radius $A(n, L_k) \leq 3nL_k$ (concentric with the $(E - E_a'(\mathbf{H}'), m)$-S cube $\mathbf{\Lambda}_{L_k}^{(n'')}(\mathbf{w})$). Since $\mathbf{\Lambda}_{L_{k+1}}^{(n'')}(\mathbf{u}'')$ is $(E - E_a'(\mathbf{H}'))$-NR, we can apply Lemma 2.4.4 (based in turn on the radial descent estimate).

Lemma 2.4.4 implies that $\mathbf{\Lambda}_{L_{k+1}}^{(n'')}(\mathbf{u}'')$ is also $(E - E_a'(\mathbf{H}'), m)$-NS.

Similarly, we conclude that the cube $\mathbf{\Lambda}_{L_{k+1}}^{(n')}(\mathbf{u}')$ is $(E - E_b''(\mathbf{H}''), m)$-NS.
Now the assertion follows directly from Lemma 4.3.5. □

Lemma 4.4.2. *Given integers $N \geq 2$ and $n \in \{2, \dots, N\}$, suppose that for some $m \geq 1$ and $k \in \mathbb{N}$, for all $\tilde{n} \in \{1, \dots, n-1\}$ and all $\tilde{n}$-particle cubes $\mathbf{\Lambda}_{L_{k+1}}^{(\tilde{n})}(\mathbf{u})$, the following bound holds true:*

$$\mathbb{P}\left\{\mathbf{\Lambda}_{L_{k+1}}^{(\tilde{n})}(\mathbf{u}) \textit{ is } m\textit{-}T\right\} \leq L_{k+1}^{-2p_{k+1}(\tilde{n})}.$$

Then for any n-particle PI *cube* $\mathbf{\Lambda}_{L_{k+1}}^{(n)}(\mathbf{x})$,

$$\mathbb{P}\left\{\mathbf{\Lambda}_{L_{k+1}}^{(n)}(\mathbf{x}) \textit{ is } m\textit{-}PT\right\} \leq 2L_{k+1}^{-2p_{k+1}(n-1)} = 2L_{k+1}^{-8p_{k+1}(n)}. \tag{4.62}$$

Proof. By Lemma 4.2.2, $\mathbf{\Lambda}_{L_{k+1}}^{(n)}(\mathbf{x})$ admits a canonical decomposition

$$\mathbf{\Lambda}_{L_{k+1}}^{(n)}(\mathbf{x}) = \mathbf{\Lambda}_{L_{k+1}}^{(n')}(\mathbf{x}') \times \mathbf{\Lambda}_{L_{k+1}}^{(n'')}(\mathbf{x}''), \tag{4.63}$$

with some $n', n'' \leq n - 1$. By Definition 4.4.2, $\mathbf{\Lambda}_{L_{k+1}}^{(n)}(\mathbf{x})$ is m-PT iff one of the cubes $\mathbf{\Lambda}_{L_{k+1}}^{(n')}(\mathbf{x}')$, $\mathbf{\Lambda}_{L_{k+1}}^{(n'')}(\mathbf{x}'')$ is m-T. Hence,

$$\begin{aligned}
&\mathbb{P}\left\{\mathbf{\Lambda}_{L_{k+1}}^{(n)}(\mathbf{x}) \text{ is } m\text{-PT}\right\} \\
&\quad \leq \mathbb{P}\left\{\mathbf{\Lambda}_{L_{k+1}}^{(n')}(\mathbf{x}') \text{ is } m\text{-T}\right\} + \mathbb{P}\left\{\mathbf{\Lambda}_{L_{k+1}}^{(n'')}(\mathbf{x}'') \text{ is } m\text{-T}\right\} \\
&\quad \leq 2L_{k+1}^{-2p_{k+1}(n-1)} = 2L_{k+1}^{-8p_{k+1}(n)}.
\end{aligned}$$

□

4.4.5 Analysis of Partially Interactive Pairs

The arguments developed in the previous subsection will be now used for analyzing inductive properties of separable PI pairs of multi-particle cubes. The main outcome of this subsection is Lemma 4.4.5 guaranteeing the transition $n-1 \rightsquigarrow n$ for PI cubes of radius L_{k+1}. Since the validity of property $(\widetilde{\mathbf{DS}}.N, 1, m, p_k, L_k)$ for all $k \in \mathbb{N}$ is guaranteed by the single-particle MSA, Lemma 4.4.3 allows us to make the passage $k \rightsquigarrow k+1$ for separable PI pairs of n-particle cubes for any $n = 2, \dots, N$.

At this point we need to extend the formal definition of a resonance to the multi-particle setting; here we give it in the context of the scale induction, addressing the resolvent $\mathbf{G}_{\mathbf{\Lambda}_{L_{k+1}}(\mathbf{u})}(E,\omega) = \left(\mathbf{H}_{\mathbf{\Lambda}_{L_{k+1}}(\mathbf{u})}(\omega) - E\right)^{-1}$; cf. Definition 2.2.3. Again we will operate with $\beta = 1/2$.

Lemma 4.4.3. *Suppose that for all $\tilde{n} \in \{2, \dots, n-1\}$ and any $\tilde{n}$-particle cube $\mathbf{\Lambda}^{(\tilde{n})}_{L_{k+1}}(\mathbf{u})$ the following bound holds true:*

$$\mathbb{P}\left\{ \mathbf{\Lambda}^{(\tilde{n})}_{L_{k+1}}(\mathbf{u}) \text{is } (m, \mathbb{R})\text{-}PT \right\} \le L_{k+1}^{-2p_{k+1}(\tilde{n})}.$$

Then for any pair of distant n-particle PI cubes, $\mathbf{\Lambda}^{(n)}_{L_{k+1}}(\mathbf{x})$ and $\mathbf{\Lambda}^{(n)}_{L_{k+1}}(\mathbf{y})$,

$$\begin{aligned} &\mathbb{P}\left\{ \exists\, E \in \mathbb{R} : \mathbf{\Lambda}^{(n)}_{L_{k+1}}(\mathbf{x}) \text{ and } \mathbf{\Lambda}^{(n)}_{L_{k+1}}(\mathbf{y}) \text{ are } (E,m)\text{-S} \right\} \\ &\quad \le 5L_{k+1}^{-2p_{k+1}(n-1)} < L_{k+1}^{-2p_{k+1}(n)}. \end{aligned} \tag{4.64}$$

Proof. Given $E \in \mathbb{R}$, consider the following events:

$$\begin{aligned} \mathcal{S} &= \{\exists\, E \in \mathbb{R} :\ \mathbf{\Lambda}^{(n)}_{L_{k+1}}(\mathbf{x}) \text{ and } \mathbf{\Lambda}^{(n)}_{L_{k+1}}(\mathbf{y}) \text{ are } (E,m)\text{-S}\} \\ \mathcal{P}(\mathbf{x}) &= \{\mathbf{\Lambda}^{(n)}_{L_{k+1}}(\mathbf{x}) \text{ is } (m, \mathbb{R})\text{-PT}\} \\ \mathcal{P}(\mathbf{y}) &= \{\mathbf{\Lambda}^{(n)}_{L_{k+1}}(\mathbf{y}) \text{ is } (m, \mathbb{R})\text{-PT}\} \\ \mathcal{R} &= \{\exists\, E \in \mathbb{R} :\ \mathbf{\Lambda}^{(n)}_{L_{k+1}}(\mathbf{x}) \text{ and } \mathbf{\Lambda}^{(n)}_{L_{k+1}}(\mathbf{y}) \text{ are } E\text{-R}\} \end{aligned}$$

By Lemma 4.4.1, if the cube $\mathbf{\Lambda}^{(n)}_{L_{k+1}}(\mathbf{x})$ is $(m, \mathbb{R})$-NPT and E-NR, with some $E \in \mathbb{R}$, then, for L_0 sufficiently large, $\mathbf{\Lambda}^{(n)}_{L_{k+1}}(\mathbf{x})$ is (E,m)-NS; the same is true, of course, for $\mathbf{\Lambda}^{(n)}_{L_{k+1}}(\mathbf{y})$. Therefore, $\mathcal{S} \subset \mathcal{P}(\mathbf{x}) \cup \mathcal{P}(\mathbf{y}) \cup \mathcal{R}$. Further, by virtue of Lemma 4.4.2, we have that

$$\max\left[\mathbb{P}\{\mathcal{P}(\mathbf{x})\}, \mathbb{P}\{\mathcal{P}(\mathbf{y})\}\right] \le 2L_{k+1}^{-2p_{k+1}(n-1)}. \tag{4.65}$$

Next, by the EVC bound (3.52) (cf. Corollary 3.1) with $Q = 2 \cdot 4^{N-n+1} p_0$ and L_0 large enough, we also have

$$\mathbb{P}\{\mathcal{R}\} < L_{k+1}^{-Q(1+\theta)^{k+1}} = L_{k+1}^{-2p_{k+1}(n-1)}.$$

By summing up, we finally obtain

$$\begin{aligned}\mathbb{P}\{\,\mathcal{S}\,\} &\le \mathbb{P}\{\,\mathcal{P}(\mathbf{x})\,\} + \mathbb{P}\{\,\mathcal{P}(\mathbf{y})\,\} + \mathbb{P}\{\,\mathcal{R}\,\}\\ &\le 2L_{k+1}^{-2p_{k+1}(n-1)} + 2L_{k+1}^{-2p_{k+1}(n-1)} + L_{k+1}^{-2p_{k+1}(n-1)}\\ &\le 5L_{k+1}^{-2p_{k+1}(n-1)} = 5L_{k+1}^{-8p_{k+1}(n)} = L_{k+1}^{-2p_{k+1}(n)} \cdot 5L_{k+1}^{-6p_{k+1}(n)}\\ &< L_{k+1}^{-2p_{k+1}(n)},\end{aligned}$$

since $L_{k+1} \ge 2$, $p_{k+1} \ge p_0 > 1$, so $L_{k+1}^{6p_{k+1}(n)} > 2^6 > 5$. □

Lemma 4.4.4. *Suppose that for some* $2 \le n \le N$ *and* $k \ge 1$, *the property* $(\widetilde{\mathbf{DS}}.N, \tilde{n}, m, p_k, L_k)$ *is fulfilled for all* $\tilde{n} \in \{1, \dots, n-1\}$. *Then,* $\forall$ *n-particle* PI *cube* $\mathbf{\Lambda}_{L_{k+1}}^{(n)}(\mathbf{x})$,.

$$\begin{aligned}\mathbb{P}\left\{\ \mathbf{\Lambda}_{L_{k+1}}^{(n)}(\mathbf{x})\ is\ (m,\mathbb{R})\text{-}PT\right\} &< 3^{2nd} L_{k+1}^{-\frac{30}{7}p_{k+1}(n)}\\ &< 3^{2nd} L_{k+1}^{-4p_{k+1}(n)}.\end{aligned} \tag{4.66}$$

Proof. As before, the PI cube $\mathbf{\Lambda}_{L_{k+1}}^{(n)}(\mathbf{x})$ with the canonical decomposition

$$\mathbf{\Lambda}_{L_{k+1}}^{(n)}(\mathbf{x}) = \mathbf{\Lambda}_{L_{k+1}}^{(n')}(\mathbf{x}') \times \mathbf{\Lambda}_{L_{k+1}}^{(n'')}(\mathbf{x}'')$$

is $(m, \mathbb{R})$-PT if at least one of the following events occurs:

$$\mathcal{T}(\mathbf{x}') := \{\mathbf{\Lambda}_{L_{k+1}}^{(n')}(\mathbf{x}') \text{ is } (m,\mathbb{R})\text{-T}\}$$

or

$$\mathcal{T}(\mathbf{x}'') := \{\mathbf{\Lambda}_{L_{k+1}}^{(n'')}(\mathbf{x}'') \text{ is } (m,\mathbb{R})\text{-T}\}.$$

Therefore,

$$\mathbb{P}\left\{\ \mathbf{\Lambda}_{L_{k+1}}^{(n)}(\mathbf{x}) \text{ is } (m,\mathbb{R})\text{-PT}\right\} \le 2\max\left[\,\mathbb{P}\{\,\mathcal{T}(\mathbf{x}')\,\}, \mathbb{P}\{\,\mathcal{T}(\mathbf{x}'')\,\}\,\right]. \tag{4.67}$$

We will now focus on assessing $\mathbb{P}\{\,\mathcal{T}(\mathbf{x}')\,\}$; the probability $\mathbb{P}\{\,\mathcal{T}(\mathbf{x}'')\,\}$ is bounded in the same way.

The cube $\mathbf{\Lambda}_{L_{k+1}}^{(n')}(\mathbf{x}')$ is $(m, \mathbb{R})$-T iff for some $E \in \mathbb{R}$ it contains two distant (E, m)-S cubes $\mathbf{\Lambda}_{L_k}^{(n')}(\mathbf{u})$, $\mathbf{\Lambda}_{L_k}^{(n')}(\mathbf{v})$. The number of such pairs is bounded by

$$\frac{1}{2}|\mathbf{\Lambda}_{L_{k+1}}^{(n')}(\mathbf{x}')|^2 \le \frac{1}{2}(2L_{k+1}+1)^{2dn'} \le \frac{3^{2dn'}}{2}L_{k+1}^{2dn'}.$$

Next, for fixed centers $\mathbf{u}$, $\mathbf{v}$ property $(\widetilde{\mathbf{DS}}.N, n', m, p_k, L_k)$ states that

$$\mathbb{P}\left\{\exists\, E \in \mathbb{R}:\ \mathbf{\Lambda}_{L_k}^{(n')}(\mathbf{u}) \text{ and } \mathbf{\Lambda}_{L_k}^{(n')}(\mathbf{v}) \text{ are } (E,m)\text{-S}\right\}$$
$$\le L_k^{-2p_k(n')} \le L_k^{-2p_k(n-1)} \le L_{k+1}^{-\frac{2p_{k+1}(n)\cdot 4}{\alpha(1+\theta)}}$$
$$= L_{k+1}^{-\frac{2p_{k+1}(n)\cdot 8}{3(1+\theta)}}.$$

Therefore,

$$\mathbb{P}\{\mathcal{T}'\} \le \frac{3^{2dn'}}{2} L_{k+1}^{-\frac{16p_{k+1}(n)}{3(1+\theta)}+2dn'}.$$

Recall that

$$p_{k+1}(n) = 4^{N-n} p_0 (1+\theta)^{k+1} > p_0 > 6Nd \text{ and } 0 < \theta < 1/6, \tag{4.68}$$

implying that

$$\begin{aligned}
\frac{16p_{k+1}(n)}{3(1+\theta)} - 2n'd &\ge \frac{2p_{k+1}(n)}{1+\theta}\left(\frac{8}{3} - \frac{Nd(1+\theta)}{p_{k+1}(n)}\right) \\
&= \frac{2p_{k+1}(n)}{1+\theta}\left(\frac{8}{3} - \frac{Nd(1+\theta)}{p_k(n)(1+\theta)}\right) \\
&> \frac{2p_{k+1}(n)}{1+\theta}\left(\frac{8}{3} - \frac{1}{6}\right) = \frac{2p_{k+1}(n)}{1+\theta}\cdot\frac{5}{2} \\
&> \frac{30}{7}\, p_{k+1}(n) > 4p_{k+1}(n).
\end{aligned}$$

Consequently, with $n' \le N-1 < N$,

$$\mathbb{P}\{\mathcal{T}'\} < \frac{3^{2dN}}{2} L_{k+1}^{-\frac{30}{7}p_{k+1}(n)} < \frac{3^{2dN}}{2} L_{k+1}^{-4p_{k+1}(n)}.$$

Similarly,

$$\mathbb{P}\{\mathcal{T}''\} < \frac{3^{2dn'}}{2} L_{k+1}^{-\frac{30}{7}p_{k+1}(n)} < \frac{3^{2dN}}{2} L_{k+1}^{-4p_{k+1}(n)}.$$

Taking into account Eq. (4.67), the assertion of the lemma follows. □

Lemma 4.4.5. *Suppose that the property* $(\widetilde{\mathbf{DS}}.N, \tilde{n}, m, p_k, L_k)$ *is fulfilled for all* $\tilde{n} \in [[1, n-1]]$, *and that the EVC bounds* (3.52) *holds true with* $Q = 4^N p_0$ *for all* $k \ge 0$. *Let* $\mathbf{\Lambda}_{L_{k+1}}^{(n)}(\mathbf{x})$ *and* $\mathbf{\Lambda}_{L_{k+1}}^{(n)}(\mathbf{y})$ *be distant* PI *cubes. Then*

$$\mathbb{P}\left\{\exists\, E \in \mathbb{R}:\ \mathbf{\Lambda}_{L_{k+1}}^{(n)}(\mathbf{x}) \text{ and } \mathbf{\Lambda}_{L_{k+1}}^{(n)}(\mathbf{y}) \text{ are } (E,m)\text{-S}\right\} \le C\, L_{k+1}^{-\frac{30}{7}p_{k+1}(n)} \tag{4.69}$$

with some $C = C(d, N)$.

Proof. As in the proof of Lemma 4.4.3, we work with the events

$$\begin{aligned}
\mathcal{S} &= \{\exists\, E \in \mathbb{R} :\ \mathbf{\Lambda}^{(n)}_{L_{k+1}}(\mathbf{x}) \text{ and } \mathbf{\Lambda}^{(n)}_{L_{k+1}}(\mathbf{y}) \text{ are } (E,m)\text{-S}\},\\
\mathcal{P}(\mathbf{x}) &= \{\mathbf{\Lambda}^{(n)}_{L_{k+1}}(\mathbf{x}) \text{ is } (m,\mathbb{R})\text{-PT}\},\\
\mathcal{P}(\mathbf{y}) &= \{\mathbf{\Lambda}^{(n)}_{L_{k+1}}(\mathbf{y}) \text{ is } (m,\mathbb{R})\text{-PT}\},\\
\mathcal{R} &= \{\exists\, E \in \mathbb{R} :\ \mathbf{\Lambda}^{(n)}_{L_{k+1}}(\mathbf{x}) \text{ and } \mathbf{\Lambda}^{(n)}_{L_{k+1}}(\mathbf{y}) \text{ are } E\text{-R}\}.
\end{aligned}$$

By Lemma 4.4.1, if $\mathbf{\Lambda}^{(n)}_{L_{k+1}}(\mathbf{x})$ is $(m,\mathbb{R})$-NPT and E-NR, with some $E \in \mathbb{R}$, then it is (E,m)-NS; the same is true, of course, for $\mathbf{\Lambda}^{(n)}_{L_{k+1}}(\mathbf{y})$. Therefore, $\mathcal{S} \subset \mathcal{P}(\mathbf{x}) \cup \mathcal{P}(\mathbf{y}) \cup \mathcal{R}$. Further, by our assumption $(\widetilde{\mathbf{DS}}.N, \tilde{n}, m, p_k, L_k)$, Lemma 4.4.4 applies:

$$\mathbb{P}\{\mathcal{P}(\mathbf{x})\} \le C'\, L_{k+1}^{-\frac{30}{7}p_{k+1}(n)}, \quad \mathbb{P}\{\mathcal{P}(\mathbf{y})\} \le C'\, L_{k+1}^{-\frac{30}{7}p_{k+1}(n)},$$

with $C' = C'(d,N)$. Next, by the EVC bound (3.52) with $Q = \frac{30}{7}\cdot 4^N p_0$, $p_0 > 6Nd$ and $L_0 \ge 2$, one also has

$$\mathbb{P}\{\mathcal{R}\} \le L_{k+1}^{-Q(1+\theta)^{k+1}} \le L_{k+1}^{-\frac{30}{7}p_{k+1}(n)}.$$

Finally, we obtain, for some $C = C(d,N) < \infty$,

$$\mathbb{P}\{\mathcal{S}\} \le \mathbb{P}\{\mathcal{P}(\mathbf{x})\} + \mathbb{P}\{\mathcal{P}(\mathbf{y})\} + \mathbb{P}\{\mathcal{R}\} \le C\, L_{k+1}^{-\frac{30}{7}p_{k+1}(n)}. \qquad \square$$

Let $K_{\mathrm{PI}}(\mathbf{\Lambda}_{L_{k+1}}(\mathbf{x}), E)$ denote the largest cardinality[1] of a collection of pairwise distant, partially interactive (E,m)-S cubes of radius L_k contained in the cube $\mathbf{\Lambda}_{L_{k+1}}(\mathbf{x})$. Further, set:

$$K_{\mathrm{PI}}\big(\mathbf{\Lambda}_{L_{k+1}}(\mathbf{x})\big) = \sup_{E\in\mathbb{R}} K_{\mathrm{PI}}\big(\mathbf{\Lambda}_{L_{k+1}}(\mathbf{x}), E\big).$$

Then Lemma 4.4.5 leads directly to the following

Corollary 4.4. *Suppose that the property* $(\widetilde{\mathbf{DS}}.N, \tilde{n}, m, p_{k-1}, L_{k-1})$ *is fulfilled for all* $\tilde{n} \in [[1, n-1]]$. *Then for any n-particle cube* $\mathbf{\Lambda}^{(n)}_{L_{k+1}}(\mathbf{x})$ *and some* $C' = C'(N,d) < \infty$,

$$\mathbb{P}\Big\{ K_{\mathrm{PI}}\big(\mathbf{\Lambda}^{(n)}_{L_{k+1}}(\mathbf{x})\big) \ge 2 \Big\} \le C'\, L_{k+1}^{-\frac{83}{21}p_{k+1}(n)}. \tag{4.70}$$

---

[1]Note that for a given E, there may be several maximal families of pairwise distant partially interactive (E,m)-S cubes inside $\mathbf{\Lambda}_{L_{k+1}}(\mathbf{x})$, but their maximal cardinality is well-defined, since $\mathbf{\Lambda}_{L_{k+1}}(\mathbf{x})$ is a finite set. The same is true for the maximal families of distant FI cubes; see below.

Proof. By our assumption ($\widetilde{\mathbf{DS}}.N, \tilde{n}, m, p_{k-1}, L_{k-1}$) for $\tilde{n} = 1, \ldots, n-1$, Lemma 4.4.5 applies. Then, for any distant pair of PI cubes $\mathbf{\Lambda}_{L_k}(\mathbf{u})$, $\mathbf{\Lambda}_{L_k}(\mathbf{v})$ inside $\mathbf{\Lambda}_{L_{k+1}}(\mathbf{x})$, the estimate of the form (4.69) holds true:

$$\mathbb{P}\left\{\exists\, E \in \mathbb{R} : \mathbf{\Lambda}_{L_k}^{(n)}(\mathbf{u}) \text{ and } \mathbf{\Lambda}_{L_k}^{(n)}(\mathbf{v}) \text{ are } (E,m)\text{-}S\right\} \leq C\, L_{k+1}^{-\frac{30}{7}\, p_{k+1}(n)}.$$

The number of pairs of cubes of radius L_k inside $\mathbf{\Lambda}_{L_{k+1}}(\mathbf{x})$ is bounded by

$$\frac{1}{2}(2L_{k+1} + 1)^{2nd} < 3^{2nd} L_{k+1}^{2nd},$$

so we finally get the inequality, for some $C' = C'(N,d) < \infty$,

$$\mathbb{P}\left\{ K_{\mathrm{PI}}\big(\mathbf{\Lambda}_{L_{k+1}}^{(n)}(\mathbf{x})\big) \geq 2 \right\} < C'\, L_{k+1}^{-\frac{30}{7}\, p_{k+1}(n)+2nd}.$$

Finally, note that

$$\begin{aligned}\frac{30}{7}\, p_{k+1}(n) - 2nd &= 2p_{k+1}(n)\left(\frac{15}{7} - \frac{nd}{p_{k+1}(n)}\right)\\ &> 2p_{k+1}(n)\left(\frac{15}{7} - \frac{1}{6}\right) = \frac{83}{21}\, p_{k+1}(n).\end{aligned}$$

□

4.4.6 Analysis of Fully Interactive Pairs

The analysis of a pair of fully interactive cubes $\mathbf{\Lambda}_{L_{k+1}}(\mathbf{x})$ and $\mathbf{\Lambda}_{L_{k+1}}(\mathbf{y})$ results in Lemma 4.4.7 below.

Let $K_{\mathrm{FI}}(\mathbf{\Lambda}_{L_{k+1}}^{(n)}(\mathbf{x}), E)$ stand for the largest cardinality of a collection of pairwise distant FI cubes of radius L_k inside a cube $\mathbf{\Lambda}_{L_{k+1}}(\mathbf{x})$. Further, set:

$$K_{\mathrm{FI}}\big(\mathbf{\Lambda}_{L_{k+1}}^{(n)}(\mathbf{x})\big) = \sup_{E \in \mathbb{R}} K_{\mathrm{FI}}\big(\mathbf{\Lambda}_{L_{k+1}}^{(n)}(\mathbf{x}), E\big).$$

Lemma 4.4.6. *Suppose that for some $n \in \{1, \ldots, N\}$ and any distant pair of n-particle* FI *cubes $\mathbf{\Lambda}_{L_k}^{(n)}(\mathbf{x})$ and $\mathbf{\Lambda}_{L_k}^{(n)}(\mathbf{y})$, the following bound holds true:*

$$\mathbb{P}\left\{\exists\, E \in \mathbb{R} :\ \mathbf{\Lambda}_{L_k}^{(n)}(\mathbf{x}) \textit{ and } \mathbf{\Lambda}_{L_k}^{(n)}(\mathbf{y}) \textit{ are } (E,m)\text{-S}\right\} \leq L_k^{-2p_k(n)}.$$

(In other words, the property ($\widetilde{\mathbf{DS}}.N, n, m, p_k, L_k$) holds true for pairs of n-particle FI *cubes.) Then for any n-particle cube $\mathbf{\Lambda}_{L_{k+1}}^{(n)}(\mathbf{u})$ and some $C'' = C''(N,d) < \infty$,*

$$\mathbb{P}\left\{ K_{\mathrm{FI}}\big(\mathbf{\Lambda}^{(n)}_{L_{k+1}}(\mathbf{u})\big) \geq 6 \right\} \leq C'' L_{k+1}^{-\frac{18}{7} p_{k+1}(n)}. \tag{4.71}$$

Proof. The number of all possible collections of 6 cubes of radius L_k inside $\mathbf{\Lambda}_{L_{k+1}}(\mathbf{u})$, including fully interactive ones, is bounded by $(2L_{k+1} + 1)^{6nd}/6!$. Dividing 6 cubes into 3 pairs and applying the inductive assumption of the lemma, we obtain

$$\mathbb{P}\left\{ K_{\mathrm{FI}}\big(\mathbf{\Lambda}^{(n)}_{L_{k+1}}(\mathbf{u})\big) \geq 6 \right\} \leq \frac{1}{6!} L_{k+1}^{-\frac{6p_k(n)}{\alpha}} (2L_{k+1} + 1)^{6nd}. \tag{4.72}$$

With $\alpha = 3/2$, $p_k(n) = 4^{N-n} p_0(1+\theta)^k$ $p_{k+1}(n) = p_k(n)(1+\theta)$, $p_0 > 6Nd$, $0 < \theta < 1/6$, one has

$$\begin{aligned} \frac{6p_k(n)}{\alpha} - 6nd &= 4p_k(n) - 6nd \\ &> \frac{3\,p_k(n)\cdot(1+\theta)}{(1+\theta)} \\ &> \frac{3\,p_{k+1}(n)}{\frac{7}{6}} = \frac{18\,p_{k+1}(n)}{7}. \end{aligned} \tag{4.73}$$

□

Lemma 4.4.7. *Under the assumptions of Lemma 4.4.6, assume also that the property* $(\widetilde{\mathbf{DS}}.N, \tilde{n}, m, p_{k-1}, L_{k-1})$ *holds true for any* $1 \leq \tilde{n} < n$. *Then for any pair of distant n-particle* FI *cubes* $\mathbf{\Lambda}^{(n)}_{k+1}(\mathbf{x})$, $\mathbf{\Lambda}^{(n)}_{k+1}(\mathbf{y})$, *the following bound holds true:*

$$\mathbb{P}\left\{ \exists\, E \in \mathbb{R} : \mathbf{\Lambda}^{(n)}_{k+1}(\mathbf{x}) \text{ and } \mathbf{\Lambda}^{(n)}_{k+1}(\mathbf{y}) \text{ are } (E,m)\text{-S} \right\} \leq L_{k+1}^{-2p_{k+1}(n)}. \tag{4.74}$$

Proof. Consider the following events:

$$\begin{aligned} \mathcal{S} &= \left\{ \exists\, E \in \mathbb{R} : \mathbf{\Lambda}^{(n)}_{L_{k+1}}(\mathbf{x}) \text{ and } \mathbf{\Lambda}^{(n)}_{L_{k+1}}(\mathbf{y}) \text{ are } (E,m)\text{-S} \right\}, \\ \mathcal{B}_{\mathrm{FI}} &= \left\{ \max\left[K_{\mathrm{FI}}\big(\mathbf{\Lambda}^{(n)}_{L_{k+1}}(\mathbf{x})\big), K_{\mathrm{FI}}\big(\mathbf{\Lambda}_{L_{k+1}}(\mathbf{y})\big) \right] \geq 6 \right\}, \\ \mathcal{B}_{\mathrm{PI}} &= \left\{ \max\left[K_{\mathrm{PI}}\big(\mathbf{\Lambda}^{(n)}_{L_{k+1}}(\mathbf{x})\big), K_{\mathrm{PI}}\big(\mathbf{\Lambda}^{(n)}_{L_{k+1}}(\mathbf{y})\big) \right] \geq 2 \right\}, \\ \mathcal{R} &= \left\{ \exists\, E \in \mathbb{R} : \mathbf{\Lambda}^{(n)}_{L_{k+1}}(\mathbf{x}) \text{ and } \mathbf{\Lambda}^{(n)}_{L_{k+1}}(\mathbf{y}) \text{ are } E\text{-PR} \right\}. \end{aligned}$$

With the help of Theorem 2.6.3 we will deduce that $\mathcal{S} \subset \mathcal{B}_{\mathrm{FI}} \cup \mathcal{B}_{\mathrm{PI}} \cup \mathcal{R}$. Indeed, suppose that $\omega \notin (\mathcal{B}_{\mathrm{FI}} \cup \mathcal{B}_{\mathrm{PI}} \cup \mathcal{R})$. Then, for any $E \in \mathbb{R}$, neither of the cubes $\mathbf{\Lambda}^{(n)}_{L_{k+1}}(\mathbf{x})$, $\mathbf{\Lambda}^{(n)}_{L_{k+1}}(\mathbf{y})$ can contain

- Six or more distant FI (E,m)-S cubes of radius L_k,
- Two or more distant PI (E,m)-S cubes of radius L_k.

This gives at most 5 pairwise distant FI cubes and at most one PI cubes which can be (E,m)-S. Taking into account Lemma 3.3.2, all PI (E,m)-S cubes of radius L_k in $\mathbf{\Lambda}^{(n)}_{L_{k+1}}(\mathbf{x})$ can be covered by at most $K(n)+1$ cubes of radius $2n(L_k+r_0)$, and all FI (E,m)-S cubes of radius L_k can be covered by at most 5 cubes of radius $11nL_k$. Thus all (E,m)-S cubes of radius L_k in $\mathbf{\Lambda}^{(n)}_{L_{k+1}}(\mathbf{x})$ can be covered by $\tilde{K} \le 5+1 < \infty$ cubes of radius $\tilde{C}L_k$, $\tilde{C} = \tilde{C}(n) < \infty$ and, therefore, by a family of annuli concentric with $\mathbf{\Lambda}^{(n)}_{L_{k+1}}(\mathbf{x})$ of total width $\le 2\tilde{C}(n)\tilde{K}$.

The same is true, of course, for $\mathbf{\Lambda}^{(n)}_{L_{k+1}}(\mathbf{y})$.

Since $\omega \notin \mathcal{R}$, at least one of cubes $\mathbf{\Lambda}^{(n)}_{L_{k+1}}(\mathbf{x})$, $\mathbf{\Lambda}^{(n)}_{L_{k+1}}(\mathbf{y})$ must be E-CNR. Now Theorem 2.6.3 applies to n-particle E-CNR cubes; this shows that either $\mathbf{\Lambda}^{(n)}_{L_{k+1}}(\mathbf{x})$ or $\mathbf{\Lambda}^{(n)}_{L_{k+1}}(\mathbf{y})$ is (E,m)-NS. Hence, the assumption that $\omega \notin (\mathcal{B}_{\rm FI} \cup \mathcal{B}_{\rm PI} \cup \mathcal{R})$ implies that $\omega \notin \mathcal{S}$, i.e., $\mathcal{S} \subset \mathcal{B}_{\rm FI} \cup \mathcal{B}_{\rm PI} \cup \mathcal{R}$.

Therefore, it suffices now to assess the probability

$$\mathbb{P}\{\mathcal{S}\} \le \mathbb{P}\{\mathcal{B}_{\rm FI}\} + \mathbb{P}\{\mathcal{B}_{\rm PI}\} + \mathbb{P}\{\mathcal{R}\}.$$

As in the proof of Lemma 4.4.6, owing to the EVC bound (3.52) with $Q = 4 \cdot 4^{N-n} p_0$ (cf. Corollary 3.1), we can write:

$$\mathbb{P}\{\mathcal{R}\} < L_{k+1}^{-Q(1+\theta)^{k+1}} = L_{k+1}^{-4p_{k+1}(n)} \le \frac{1}{2} L_{k+1}^{-2p_{k+1}(n)},$$

provided that L_0 is large enough.

Further, by Corollary 4.4 (cf. Eq. (4.70))

$$\mathbb{P}\{\mathcal{B}_{\rm PI}\} \le C' L_{k+1}^{-\frac{83}{21}p_{k+1}(n)}, \quad \frac{83}{21} > 2,$$

and by Lemma 4.4.6,

$$\mathbb{P}\{\mathcal{B}_{\rm FI}\} \le C'' L_{k+1}^{-\frac{18}{7}p_{k+1}(n)}, \quad \frac{18}{7} > 2.$$

Taking into account these bounds, we conclude that, for L_0 large enough,

$$\mathbb{P}\{\mathcal{S}\} < L_{k+1}^{-2p_{k+1}(n)}. \qquad \square$$

4.4.7 *Analysis of Mixed Pairs*

As we mentioned earlier, the analysis of a mixed pair of cubes $\mathbf{\Lambda}_{L_{k+1}}(\mathbf{x})$, $\mathbf{\Lambda}_{L_{k+1}}(\mathbf{y})$ requires a combination of methods used in the analysis of PI and FI pairs. The main outcome of this section is Lemma 4.4.8 followed by Theorem 4.4.9 adding up the information obtained in Sects. 4.4.5–4.4.7.

Lemma 4.4.8. *Suppose that L_0 is large enough and assume that the following properties are fulfilled:*

- $(\widetilde{\mathbf{DS}}.N, \tilde{n}, m, p_k, L_k)$ *for all* $\tilde{n} \in \{1, \dots, n-1\}$;
- $(\widetilde{\mathbf{DS}}.N, \tilde{n}, m, p_{k-1}, L_{k-1})$ *for all* $\tilde{n} \in \{1, \dots, n-1\}$;
- $(\widetilde{\mathbf{DS}}.N, n, m, p_k, L_k)$.

Let $\mathbf{\Lambda}^{(n)}_{L_{k+1}}(\mathbf{x})$ *and* $\mathbf{\Lambda}^{(n)}_{L_{k+1}}(\mathbf{y})$ *be distant n-particle cubes, where* $\mathbf{\Lambda}^{(n)}_{L_{k+1}}(\mathbf{x})$ *is* FI *and* $\mathbf{\Lambda}^{(n)}_{L_{k+1}}(\mathbf{y})$ PI. *Then*

$$\mathbb{P}\left\{\exists\, E \in \mathbb{R} :\ \mathbf{\Lambda}^{(n)}_{k+1}(\mathbf{x}) \text{ and } \mathbf{\Lambda}^{(n)}_{k+1}(\mathbf{y}) \text{ are } (E,m)\text{-}S\right\} \le L_{k+1}^{-2p_{k+1}(n)}. \tag{4.75}$$

Proof. Once more, consider the events

$$\begin{aligned}
\mathcal{S} &= \{\exists\, E \in \mathbb{R} : \mathbf{\Lambda}^{(n)}_{k+1}(\mathbf{x}) \text{ and } \mathbf{\Lambda}^{(n)}_{k+1}(\mathbf{y}) \text{ are } (E,m)\text{-S}\},\\
\mathcal{B}_{\mathrm{FI}}(\mathbf{x}) &= \{K_{\mathrm{FI}}(\mathbf{\Lambda}^{(n)}_{L_{k+1}}(\mathbf{x})) \ge 6\},\\
\mathcal{B}_{\mathrm{PI}}(\mathbf{x}) &= \{K_{\mathrm{PI}}(\mathbf{\Lambda}^{(n)}_{L_{k+1}}(\mathbf{x})) \ge 2\},\\
\mathcal{P}(\mathbf{y}) &= \{\mathbf{\Lambda}^{(n)}_{L_{k+1}}(\mathbf{y}) \text{ is } (m,\mathbb{R})\text{-PT}\},\\
\mathcal{R} &= \{\exists\, E \in \mathbb{R} :\ \mathbf{\Lambda}^{(n)}_{L_{k+1}}(\mathbf{x}) \text{ and } \mathbf{\Lambda}^{(n)}_{L_{k+1}}(\mathbf{y}) \text{ are } E\text{-PR}\}.
\end{aligned}$$

Due to assumptions $(\widetilde{\mathbf{DS}}.N, \tilde{n}, m, p_{k-1}, L_{k-1})$ and $(\widetilde{\mathbf{DS}}.N, \tilde{n}, m, p_k, L_k)$, Corollary 4.4 and Lemma 4.4.6 apply and give

$$\begin{aligned}
\mathbb{P}\{\mathcal{B}_{\mathrm{FI}}(\mathbf{x}) \cup \mathcal{B}_{\mathrm{PI}}(\mathbf{y})\} &< \frac{1}{8} L_{k+1}^{-2p_{k+1}(n)} + \frac{1}{8} L_{k+1}^{-2p_{k+1}(n)}\\
&= \frac{1}{4} L_{k+1}^{-2p_{k+1}(n)},
\end{aligned} \tag{4.76}$$

and if $\omega \notin \mathcal{B}_{\mathrm{FI}}(\mathbf{x}) \cup \mathcal{B}_{\mathrm{PI}}(\mathbf{x})$ then

$$K_{\mathrm{FI}}(\mathbf{\Lambda}^{(n)}_{L_{k+1}}(\mathbf{x})) + K_{\mathrm{PI}}(\mathbf{\Lambda}^{(n)}_{L_{k+1}}(\mathbf{x})) \le 5 + 1 = 6.$$

Therefore, owing to Theorem 2.6.3, if it is E-CNR, then it cannot be (E,m)-S.

Using once again the EVC bound (3.52), with $Q = 4 \cdot 4^{N-n} p_0$ and L_0 large enough, we can write

$$\mathbb{P}\{\mathcal{R}\} \le \frac{1}{4} L_{k+1}^{-2p_{k+1}(n)}.$$

Further, by Lemma 4.4.1, if $\mathbf{\Lambda}^{(n)}_{L_{k+1}}(\mathbf{y})$ is $(m, \mathbb{R})$-NPT and E-CNR for some $E \in \mathbb{R}$, then it is (E, m)-NS. Therefore,

$$\mathcal{S} \subset \mathcal{B}_{\mathrm{FI}}(\mathbf{x}) \cup \mathcal{B}_{\mathrm{PI}}(\mathbf{y}) \cup \mathcal{P}(\mathbf{y}) \cup \mathcal{R}$$

yielding

$$\begin{aligned}\mathbb{P}\{\mathcal{S}\} &\leq \mathbb{P}\{\mathcal{B}_{\mathrm{FI}}(\mathbf{x}) \cup \mathcal{B}_{\mathrm{PI}}(\mathbf{x})\} + \mathbb{P}\{\mathcal{P}(\mathbf{y})\} + \mathbb{P}\{\mathcal{R}\} \\ &\leq \frac{1}{4}L_{k+1}^{-2p_{k+1}(n)} + \frac{1}{4}L_{k+1}^{-2p_{k+1}(n)} + \frac{1}{4}L_{k+1}^{-2p_{k+1}(n)} \\ &< L_{k+1}^{-2p_{k+1}(n)},\end{aligned}$$

where $\mathbb{P}\{\mathcal{P}(\mathbf{y})\}$ is estimated with the help of Lemma 4.4.4 due to assumption $(\widetilde{\mathbf{DS}}.N, \tilde{n}, m, p_k, L_k)$. □

The arguments developed in Sects. 4.4.5–4.4.7 are summarized in Theorem 4.4.9 below.

Theorem 4.4.9. *Assume L_0 is chosen sufficiently large and suppose that properties $(\widetilde{\mathbf{DS}}.N, \tilde{n}, m, p_k, L_k)$ and $(\widetilde{\mathbf{DS}}.N, \tilde{n}, m, p_{k-1}, L_{k-1})$ is fulfilled for all $\tilde{n} \in [[1, n]]$. Then property $(\widetilde{\mathbf{DS}}.N, n, m, p_{k+1}, L_{k+1})$ also holds true for all $\tilde{n} \in [[1, n]]$.*

Consequently, if $(\widetilde{\mathbf{DS}}.N, \tilde{n}, m, p_0, L_0)$ is satisfied for all $\tilde{n} \in [[1, n]]$, then property $(\widetilde{\mathbf{DS}}.N, \tilde{n}, m, p_k, L_k)$ is fulfilled for all $k \geq 1$ and $\tilde{n} \in [[1, n]]$.

4.4.8 Conclusion of the Variable-Energy MPMSA

Theorems 4.3.3 and 4.4.9 lead, by induction in $k = 0, 1, \ldots$, to the main result of the variable-energy multi-particle multi-scale analysis.

Theorem 4.4.10. *For any integers $N \geq 2$, $d \geq 1$ and real numbers $m \geq 1$, $n \in \{1, \ldots, N\}$ there exists $L_0^* \in \mathbb{N}$ such that for any $L_0 \geq L_0^*$ and $|g| \geq g^* = g^*(N, d, m, p, L_0) \in (0, \infty)$, $(\widetilde{\mathbf{DS}}.N, n, m, p_k, L_k)$ holds true $\forall\, k \geq 0$.*

Proof. Fix arbitrary integers $d \geq 1$, $N \geq 2$ and real numbers $m \geq 1$ and $p > 6Nd$. By Theorem 4.3.3, for any $L_0 \in \mathbb{N}$ there is $g^* > 0$ such that for all $|g| \geq g^*$, the initial length scale bound $(\widetilde{\mathbf{DS}}.N, n, m, p_0, L_0)$ holds true for all $n \in \{1, \ldots, N\}$. Pick L_0^* large enough and let $L_0 \geq L_0^*$, so as to satisfy all numeric assumptions on L_0 specified in all statements used in the proof of Corollary 4.4 (inductive step for partially interactive pairs of cubes), Lemma 4.4.7 (inductive step of fully interactive pairs) and Lemma 4.4.8 (inductive step for mixed pairs).

Using induction in $k = 0, 1, \ldots$, we proceed as follows:

- Fix the integers $N \geq 2$ and $1 \leq n \leq N - 1$.

- We are allowed to assume that the entire MPMSA procedure is successfully carried out for $\tilde{n}$-particle systems for all $1 \le \tilde{n} \le n-1$. Indeed, the base of induction in n is provided by the standard, single-particle MSA for lattice Schrödinger operators with the IID random potential.

  Moreover, our MPMSA procedure is an extension of the single-particle MSA induction, so the case $n = 1$ can be easily recovered by simplifying our proofs in this case.

  Specifically, the key bounds $(\widetilde{\mathbf{DS}}.N, \tilde{n}, m, p_k, L_k)$ can be assumed for all $k \ge 0$ and $1 \le \tilde{n} \le n-1$. Now we have to prove their counterparts for n-particle systems, for all $k \ge 0$.
- The initial length scale bound $(\widetilde{\mathbf{DS}}.N, n, m, p_0, L_0)$ is established in Theorem 4.3.3.
- Fix $k \ge 0$. The derivation of $(\widetilde{\mathbf{DS}}.N, n, m, p_{k+1}, L_{k+1})$ is divided into three separate results:

  - The estimate for the pairs of n-particle PI cubes (cf. Lemma 4.4.5);
  - The estimate for the pairs of n-particle FI cubes (cf. Lemma 4.4.7);
  - The estimate for the mixed pairs of n-particle cubes, of which one is FI and the other is PI (cf. Lemma 4.4.8).

  The combined results of the above three steps are summarized in Theorem 4.4.9. It is to be stressed that the upper bound on the number of particles, $n \ge N$, is vital for the scale induction step.
- Since the estimate $(\widetilde{\mathbf{DS}}.N, n, m, p_{k+1}, L_{k+1})$ is derived from the estimates $(\widetilde{\mathbf{DS}}.N, n, m, p_{\tilde{k}}, L_{\tilde{k}}), \tilde{k} = 0, \dots, k$, combined with all estimates $(\widetilde{\mathbf{DS}}.N, \tilde{n}, m, p_k, L_k)$, $1 \le \tilde{n} \le n-1$ $k \ge 0$, we conclude by induction (in k, i.e., by the scale induction) that $(\widetilde{\mathbf{DS}}.N, n, m, p_k, L_k)$ also holds true for all $k \ge 0$, provided that $n \le N$.
- Consequently, the finite induction in the number of particles $1 \le n \le N$, with N fixed from the beginning (and the disorder parameter chosen appropriately (viz., sufficiently large: $|g| \ge g(N) < \infty$), establishes the key MPMSA estimate $(\widetilde{\mathbf{DS}}.N, n, m, p_k, L_k)$ for all $1 \le n \le N$ and all $k \ge 0$.

This completes the proof. □

4.5 Exponential Localization: From MPMSA to Theorem 3.1.1

Under Assumption C, the key probabilistic bound of the (variable-energy) MSA for multi-particle systems,

$$\mathbb{P}\left\{ \exists\, E \in I : \ \mathbf{\Lambda}_{L_k}^{(n)}(\mathbf{u}) \text{ and } \mathbf{\Lambda}_{L_k}^{(n)}(\mathbf{v}) \text{ are } (E,m)\text{-}S \right\} \le L_k^{-2p}, \tag{4.77}$$

has been established for all $11NL_k$-distant pairs of n-particle cubes of radius L_k, $k \geq 0$, i.e., with $|\mathbf{u}-\mathbf{v}| \geq 11NL_k$. Recall that Theorem 2.7.2, formulated for lattice Schrödinger operators $H(\omega) = -\Delta + W(x;\omega)$ acting in $\ell^2(\mathbb{Z}^D)$, $1 \leq D < \infty$, infers exponential spectral localization from the estimate

$$\mathbb{P}\{\exists E \in I : \Lambda_{L_k}(u) \text{ and } \Lambda_{L_k}(v) \text{ are } (E,m)\text{-S}\} \leq L_k^{-2p}, \tag{4.78}$$

valid for cubes $\Lambda_{L_k}(u), \Lambda_{L_k}(v) \subset \mathbb{Z}^D$, $k \geq 0$, with $|u-v| \geq 2CL_k + 1$. Clearly, with $C = 6n$, the inequality $|u-v| \geq 2CL_k + 1$ implies $|u-v| \geq 11nL_k$.

Apart from the assumption that Δ is the lattice Laplacian and the hypothesis (4.78), the probabilistic properties of the random potential $W(x;\omega)$ are not used in the proof of Theorem 2.7.2. As was pointed out in Remark 2.7.2, this allows to apply Theorem 2.7.2 to n-particle Hamiltonians $\mathbf{H}^{(n)}(\omega)$ acting in $\ell^2(\mathbb{Z}^{nd})$. Indeed, the particularity of $\mathbf{H}^{(n)}(\omega)$ resides in the structure of its potential energy,

$$\mathbf{W}(x;\omega) = g\sum_{j=1}^{n} V(x_j;\omega) + \mathbf{U}(x_1,\ldots,x_n).$$

Therefore, setting $D = nd$, one can identify $\mathbf{H}^{(n)}(\omega)$ with the LSO $H(\omega) = -\Delta + W(x;\omega)$, acting in $\ell^2(\mathbb{Z}^D)$, where Δ is the nearest-neighbor lattice Laplacian and

$$W(x;\omega) = g\sum_{j=1}^{n} V(x_j;\omega) + \mathbf{U}(x_1,\ldots,x_n).$$

Finally, with $C = 6n$, Theorem 2.7.2 implies directly the following result.

Theorem 4.5.1. *Fix an interval $I \subset \mathbb{R}$ and suppose that, for some $L_0 \geq 1$, $m > 0$ and*[2] *$p > Nd\alpha$ and all $n = 1,\ldots,N$ and $k \geq 0$ the following probabilistic bound holds true: For any distant pair of n-particle cubes $\mathbf{\Lambda}^{(n)}_{L_k}(\mathbf{u})$, $\mathbf{\Lambda}^{(n)}_{L_k}(\mathbf{v})$,*

$$\mathbb{P}\left\{\exists E \in I : \mathbf{\Lambda}^{(n)}_{L_k}(\mathbf{u}) \text{ and } \mathbf{\Lambda}^{(n)}_{L_k}(\mathbf{v}) \text{ are } (E,m)\text{-S}\right\} \leq L_k^{-2p}. \tag{4.79}$$

Then $\exists$ an event $\Omega^* = \Omega^*(I) \subset \Omega$, such that $\mathbb{P}\{\Omega^*\} = 1$ and $\forall\, \omega \in \Omega^*$ and $E \in I$, if $\boldsymbol{\psi} = \boldsymbol{\psi}_\omega$ is a polynomially bounded non-zero solution to the equation $\mathbf{H}^{(n)}(\omega)\boldsymbol{\psi} = E\boldsymbol{\psi}$ then

$$\limsup_{|\mathbf{x}|\to\infty} \frac{\ln|\boldsymbol{\psi}_\omega(\mathbf{x})|}{|\mathbf{x}|} \leq -m. \tag{4.80}$$

[2]Recall: we always assume that $p_0 > 6Nd = 4Nd\alpha$, $\alpha = 3/2$, yielding $p > Nd\alpha$.

Consequently, for all $\omega \in \Omega^*$ the operator $\mathbf{H}^{(n)}(\omega)$ has p.p. spectrum in I, and any normalized eigenfunction $\boldsymbol{\psi} = \boldsymbol{\psi}_\omega$ of $\mathbf{H}^{(n)}(\omega)$ with eigenvalue $E(\omega) \in I$, satisfies

$$\forall\, \mathbf{x} \in \left(\mathbb{Z}^d\right)^n \quad |\boldsymbol{\psi}_\omega(\mathbf{x})| \le C(\boldsymbol{\psi}_\omega)\, \mathrm{e}^{-m|\mathbf{x}|} \tag{4.81}$$

with some random constant $C\left(\boldsymbol{\psi}_\omega\right) < \infty$.

As was mentioned in Chap. 2, owing to Shnol–Simon-type theorems (cf. [150, 151]), for spectrally a.e. $E \in \sigma(\mathbf{H})$ there exists a generalized eigenfunction $\boldsymbol{\psi}_E$ which is polynomially bounded. (cf. [36, 151]), so that Theorem 4.5.1 implies that, under the assumptions of the theorem, the spectrum of the random N-particle operator $\mathbf{H}(\omega)$ is pure point,[3] and that all ℓ^2-eigenfunctions decay exponentially, thus proving Theorem 3.1.1.

Remark 4.5.1. Assertion (C) of Lemma 3.3.2, hides an important distinction of the multi-particle eigenfunction localization bound from its conventional, single-particle analog.

Namely, the asymptotic exponential decay for a given square-summable eigenfunction $\boldsymbol{\psi}$ guaranteed by our argument depends, in fact, upon the location of the site $\mathbf{u} \in \left(\mathbb{Z}^d\right)^N$ (actually, upon diam $\Pi\mathbf{u}$). More to the point, had we had to establish an efficient, physically relevant bound of the decay, this site $\mathbf{u}$ should be chosen in the area where $\boldsymbol{\psi}$ is not too small. The above proof, based on the EVC bound from Theorem 3.4.2, leaves an uncomfortable feeling that the decay properties are not homogeneous in the multi-particle configuration space. This raises a legitimate question[4] of whether it is an artefact of the method used in the proof or a physical phenomenon. Replacing Theorem 3.4.2 by a more optimal one, stemming from Corollary 3.3 (if and where it is applicable) solves this mystery and provides decay bounds on eigenfunctions $\boldsymbol{\psi}$ which do not depend upon the position of their respective centers of localization.

4.6 Strong Dynamical Localization: From MPMSA to Theorem 3.1.2

The proof of Theorem 4.3.13, leading to the n-particle strong dynamical localization in $\mathbb{R}$, is based on a bound of the following form (cf. (4.53)):

$$\mathbb{P}\left\{\exists\, E \in I : \min\left(\mathrm{M}_{\mathbf{x},L}(E), \mathrm{M}_{\mathbf{y},L}(E)\right) > a(L)\right\} \le h(L), \tag{4.82}$$

[3]This fact will also follow from the results of Sect. 4.6 below, combined with RAGE theorems. However, RAGE-type theorems do not imply directly that the square-summable eigenfunctions decay exponentially.

[4]See the discussion in [9] and in [57].

where

$$\mathrm{M}_{\mathbf{u},L}(E) := \max_{\mathbf{v}\in\partial^{-}\mathbf{\Lambda}_L^{(n)}(\mathbf{u})} |G_{\mathbf{\Lambda}_L^{(n)}}(\mathbf{u},\mathbf{v};E)|, \quad \mathbf{u}\in(\mathbb{Z}^d)^n, \tag{4.83}$$

and $h(L)$ decays sufficiently fast as $L \to \infty$. Such a bound is obtained directly in the course of the variable-energy MPMSA inductive procedure (in the fixed-energy MPMSA, it is derived from its fixed-energy counterpart). Therefore, the variable-energy MPMSA also implies multi-particle strong dynamical localization.

A more traditional derivation of the (single-particle) strong dynamical localization from the probabilistic estimates established in the (single-particle) MSA induction, presented in Chap. 2 (cf. Sect. 2.7.2) and going back following the original works by Germinet–De Bièvre [98] and Damanik–Stollmann [78], can also be adapted to multi-particle systems. The only (and minor) modification of the geometrical arguments required for such an adaptation is replacing pairs of *disjoint* cubes by *distant* pairs of cubes for which a two-volume EVC bound can be proven.

Such a modification was made in the formulation and the proof of Theorem 2.7.2. However, taking into account the complexity of the proof of Theorem 2.7.4, we do not provide a similar (elementary, but rather tedious) adaptation to the case where the key bound (4.78) holds true for $O(L)$-distant pairs of cubes of radius L.

Recall that, as we have explained in Chap. 3 (see Remark (3.1.3) in Sect. 3.1.2, Eq. (3.26) and (3.27)), the fast decay of the eigenfunction correlators implies strong dynamical localization stated in Theorem 3.1.2.

Comparing the fixed-energy and variable-energy (MP)MSA, we would like to stress the following. The derivation of variable-energy MSA bounds from their fixed-energy counterparts (cf. Theorem 2.5.2 and Theorem 4.3.10) is adapted to probabilistic estimates in a bounded interval $I \subset \mathbb{R}$. Such a situation is natural in the MSA for differential (hence unbounded) random Schrödinger operators in $\mathbb{R}^d$ (or in quantum graphs), where localization is usually proven in a bounded, or even very small, energy interval near the lower edge of the spectrum. (A notable exception is the one-dimensional case, where one can use a non-perturbative localization analysis.) The proof of complete localization (in $I = \mathbb{R}$) in the lattice models required, as we have seen in Sect. 2.5 and in Sect. 4.3.6, some additional arguments. In the simplest situation, where the random potential is a.s. bounded, and so is, therefore, the spectrum, it suffices to take a bounded interval I containing the a.s. spectrum. While the variable-energy MPMSA procedure is considerably more complex than its fixed-energy variant, it has, for example, the advantage to provide naturally two-volume estimates of the form $(\widetilde{\mathbf{DS}}.N, n, m, p_k, L_k)$,

$$\mathbb{P}\left\{ \exists\, E \in \mathbb{R} :\ \mathbf{\Lambda}_{L_k}^{(n)}(\mathbf{u}) \text{ and } \mathbf{\Lambda}_{L_k}^{(n)}(\mathbf{v}) \text{ are } (E,m)\text{-S} \right\} \le L_k^{-2p_k(n)},$$

directly in the entire real line $\mathbb{R}$ (or in an interval near the spectral bottom).

On the other hand, in the single-particle localization theory, the fixed-energy MSA proved its greater flexibility, for example, in the context of correlated random potentials (cf., e.g., [92, 162]). A traditional derivation of spectral localization from the fixed-energy MSA is based upon a deep and very general result by Simon and Wolff [153]. Unfortunately, there has been no analog of their method for multi-particle random Hamiltonians. To a certain extent, the method presented in Sect. 4.3.6, plays a role similar to that of the Simon–Wolff argument, and even allows to infer (single- or multi-particle) strong dynamical localization from the results of the fixed-energy analysis.

4.7 Further Progress in Multi-particle Localization Theory

In this section we briefly discuss some fresh results that have been achieved or are in progress, on multi-particle localization, of which the authors of this book became aware at the time of sending the book to the publishers. An exception is Sect. 4.7.2 where we discuss possible roles of the concepts of ergodicity and the integrated density of states (IDoS) in the context of multi-particle models.

4.7.1 Multi-particle Localization at Extreme Energies

As was explained in Sects. 2.2 and 2.3, in the single-particle localization theory the initial scale bound on Green's functions $G_{\Lambda_{L_0}(u)}(x, y; E, \omega)$ can be proven *for some energies* $E \in \mathbb{R}$ without assuming the modulus of the amplitude g of the disorder to be large. In fact, such bounds for energies E near a given point $E_0 \in \mathbb{R}$ are usually based on an assumption that, speaking informally, the density of states near E_0 is sufficiently low. In more precise terms, the probability of having at least one eigenvalue of the operator $H_{\Lambda_{L_0}(u)}$ in a given interval I must be sufficiently small. Naturally, multiplying a sample of the potential $\{V(x; \omega), x \in \Lambda_{L_0}(u)\}$, by g with large $|g|$ eventually leads to the required bound. On the other hand, the Lifshitz tails phenomenon provides an alternative mechanism which has a similar effect: at extreme energies, e.g., near the edge $E_0 = 0$ of the range of values of the random potential under Assumption B (see Eq. (2.4)). Indeed, it is unlikely to have even one eigenvalue of $H_{\Lambda_{L_0}(u)}$ near point 0, for such "extreme" eigenvalues are associated with samples of the potential having a very small probability.

It is a legitimate and natural question, if a similar phenomenon occurs in interacting multi-particle disordered systems. The answer is affirmative, as has been shown by Ekanga (cf. [82, 83]) who adapted to the edge-band localization the inductive procedure from [70, 71]. See below. In the case where the periodic lattice $\mathbb{Z}^d$, serving as the single-particle (physical) configuration space, is replaced by a more general countable graph (with polynomially bounded growth of balls), such an adaptation was performed in a recent preprint [62], again with the help of the Lifshitz-type asymptotic analysis (in this case, on graphs of polynomial growth; cf. [60]).

The validity of the initial-scale N-particle MSA bound at low energies can be proven with help of the following argument. For the sake of brevity, we assume in this section that the random potential $V(x;\cdot)$, $x \in \mathbb{Z}^d$, satisfies Assumptions A and B (Eqs. (2.3) and (2.4), respectively), while the interaction potentials $U^{(n)}$, $n = 2, \dots, N$, are non-negative and of a finite range (i.e., obeys Assumption 3.1.1: cf. (3.5)–(3.7)).

For the purposes of this subsection introduce an analog of the property $(\mathbf{DS}.N, n, m, p_k, L_k)$ for energies in a bounded interval $I \subset \mathbb{R}$:

$(\mathbf{DS}.I, N, n, m, p_k, L_k)$ *For any pair of distant n-particle cubes* $\mathbf{\Lambda}^{(n)}_{L_k}(\mathbf{u})$, $\mathbf{\Lambda}^{(n)}_{L_k}(\mathbf{v})$ *the following bound holds true:*

$$\mathbb{P}\left\{\exists\, E \in I:\ \mathbf{\Lambda}^{(n)}_{L_k}(\mathbf{u}) \text{ and } \mathbf{\Lambda}^{(n)}_{L_k}(\mathbf{v}) \text{ are } (E,m)\text{-S}\right\} \le L_k^{-2p_k(n)}. \tag{4.84}$$

Lemma 4.7.1 (Cf. [83]). *Let Assumptions A and B hold true, fix an integer $N \ge 2$ and suppose that the interaction $\mathbf{U}^{(N)}(\mathbf{x})$ is non-negative and has finite range.*

Then for any $g \neq 0$ and $p > 0$ there exist $\eta = \eta(g,p) \in (0,\infty)$ and $C = C(\eta,p) \in (0,+\infty)$ such that $(\mathbf{DS}.I, N, n, m, p, L_0)$ holds true for the interval $I = [0,\eta]$ and any $n = 1,\dots,N$, provided, L_0 is sufficiently large and with $m = m(\eta) > L_0^{-1/2}$.

Sketch of the proof. Fix a cube $\mathbf{\Lambda}_{L_0}(\mathbf{u})$ where $\mathbf{u} = (u_1,\dots,u_N) \in \left(\mathbb{Z}^d\right)^N$. Since $\mathbf{U}_{\mathbf{\Lambda}_{L_0}(\mathbf{u})}$ is a non-negative operator, the lowest eigenvalue of $\mathbf{H}_{\mathbf{\Lambda}_{L_0}(\mathbf{u})}(\omega)$ is bounded from below by the lowest eigenvalue of the Hamiltonian of the non-interacting N-particle system $\mathbf{U}_{\mathbf{\Lambda}_L(\mathbf{u})}$:

$$\mathbf{H}_{0,\mathbf{\Lambda}_{L_0}(\mathbf{u})} + g\mathbf{V}_{\mathbf{\Lambda}_{L_0}(\mathbf{u})}(\omega).$$

The eigenvalues of the operator $\mathbf{H}_{0,\mathbf{\Lambda}_{L_0}(\mathbf{u})} + g\mathbf{V}_{\mathbf{\Lambda}_{L_0}(\mathbf{u})}(\omega)$ are the sums $\sum_{j=1}^{N} E^{(j)}_{i_j}$ where $i_j \in \{1,\dots,2L_0+1\}$ and for each $J = 1,\dots N$, $E^{(j)}_{i_j}$ are the eigenvalues of the single-particle LSO

$$H_{\Lambda_{L_0}(u_j)} = H_{0,\Lambda_{L_0}(u_j)} + gV_{\Lambda_{L_0}(u_j)}(\omega).$$

Every eigenvalue $E^{(j)}_{i_j}$ is non-negative, due to non-negativity of the random potential V. Therefore, owing to the fact that the values $V(x;\omega)$, $x \in \mathbb{Z}^d$, are IID, the spectrum of $\mathbf{H}_{\mathbf{\Lambda}_{L_0}(\mathbf{u})}$ satisfies

$$\mathbb{P}\left\{\sigma\big(\mathbf{H}_{\mathbf{\Lambda}_{L_0}(\mathbf{u})}(\omega)\big) \cap [0,\eta] \neq \varnothing\right\} \le \mathbb{P}\left\{E_1(H_{\Lambda_{L_0}(u_1)}) \le \eta\right\}$$

where $E_1^{(1)}(H_{\Lambda_{L_0}(u_1)})$ is the lowest eigenvalue of $H_{\Lambda_{L_0}(u_j)}$. Now the problem is reduced to the single-particle estimate near a spectral edge, so the required upper bound on the probability can be bounded with the help of Theorem 2.3.5. We skip further details. □

As was mentioned in Sect. 3.1.2, methods developed in Chaps. 3 and 4 can be adapted to the proof of the multi-particle spectral and dynamical localization for the N-particle Hamiltonian $H(\omega)$ in interval $[0, \eta)$. One of the main ingredients of such an adaptation is the replacement of the main bound (**DS**.N, n, m, p_k, L_k)—stated for *all* $E \in \mathbb{R}$—by its analog for energies $E \in I^{(n)}$, with properly chosen energy bands $I^{(n)}$, $1 \leq n \leq N$. See the details in [82, 83].

Remark 4.7.1. One can prove[5] that if

$$\inf\left\{\lambda \in \mathbb{R} : \mathbb{P}\{\, V(0,\omega) \leq \lambda \,\} > 0\right\} = 0, \qquad (4.85)$$

then with $\mathbb{P}$-probability one, the spectrum of $\mathbf{H}(\omega)$ in any interval $[0, \eta]$ with $\eta > 0$ is non-empty. The idea of the proof is as follows. The Hamiltonian without interaction is a direct sum of 1-particle operators $H^{1,j}(\omega) = -\Delta + V(x_j;\omega)$, each acting on the respective variable x_j, $1 \leq j \leq N$. It is well-known (cf., e.g., [113]) that under the assumption (4.85), $\inf \sigma\left(H^{(1,j)}(\omega)\right) = 0$. The usual proof is based on the Weyl argument (see Sect. 3.4 in [113]), which requires only the existence of arbitrarily large cubes with *all* values of the potential energy arbitrarily close to 0.

Observe that in the N-particle lattice there are arbitrarily large cubes on which the interaction potential vanishes: it suffices to consider the cubes of the form

$$\mathbf{\Lambda}^{(N)}(\mathbf{u}) = \times_{j=1}^{N} \Lambda_L(u_j)$$

such that for all $i \neq j$, $\mathrm{dist}(\Lambda_L(u_i), \Lambda_L(u_i)) > r_0$, where r_0 is the range of interaction. Now it suffices to apply the Weyl argument to such cubes to show that the spectrum of $\mathbf{H}^{(N)}(\omega)$, too, is a.s. nonempty in any interval $[0, \eta]$, $\eta > 0$. It is worth mentioning that this property holds true for arbitrary interactions of finite range. However, for non-negative interactions this implies also that $\inf \sigma\left(\mathbf{H}^{(N)}(\omega)\right) = 0$, since in this case $\mathbf{H}^{(N)}(\omega)$ is a.s. non-negative, hence $\inf \sigma\left(\mathbf{H}^{(N)}(\omega)\right) \geq 0$.

Remark 4.7.2. The assumption of Hölder-continuity in the multi-particle context was used in this book only in order to obtain bounds (**DS**.N, N, m, p_k, L_k) with $p_k = p_k(N) \to \infty$ as $k \to \infty$. We repeat that the spectral n-particle localization follows from the property (**DS**.N, N, m, p_k, L_k) with a fixed $p > 6Nd$.

Also, the assumption of non-negativity of the potential V can be replaced by a lower bound $V(x;\omega) \geq E_*$ where $E_* > -\infty$. Indeed, it is not difficult to see that this, more general case is reduced to that of $E_* = 0$ by replacing the random potential field: $\tilde{V}(x;\omega) := V(x;\omega) - E_* \geq 0$. This results in a shifted Hamiltonian $\tilde{\mathbf{H}}(\omega) = \mathbf{H}(\omega) - NE_*\mathbf{1}$, sharing with $\mathbf{H}(\omega)$ the localized eigenfunctions.

---

[5] See [83].

4.7.2 The Role of Ergodicity and the Integrated Density of States for Multi-particle Systems

In the single-particle localization theory, ergodicity and related properties of the operator ensemble $\{H(\omega), \omega \in \Omega\}$ often play an important role, for instance, in the proof of localization at extreme energies. See Sect. 2.2.4. (Of course, these properties are guaranteed when the external potential $V(x;\omega)$ is IID.) Indeed, the asymptotic analysis of "Lifshitz tails," based on large deviations estimates, requires the knowledge of the a.s. location of spectral edges (the lower edge of the spectrum, in the simplest case). The location of the spectrum is indicated (with probability one) by the support of the IDoS measure, and the existence (and desired properties) of the latter is usually established with the help of ergodicity arguments.

In the previous subsection, we reduced the initial scale bound for a multi-particle system to that for its single-particle counterpart; in a sense, ergodicity of the (IID) random potential $V : \mathbb{Z}^d \times \Omega \to \mathbb{R}$ has been implicitly used in this argument. However, we would like to stress that the role of ergodicity arguments and of the IDoS for the multi-particle systems seems quite different.

It is still true that the operator ensemble $\{\mathbf{H}^{(N)}(\omega), \omega \in \Omega\}$, with $N > 1$, is an ergodic operator ensemble, as defined, e.g., in [158]. (cf. Definition 1.2.3 in [158]). Since we did not discuss this notion in Chap. 2, we briefly define it now, for the reader's convenience.

- Let T be a dynamical system on some probability space $(\Omega, \mathfrak{F}, \mathbb{P})$ with time given by a commutative group A, i.e., a mapping $T : A \times \Omega \to \Omega$ measurable with respect to the sigma-algebra $\mathfrak{F}$, preserving $\mathbb{P}$,

$$\forall\, a \in A \;\; \forall\, \mathcal{B} \in \mathfrak{F} \qquad \mathbb{P}\big(T^{-a}\mathcal{B}\big) = \mathbb{P}(\mathcal{B}),$$

  and obeying the identity $T^a T^b = T^{a+b}$ (here "+" stands for the group operation in A, and $(-a)$ for the inverse of the element a).
- The dynamical system T is assumed ergodic, i.e., any T-invariant $\mathfrak{F}$-measurable subset $\mathcal{B} \subset \Omega$ must be trivial: $\mathbb{P}(\mathcal{B}) = 0$ or $\mathbb{P}(\mathcal{B}) = 1$.
- Next, it is assumed that there is a measurable mapping $H : \Omega \to \mathcal{S}(\mathcal{H})$ to the set of self-adjoint operators in $\mathcal{H}$, not necessarily bounded; the measurability (for possibly unbounded operators) can be defined in several equivalent ways (cf. [158]); e.g., it suffices to assume that for any $t \in \mathbb{R}$ and any vectors $\varphi, \psi \in \mathcal{H}$, the mapping

$$\Omega \ni \omega \mapsto \langle \varphi, e^{itH(\omega)} \psi \rangle \in \mathbb{C}$$

  is $\mathfrak{F}$-measurable.
- Finally, it is assumed that there is a representation $\mathcal{U}$ of the dynamical system T into the group of unitary operators in $\mathcal{H}$, such that

$$\forall\, a \in A \qquad H(T^a\omega) = \mathcal{U}^{-a} H(\omega)\,\mathcal{U}^a .$$

(This property is often called the covariance of the mapping H with respect to T.)

Using the expression "ergodic ensemble of [self-adjoint] operators" assumes that all the above-mentioned objects are defined and fulfill the listed conditions.

In the case where $\mathcal{H}$ is the Hilbert space of square-summable functions on a periodic lattice $\mathbb{Z}^d$ (resp, on a Euclidean space $\mathbb{R}^d$), a natural subgroup of unitary operators is given by the shifts defined by $(\mathcal{U}^a f)(x) := f(x-a)$, with arbitrary $a \in \mathbb{Z}^d$ (resp., $a \in \mathbb{R}^d$). A subtlety of this construction is that, in contrast with an earlier approach developed back in the 1970s, the representation $\mathcal{U}$ is not required to be *surjective* onto the group of the shift operators. This turns out to be an advantage, allowing to apply the above construction to the multi-particle operators $\mathbf{H}(\omega)$.

First, note that the potential energy field $\mathbf{W} : (\mathbb{Z}^d)^N \times \Omega \to \mathbb{R}$ defined by

$$\mathbf{W}(\mathbf{x}, \omega) = \sum_{j=1}^{N} V(x_j; \omega) + \mathbf{U}^{(N)}(\mathbf{x})$$

is not necessarily ergodic, and even not stationary with respect to the entire group of translations in $(\mathbb{Z}^d)^N$, due to the presence of the interaction $\mathbf{U}^{(N)}$ that hasn't been assumed shift-invariant in $(\mathbb{Z}^d)^N$. However, if the interaction *is translation invariant* with respect to the shifts in the "physical space" $\mathbb{Z}^d$, then the random field $\mathbf{W}$ is stationary with respect to the subgroup $\cong \mathbb{Z}^d$ of diagonal translations

$$(x_1, \dots, x_N) \mapsto (x_1 + t, \dots, x_N + t), \quad t \in \mathbb{Z}^d .$$

This would allow to conclude that the spectrum $\sigma\big(\mathbf{H}(\omega)\big)$—as a set—is a.s. constant with respect to $\omega \in \Omega$. This phenomenon, although important from the mathematical and physical points of view, is actually irrelevant for the proof of single- and multi-particle Anderson localization in the case of strong disorder ($|g| \gg 1$). In the proof of localization at low energies, the exact location of the lower edge of the spectrum, $\inf \sigma\big(\mathbf{H}(\omega)\big)$, can also be established without using explicitly the ergodicity arguments. For this reason, we did not discuss this notion in Chap. 2 and earlier in this chapter.

At the same time, without additional considerations and conditions, this is insufficient for the existence of the IDoS. There is a certain similarity between this situation and the case of a single-particle model in $\mathbb{Z}^d$ with a "background" potential invariant only with respect to translations forming a sub-lattice of nonzero co-dimension. In the latter case, it is easy to construct examples where the IDoS does not exist.

The existence of the IDoS for multi-particle lattice models with decaying interaction (not necessarily of finite range) can apparently be established without using ergodicity arguments; cf. [120]. However, another subtlety of multi-particle models is that the support of the IDoS does not necessarily give the a.s. support of the spectral measure, as suggests the following simple example.

Take $d = 1$ and $N = 2$. Let the random potential $V(x;\omega)$, $x \in \mathbb{Z}^1$, be IID with uniform marginal distribution on [0, 1]. Since the two-particle lattice Laplacian $(-\mathbf{\Delta}^{(2)})$ is non-negative, the Hamiltonian without interaction

$$-\mathbf{\Delta}^{(2)} + V(x_1;\omega) + V(x_2;\omega)$$

is also non-negative (as a quadratic form).

Now let the interaction potential $a\mathbf{U}$ have the following form:

$$aU^{(2)}(x_1, x_2) = a\delta_{x_1,x_2}, \quad \text{where } a < 0.$$

Arguing as in [120], one can conclude[6] that the IDoS for the non-interacting system is supported by (a subset of) the positive half-line $[0, +\infty)$, and the same is true, therefore, for the IDoS of the interacting system. However, it is readily seen that the lower edge of the spectrum for the Hamiltonian with interaction,

$$\mathbf{H}_a^{(2)} = -\mathbf{\Delta}^{(2)} + V(x_1;\omega) + V(x_2;\omega) + aU^{(2)}(x_1, x_2),$$

satisfies

$$\inf \sigma\left(\mathbf{H}_a^{(2)}\right) \underset{a\to-\infty}{\longrightarrow} -\infty.$$

To this end, it suffices to evaluate the quadratic form $\langle \boldsymbol{\phi}, \mathbf{H}_a^{(2)} \boldsymbol{\phi} \rangle$ for $\boldsymbol{\phi} = \delta_{(0,0)}$ and make use of the norm-boundedness of the non-interacting component of the Hamiltonian.

Physically speaking, the IDoS is connected to the bulk spectrum, while the non-translation invariant random field $\mathbf{W}(\mathbf{x};\omega)$ gives rise to an energy band corresponding to surface states.

In the case (considered in this book) of tight-binding Anderson Hamiltonians with an IID external potential field $\{V(x;\omega)\}$, the qualitative properties of the operators $+\mathbf{H}(\omega)$ and $-\mathbf{H}(\omega)$ are similar. In particular, the above example, referring to a negative interaction, can be transformed into an example with a positive interaction, creating a surface energy band *above* the spectrum of the non-interacting component $-\mathbf{\Delta}^{(2)} + V(x_1;\omega) + V(x_2;\omega)$. This band is also undetectable by means of the IDoS of the operator $\mathbf{H}(\omega)$.

---

[6]Formally speaking, Klopp and Zenk considered in [120] a continuous model, in $\mathbb{R}^d$. However, we believe that the main strategy can be probably adapted to (and even made simpler for) lattice models.

4.7.3 Correlated Potentials and Infinite-Range Interactions

We begin with the remark that in the single-particle context, the correlated external potentials were considered long ago by von Dreifus and Klein [162]. For multi-particle systems, in a recent work [58] (see also [55]) the MPMSA scheme presented in this book has been extended in the following directions:

(a) The external random potential field $V(x;\omega)$, $x \in \mathbb{Z}^d$, can be allowed to be correlated, albeit with a strong mixing;
(b) The interaction can have infinite range, but must decay exponentially (or at least sub-exponentially) fast at infinity.

To be more precise, let us define the rate $\varpi(r) = \varpi^{(N)}(r)$, $r > 0$, of decay of interaction as follows. Recall, a finite-range interaction energy $\underline{U}^{(N)}$ was specified in Assumption 3.1.1 in terms of differences $U^{(n',n'')}\left(\mathbf{x}^{\mathcal{J}}, \mathbf{x}^{\mathcal{J}^c}\right)$ (see Eqs. (3.5)–(3.7)). Next, set

$$\begin{aligned}\varpi(r) = \sup\Big[& U^{(n',n'')}\left(\mathbf{x}^{\mathcal{J}}, \mathbf{x}^{\mathcal{J}^c}\right) : \mathbf{x} \in \left(\mathbb{Z}^d\right)^n, \ \mathcal{J} \subset \{1,\dots,n\}, \\ & \mathcal{J}^c = \{1,\dots,n\} \setminus \mathcal{J}, \ \operatorname{dist}(\mathbf{x}^{\mathcal{J}}, \mathcal{J}^c) \geq r, \\ & n' = |\mathcal{J}| \geq 1, \ n'' = |\mathcal{J}^c| \geq 1, \ n = 2,\dots,N \Big]\end{aligned} \tag{4.86}$$

The precise form of the condition used in [58] instead of Assumption 3.1.1 is:

Assumption U1. $\exists\, \theta \in (0,1] : \ \forall\, r \geq 1 \ \ \varpi(r) \leq \mathrm{e}^{-r^{\theta}}.$ (4.87)

The strong mixing condition on the potential $V(x;\cdot)$ suitable for the MPMSA may be expressed in the following form. Given $L > 1$, consider two disjoint cubes $\Lambda_L(u), \Lambda_L(u') \subset \mathbb{Z}^d$ with $|x - y| > 2L$. As before, let $\mathfrak{B}\big(\Lambda_L(u)\big)$ be the sigma-algebra generated by the values $V(x;\omega)$ with $x \in \Lambda_L(u)$ and $\mathfrak{B}\big(\Lambda_L(u')\big)$ be that generated by $V(x;\omega)$ with $x \in \Lambda_L(u')$. It is demanded that for any events $\mathcal{E} \in \mathfrak{B}\big(\Lambda_L(u)\big)$, $\mathcal{E}' \in \mathfrak{B}\big(\Lambda_L(u')\big)$, the Rosenblatt mixing condition holds true:

$$\left|\mathbb{P}\left\{\mathcal{E} \cap \mathcal{E}''\right\} - \mathbb{P}\{\mathcal{E}\}\mathbb{P}\left\{\mathcal{E}'\right\}\right| \leq \mathrm{e}^{-C \ln^2 L} \tag{4.88}$$

where $C \in (0,\infty)$ is a constant. This allows one to prove the dynamical localization with the EF correlators decaying faster than any power law. For a more traditional, power-law decay of EF correlators, it suffices to require a power-law Rosenblatt mixing. For details, cf. [58].

4.7.4 *Weak Perturbations of Localized Non-interacting Systems*

This section focuses on one-dimensional multi-particle systems which is, apparently, the most interesting area where the idea of weak perturbations of localized non-interacting systems can be used. In the introductory Chap. 1 of this book it has been stressed that for a single-particle system in one dimension, Anderson localization occurs for any nonzero amplitude of a "sufficiently random" potential $V(\cdot\,;\omega)$. See [101, 102] (continuous models) and [123] (lattice models).

Unlike the higher-dimensional models, the proof of localization in one dimension is non-perturbative and based on deep results of the theory of random matrices. There is, however, a price to pay for allowing the amplitude $|g|$ of the random potential in a single-particle one-dimensional LSO $H(\omega) = H^0 + gV(\omega)$ to be arbitrarily small: the rate of exponential decay of the eigenfunctions vanishes as $|g| \to 0$.

In fact, the localization analysis of one-dimensional disordered systems usually starts with the study of the exponential growth of a solution $\psi = \psi_E(\omega)$ to the equation $H(\omega)\psi = E\psi$. We sketch a standard argument relating the growth of a typical solution to decay properties of the Green's functions and of eigenfunctions in a finite (but large) interval $\Lambda_L(0) = [-L, L] \cap \mathbb{Z}^1$, with Dirichlet's boundary conditions outside Λ_L. (The choice of (non-random) self-adjoint boundary conditions is actually irrelevant for the argument.)

Suppose that, with probability one, the following limit, called the upper Lyapunov exponent for the solutions of the equation $H(\omega)\psi = E\psi$, exists and is non-random:

$$\lambda(E, g) = \lim_{|x|\to\infty} |x|^{-1} \ln \sqrt{\psi_E^2(x) + \psi_E^2(x+1)}. \tag{4.89}$$

In fact, the almost-sure existence and non-randomness of the above limit can be established in a general case where the random potential $V(\cdot\,;\omega)$ is an ergodic random process on $\mathbb{Z}^1$; see, e.g., [53, 144].

Further, fix an energy interval $I \subset \mathbb{R}$ and suppose also that

$$\forall\, E \in I \quad \lambda(E, g) \ge \lambda^* > 0. \tag{4.90}$$

Consider the solutions $\phi_{E,\pm}$ to the problems

$$\begin{cases} \big(H(\omega)\phi_{E,+}\big)(x;\omega) = E\phi_{E,+}(x), & x \in [-L, +\infty) \cap \mathbb{Z}^1, \\ \phi_{E,+}(-L-1) = 0 \end{cases}$$

and, respectively,

$$\begin{cases} \big(H(\omega)\phi_{E,-}\big)(x;\omega) = E\phi_{E,-}(x;\omega), & x \in (-\infty, +L] \cap \mathbb{Z}^1, \\ \phi_{E,-}(L+1) = 0. \end{cases}$$

Choose L large enough, so that, with high probability,

$$\ln \phi_{E,\pm}(0) \approx \lambda(E, g)L \geq \lambda^* L.$$

Observe that, with the boundary conditions fixed at $\pm(L + 1)$, the values $\{\phi_{E,+}(x; \omega),\ x \in [-L, 0)\}$ and $\{\phi_{E,-}(x; \omega),\ x \in (0, +L]\}$, form two independent collections of random variables, due to the recurrence equations

$$\phi_{E,+}(x + 1) = \big(E - V(x; \omega)\big)\phi_{E,+}(x) - \phi_{E,+}(x - 1),\ x \geq -L + 1,$$

and

$$\phi_{E,-}(x - 1) = \big(E - V(x; \omega)\big)\phi_{E,-}(x) - \phi_{E,-}(x + 1),\ x \leq L - 1.$$

An eigenfunction $\psi_j = \psi_{E_j,\Lambda_L(0)}(\omega)$ of the operator $H_{\Lambda_L(0)}(\omega)$, with eigenvalue $E_j = E_j\big(\Lambda_L(0)(\omega)\big)$, must obey the boundary conditions at both endpoints $]\pm(L + 1)$, thus it must coincide with $\phi_{E_j,+}$ on $[-L, 0]$ and with $\phi_{E_j,-}$ on $[0, L]$. Therefore, if both solutions $\phi_{E_j,\pm}$ are growing exponentially at rate $\approx \lambda(E_j, g) \geq \lambda^* > 0$, in the directions from the boundary to the center 0 of $\Lambda_L(0)$, then the eigenfunction ψ_j is *exponentially decaying* from the center to the boundary.

However, the above argument is incomplete: while under condition (4.90) the exponential growth holds with high probability for any fixed energy $E \in I$, it is far from obvious that it still occurs with high probability for (random) eigenvalues of the operator $H_{\Lambda_L(0)}(\omega)$. Yet, it is the case, under suitable technical hypotheses on the random potential $V(\cdot\,; \omega)$. For example, this is true when IID random variables satisfy Assumption A.

It has to be said that the derivation of the spectral (and dynamical) localization from the positivity of the upper Lyapunov exponents in a given interval $I \subset \mathbb{R}$ (eventually, with $I = \mathbb{R}$) is not a straightforward task. An interested reader can find a detailed presentation of related techniques and results in [53, 144]. Here we mention a general fact proved by Simon and Wolff [153]. In the context of an one-dimensional LSO $H(\omega)$ with an IID random potential, the main result of [153] asserts that positivity of upper Lyapunov exponents $\lambda(E, g)$ in an interval $I \subset \mathbb{R}$ implies that, with probability one, the spectrum of $H(\omega)$ in the interval I is pure point. Unfortunately, the functional-analytic techniques of [153] do not imply directly exponential decay of the ℓ^2-eigenfunctions.

The proofs of positivity of upper Lyapunov exponents for a large class of "sufficiently random" potentials $V(\cdot\,; \omega)$ are often based on the theory of (products of) random matrices. Indeed, the second-order recurrence

$$\phi(x + 1) = \big(E - V(x)\big)\phi(x) - \phi(x - 1)$$

is equivalent to the first-order recurrence for vectors

$$\begin{pmatrix} \phi(x+1) \\ \phi(x-1) \end{pmatrix} = \begin{pmatrix} E - V(x) & -1 \\ 1 & 0 \end{pmatrix} \begin{pmatrix} \phi(x) \\ \phi(x-1) \end{pmatrix}.$$

Consequently, the growth properties of the solutions $\phi_{E,\pm}$ are encrypted into those of the product of random matrices (called transfer-matrices)

$$M_L(\omega) = A_L(\omega) A_{L-1}(\omega) \cdots A_1(\omega),$$

where

$$A_x(\omega) = \begin{pmatrix} E - V(x;\omega) & -1 \\ 1 & 0 \end{pmatrix}, x \in \mathbb{Z}^1.$$

Such proofs are based on general facts from topology and algebraic geometry; as a rule, they do not provide explicit asymptotics of the upper Lyapunov exponents $\lambda(E, g)$ when $|g| \to 0$. Nevertheless, it can be shown that for a large class of random potentials the upper Lyapunov exponent $\lambda(E, g)$ admits an asymptotic formula of the form

$$\lambda(E, g) = g^2 \big(\tilde{\lambda}(E) + o(1)\big), \; |g| \ll 1,$$

with $\tilde{\lambda}(E) > 0$. See details, e.g., in [68, 144].

Let us summarize: even in a weakly disordered one-dimensional random environment, with a small value $|g|$, the eigenfunctions of a single-particle Hamiltonian almost surely decay exponentially, at rate $m(g) \sim O(g^2) > 0$. This is, of course, also true for any system of $N \geq 1$ non-interacting quantum particles subject to the same random potential.

Using this information, Aizenman and Warzel [9] proved that the phenomenon of Anderson localization, both spectral and dynamical, is preserved under weak perturbations by a short-range interaction.[7] That is, an N-particle Hamiltonian on $\mathbb{Z}^1$, of the form

$$\mathbf{H}(\omega) = \mathbf{\Delta} + g\mathbf{V}(\omega) + h\mathbf{U},$$

exhibits localization provided that the value $|h|$ is small enough, depending upon $|g|$ and N (as well as on detailed properties of the interaction $\mathbf{U}$ and the distribution function F of the on-site external potential $V(\cdot\,;\omega)$). The assumption in [9] upon the distribution function F of the IID on-site potential $V(\cdot\,;\omega)$ is the existence of a bounded probability density ρ satisfying some additional condition (see Eq. (1.6) from [9]). Recently, Ekanga [82,83] has shown, in the framework of his PhD project, that this assumption can be relaxed to a form of log-Hölder continuity of F.

[7]Note that the results of [9] on stability of Anderson localization under weak interactions are not limited to one-dimensional weakly disordered systems.

To conclude this discussion, we would like to stress once more that the specific methods of analysis of the equation $H(\omega)\psi = E\psi$ describing individual particles on the one-dimensional lattice $\mathbb{Z}^1$ are not applicable to the eigenvalue equation for $N \geq 2$ particles on $\mathbb{Z}^1$. In particular, there is no direct analog of the transfer matrices $M_L(\omega)$ of fixed dimension, independent of the size of the quantum system under consideration. It remains unclear at this moment if the multi-particle localization in one dimension occurs for all nonzero values of the disorder amplitude $|g|$, with a fixed (but arbitrary) interaction amplitude $|h|$. In our opinion, this is a challenging problem.

The (non-rigorous) results obtained by physicists (cf. [31, 105]) seem to indicate that the above mentioned difficulty is not a mere technicality, and that one probably should not expect "unconditional" complete localization in multi-particle systems with interaction even in one-dimensional physical space, i.e., for disordered systems in $(\mathbb{Z}^1)^N$.

4.7.5 *Multi-particle Localization in Euclidean Space*

The finite-difference Hamiltonians considered in this book often appear in physics in the framework of the so-called tight-binding approximation; a more precise description of quantum particles in the physical space requires differential operators. For example, a single quantum particle in $\mathbb{R}^d$ which is subject to a potential energy field $V(x)$ can be described with the help of the Schrödinger operator $-\Delta + V$, where Δ is the Laplacian in $\mathbb{R}^d$ and V is the operator of multiplication by the potential.

This operator can be properly defined as an unbounded self-adjoint operator with a dense domain $\mathcal{D} \subset L^2(\mathbb{R}^d)$. Suppose, for the sake of convenience, that function $V : x \in \mathbb{R}^d \mapsto \mathbb{R}$ is bounded from below, e.g., $V(x) \geq c > -\infty$. Then the Schrödinger operator $-\Delta + V$ is also bounded from below. Informally speaking, the tight-binding approximation consists in ignoring the spectrum of $-\Delta + V$ (and the respective eigenfunctions) above a certain threshold E^*. This gives rise to a number of qualitative changes in the nature of the resulting lattice Schrödinger operator. In particular, the spectral properties of an LSO at high energies ($E \geq \eta \gg 1$) and at low energies ($E \leq -\eta, \eta \gg 1$) may well be similar—at least, qualitatively. Also, the reduced, tight-binding kinetic energy operator H^0 (the lattice Laplacian) can be considered as a small-norm perturbation of the potential energy operator gV, when $|g|$ is large. (More precisely, it is convenient to write first $H^0 + gV = g(V + g^{-1}H^0$.) The same holds true for a system in a lattice cube $\Lambda \subset \mathbb{Z}^d$. This fact has been used in the proof of initial scale estimates of the Green's functions $G_\Lambda(x, y; E; \omega)$ in the strong disorder regime $|g| \gg 1$); cf. Sect. 2.3.2.

However, due to the unboundedness of the continuous Laplacian Δ, such an argument cannot be used for the localization analysis in Euclidean space $\mathbb{R}^d$. As a result, existing methods do not allow to prove Anderson localization in $\mathbb{R}^d$ with $d > 1$ for all energies, no matter how large the amplitude of the random potential is taken. This is not a mere technicality: there is a consensus in the physical community that in a Euclidean space $\mathbb{R}^d$ of a high enough dimension d, there exists a so-called *mobility edge* E^* separating the zone of localization $(-\infty, E^*)$ from the zone of delocalization $(E^*, +\infty)$. In absence of rigorous mathematical results in this direction, we do not want to pursue this discussion any further. However, we note that both the MSA and the FMM, adapted to Anderson-type models in $\mathbb{R}^d$, establish spectral and dynamical localization in a finite energy interval $I(g) \subset \mathbb{R}$, bounded from above and typically growing with $|g| \to \infty$.

Furthermore, when the random potential is not bounded from below, e.g., is generated by a regular Gaussian field with continuous argument in $\mathbb{R}^d$ (where regularity includes continuity of samples with probability one), the localization analysis can be easily extended to a half-line $(-\infty, E^*]$ in the energy axis. With these observations in mind, when addressing disordered systems in $\mathbb{R}^d$, we speak of localization at low energies.

It had been expected that the general MSA approach, originally developed in the framework of lattice models, may admit an extension to models in a Euclidean space of an arbitrary dimension and produce localization at low energies. For single-particle systems this has been done, e.g., in the papers [78, 98, 130] and the monograph [158], where a much more complete list of references can be found. From the technical point of view, the following key tools of the MSA require an adaptation to differential operators in $L^2(\mathbb{R}^d)$:

- The geometric resolvent inequality for the Green's functions,
- The geometric resolvent inequality for eigenfunctions,
- The EVC bounds.

As was already mentioned, the initial scale bounds at low energies for lattice systems make use of the Combes–Thomas estimate. Interestingly, the latter has been originally established for (differential) Schrödinger operators, and only later adapted to lattice models.

Once the above ingredients are at one's disposal, the general scheme of the MSA translates into the language of Anderson-type models in the Euclidean space without major obstacles; cf. [158].

Similarly, the MPMSA scheme described in Chap. 4 has been adapted in recent works [49, 50] and [72, 73] to continuous multi-particle Anderson models. Actually, the single- or multi-particle structure of the potential energy is irrelevant for a number of analytic tools such as the GRI. In fact, the latter applies to Schrödinger operators with a general potential energy, with or without the interaction terms.

As to the EVC bounds, the Stollmann's lemma (cf. 2.2.1) proved once more to be a very general and versatile analytical tool. Combined with the notion of separability of multi-particle cubes, the EVC bound presented in Chap. 3 of this book has been adapted in [50] for the so-called alloy-type Anderson models. Here the Hamiltonian $\mathbf{H}(\omega) = \mathbf{H}^{(N)}(\omega)$ is the above type

$$\mathbf{H}(\omega) = -\boldsymbol{\Delta} + \mathbf{V} + \mathbf{U},$$

with the minus-Laplacian $-\mathbf{\Delta}$ representing the N-particle kinetic energy operator $\mathbf{H}^0$.

The external potential energy operator $\mathbf{V}$ has again a summatory form: for an N-particle system in $\mathbb{R}^d$ it acts as multiplication by $\sum_{i=1}^{N} V(x_i;\omega)$, $x_1,\ldots,x_N \in \mathbb{R}^d$. The single-particle random potential $V(x;\omega)$ in an alloy-type model has the following particular form:

$$V(x;\omega) = \sum_{u\in\mathcal{Z}^d} a_u(\omega)\phi(x-u), \;\; x \in \mathbb{R}^d.$$

Here $\mathcal{Z}^d \subset \mathbb{R}^d$ is a periodic lattice with d linearly independent generators (a standard choice is $\mathcal{Z}^d = \mathbb{Z}^d$, an integer lattice). Next, the random amplitudes $a_u(\omega)$, $u \in \mathcal{Z}^d$, form an IID family, again with a Hölder-continuous marginal distribution function F. In [72] it was also assumed that the random potential is non-negative and bounded: supp $F \subset [0, v^*]$, $v^* \in (0, +\infty)$. Finally, the function ϕ (often referred as a bump function) is non-negative, with a compact support and such that

$$\sum_{u\in\mathcal{Z}} \phi(x-u) \geq C > 0, \;\; x \in \mathbb{R}^d.$$

Operator $\mathbf{U}$ describes the interaction between particles; again it is assumed to be a multiplication operator. The interaction was assumed to have a finite range r_0.

Alloy-type models are popular in physics of disordered media. We note that the finiteness of the interaction range in $\mathbf{U}$ is irrelevant for the proof of the EVC bounds, while it plays an important role in the proof of localization; see [72, 73].

In a particular case where the random variables $V(x;\omega)$ are not bounded from below, the essential spectrum of the operator $\mathbf{H}(\omega)$ on any half-axis $(-\infty, E]$ is non-empty with probability one. In addition, the decay rate of the localized eigenfunctions in this case does not tend to zero as $E \to -\infty$.

Further, an EVC bound presented in Sect. 3.5 extends easily to the situation where $V(x;\omega)$ is a regular stationary Gaussian field with continuous argument in $\mathbb{R}^d$ or a function of such a Gaussian random field. This results, in particular, in a simpler MPMSA scheme and in some more efficient decay bounds upon the localized eigenfunctions; cf. [73].

4.7.6 *Multi-particle Localization in Quantum Graphs*

Quantum systems on so-called metric (or quantum) graphs became a popular topic of research in recent times, motivated by a growing area of applications. Cf. the collection of papers [90]. In this section we discuss a class of such systems, with randomness introduced in a form of an external potential field.

Let us consider first a single-particle system. The starting point is the integer lattice $\mathbb{Z}^d \hookrightarrow \mathbb{R}^d$ embedded into the Euclidean space $\mathbb{R}^d$. We endow $\mathbb{Z}^d$ with the structure of a graph with the set of vertices $\mathbb{Z}^d$ and the set of bonds (or edges) $\mathcal{E} = \{e = (x, y)\}$ formed by pairs of nearest neighbors, i.e. lattice sites $x, y \in \mathbb{Z}^d$ with $y = x \pm e_j$, $1 \le j \le d$ where $e_j = (0, \ldots, 1, \ldots, 0)$ (the jth digit equals 1). A realistic physical system that we are going to model is composed of tubular segments in $\mathbb{R}^3$ with points of bifurcation, or contact, located at the lattice nodes. Making an idealization of a nonzero, but very narrow, cross-section of the tubular segments, we replace them by unit segments I_e labeled by edges e, each of which can be identified with an open unit interval $(0, 1) \subset \mathbb{R}$.

The kinetic energy operator H^0 is modeled by a differential operator in the direct sum of Hilbert spaces

$$\mathcal{H} = \bigoplus_{e \in \mathcal{E}} L^2(I_e), \quad \text{where } L^2(I_e) \cong L^2(0, 1).$$

More precisely, H^0 acts in each subspace $L^2(I_e)$ as the second derivative $-\mathrm{d}^2/\mathrm{d}s^2$ in the variable s varying along I_e, with properly chosen boundary conditions at the contact points. The latter are often taken as Kirchhoff-type conditions preserving the total current. That is, the core of H^0 consists of functions $\psi : \cup_{e \in \mathcal{E}} I_e \to \mathbb{C}$, with the properties that

- ψ belongs to class C^2 on each segment I_e;
- The limiting values $\psi'(x_e)$, $\psi'(y_e)$ exist at the border points of every segment I_e where $e = (x_e, y_e)$;
- For each vertex $u \in \mathbb{Z}^d$,

$$\sum_{e:\, x_e = u} \psi'(x_e) = \sum_{e:\, y_e = u} \psi'(y_e).$$

We skip on the analytic issues of the well-posedness and essential self-adjointness of H^0.

As was said above, the randomness in a system on a quantum graph can be introduced in a form of an external potential. For example, consider a bounded random field $V(s; \omega)$ with continuous samples on the configuration space $\cup_{e \in \mathcal{E}} I_e$ and form a Schrödinger-type operator $H(\omega) = H_0 + V(\omega)$. Boundedness and continuity of the potential $V(\cdot\,; \omega)$ suffice for such an operator $H(\omega)$ to be essentially self-adjoint on a suitable domain. A simple model is where

$$V(s; \omega) = \sum_e \xi_e(\omega) \mathbf{1}_{I_e}(s)$$

with IID edge amplitudes $\xi_e(\omega)$ having a (common) bounded probability density $\rho(\cdot)$.

In the single-particle context, such a model has been studied before; see, e.g., [89] and references therein. Taking into account that the kinetic energy operator H^0

is here a (special form of) one-dimensional differential Schrödinger operator, the comments made in Sect. 4.7.5 still apply, explaining why the localization can be expected only for sufficiently low energies.

An adaptation of the approach presented in this book to multi-particle systems on quantum graphs has been so far less straightforward than in the case of Anderson models in $\mathbb{R}^d$ discussed in the previous subsection. In fact, the same remark should be made here as in Sect. 4.7.4: although in a quantum graph each particle moves in a (locally) one-dimensional environment composed by intervals with contact points, already a two-particle system is described by a partial-differential operator, and the configuration space of the system is formed by squares (more generally, cubes of dimension N, for $N \geq 3$ particles). This explains why a certain number of well-known analytical tools and auxiliary results established in the analysis of single-particle systems on quantum graphs cannot be applied directly to multi-particle systems without additional work.

Such a work has been recently done by Sabri in the framework of his PhD thesis (cf. [148]). Here, exponential spectral and strong dynamical localization of all orders in the Hilbert Schmidt norm have been established in an N-particle counterpart of this model (with non-negative interaction of finite range), within an energy band near the lower edge of the spectrum. This covers the case of bounded IID random variables $\xi_e(\omega)$ with a Hölder continuous probability distribution. It was proved that the lower part of the spectrum is almost surely non-random, and that localization holds in a small interval $[E^*, E^* + \eta]$, where E^* is the almost sure bottom of the spectrum.

References

1. Anderson, P.W.: Absence of diffusion in certain random lattices. Phys. Rev. **109**, 1492–1505 (1958)
2. Abrahams, E. (ed.): 50 Years of Anderson Localization. World Scientific, Singapore (2010); reprinted in Int. J. Mod. Phys. B **24**(12–13) (2010)
3. Abrahams, E., Anderson, P.W., Licciardello, D.C., Ramakrishnan, T.C.: Scaling theory of localization: absence of quantum diffusion in two dimensions. Phys. Rev. Lett. **42**, 673–676 (1979)
4. Abu-Chacra, R., Anderson, P.W., Thouless, D.J.: A self consistent theory of localization. J. Phys. C **6**, 1734–1752 (1973)
5. Abu-Chacra, R., Anderson, P.W., Thouless, D.J.: Self consistent theory of localization. II. Localization near the band edges. J. Phys. C **7**, 65–75 (1974)
6. Aizenman, M.: Localization at weak disorder: some elementary bounds. Rev. Math. Phys. **06**(special issue), 1163–1182 (1994)
7. Aizenman, M., Molchanov, S.: Localization at large disorder and at extreme energies: an elementary derivation. Commun. Math. Phys. **157**, 245–278 (1993)
8. Aizenman, M., Warzel, S.: The canopy graph and level statistics for random operators on trees. Math. Phys. Anal. Geom. **9**(4), 291–333 (2007)
9. Aizenman, M., Warzel, S.: Localization bounds for multiparticle systems. Commun. Math. Phys. **290**, 903–934 (2009)
10. Aizenman, M., Warzel, S.: Complete dynamical localization in disordered quantum multi-particle systems. In: XVIth International Congress on Mathematical Physics, Prague, pp. 556–565. World Scientific (2010)
11. Aizenman, M., Warzel, S.: Extended states in a Lifshits tail regime for random Schrödinger operators on trees. Phys. Rev. Lett. **106**, 136801 (2011)
12. Aizenman, M., Warzel, S.: Resonant delocalization for random Schrödinger operators on tree graphs. J. Eur. Math. Soc. (2011, to appear). Preprint, `arXiv:math-ph/1104:0969`
13. Aizenman, M., Schenker, J.H., Friedrich, R.M., Hundertmark, D.: Finite-volume fractional-moment criteria for Anderson localization. Commun. Math. Phys. **224**, 219–253 (2001)
14. Aizenman, M., Elgart, A., Naboko, S., Schenker, J.H., Stoltz, G.: Moment analysis for localization in random Schrödinger operators. Invent. Math. **163**, 343–413 (2006)
15. Aizenman, M., Sims, R., Warzel, S.: Stability of the absolutely continuous spectrum of random Schrödinger operators on tree graphs. Probab. Theory Relat. Fields **136**, 363–394 (2006)
16. Aizenman, M., Sims, R., Warzel, S.: Absolutely continuous spectra of quantum tree graphs with weak disorder. Commun. Math. Phys. **264**, 371–389 (2006)

V. Chulaevsky and Y. Suhov, *Multi-scale Analysis for Random Quantum Systems with Interaction*, Progress in Mathematical Physics 65, DOI 10.1007/978-1-4614-8226-0,

17. Aizenman, M., Germinet, F., Klein, A., Warzel, S.: On Bernoulli decompositions for random variables, concentration bounds and spectral localization. Probab. Theory Relat. Fields **143**, 219–238 (2009)
18. Altshuler, B.L., Aronov, A.G., Khmelnitskii, D.E.: Effects of electron-electron collisions with small energy transfers on quantum localization. J. Phys. C **15**, 7367–7386 (1982)
19. Amrein, W., Georgescu, V.: On the characterization of bound states and scattering states in quantum mechanics. Helv. Phys. Acta **46**, 635–658 (1973)
20. Anderson, P.W.: Thoughts on localization. In: Abrahams, E. (ed.) 50 Years of Anderson Localization. World Scientific, Singapore (2010); reprinted in Int. J. Mod. Phys. B **24**, 1501–1506 (2010)
21. André, G., Aubry, S.: Analyticity breaking and Anderson localization in incommensurate lattices. Ann. Isr. Phys. Soc. **3**, 133–164 (1980)
22. Aubry, S.: The new concept of transition by breaking of analyticity. Solid State Sci. **8**, 264–277 (1978)
23. Avila, A., Damanik, D.: Absolute continuity of the integrated density of states for the almost Mathieu operator with non-critical coupling. Invent. Math. **172**(2), 439–453 (2008)
24. Avila, A., Jitomirskaya, S.: Solving the ten martini problem. In: Mathematical Physics of Quantum Mechanics. Lecture Notes in Physics, vol. 690, pp. 5–16. Springer, Berlin (2006)
25. Avila, A., Jitomirskaya, S.: The ten martini problem. Ann. Math. **170**, 303–342 (2009)
26. Avila, A., Jitomirskaya, S.: Almost localization and almost reducibility. J. Eur. Math. Soc. **12**, 93–131 (2010)
27. Avila, A., Jitomirskaya, S.: Hölder continuity of absolutely continuous spectral measures for one-frequency Schrödinger operators. Commun. Math. Phys. **301**, 563–581 (2011)
28. Avron, Y., Simon, B.: Almost periodic Schrödinger operators. I. Limit periodic potentials. Commun. Math. Phys. **82**, 101–120 (1982)
29. Avron, Y., Simon, B.: Almost periodic Schrödinger operators. II. The integrated density of states. Duke Math. J. **50**, 369–391 (1983)
30. Barbaroux, J.M., Combes, J.M., Hislop, P.D.: Localization near band edges for random Schrödinger operators. Helv. Phys. Acta **70**(1–2), 16–43 (1997)
31. Basko, D.M., Aleiner, I.L., Altshuler, B.L.: Metal–insulator transition in a weakly interacting many-electron system with localized single-particle states. Ann. Phys. **321**, 1126–1205 (2006)
32. Basko, D.M., Aleiner, L.I., Altshuler, B.L.: On the problem of many-body localization. In: Ivanov, A.L., Tikhodeev, S.G. (eds.) Problems of Condensed Matter Physics, pp. 50–70. Oxford University Press, Oxford (2008)
33. Bellissard, J., Simon, B.: Cantor spectrum for the almost Mathieu equation. J. Funct. Anal. **48**, 408–419 (1982)
34. Bellissard, J., Lima, R., Scoppola, E.: Localization in ν-dimensional incommensurate structures. Commun. Math. Phys. **88**, 465–477 (1983)
35. Bellissard, J.V., Hislop, P.D., Stolz, G.: Correlations estimates in the Anderson model. J. Stat. Phys. **129**, 649–662 (2007)
36. Berezanskii, J.M.: Expansion in Eigenfunctions of Self-Adjoint Operators. Translations of Mathematical Monographs, vol. 17. American Mathematical Society, Providence (1968)
37. Bjerklöv, K.: Positive Lyapunov exponent and minimality for a class of one-dimensional quasi-periodic Schrödinger equations. Ergod. Theory Dyn. Syst. **25**, 1015–1045 (2005)
38. Bjerklöv, K.: Positive Lyapunov exponent and minimality for the continuous 1-D quasi-periodic Schrödinger equations with two basic frequencies. Ann. Inst. Henri Poincaré **8**, 687–730 (2007)
39. Bougerol, P., Lacroix, J.: Products of Random Matrices with Applications to Schrödinger Operators. Birkhäuser, Boston (1985)
40. Bourgain, J.: Recent progress in quasi-periodic lattice Schrödinger operators and Hamiltonian partial differential equations. (Russian) Uspekhi Mat. Nauk **59**, 37–52 (2004); translation in Russ. Math. Surv. **59**, 231–246 (2004)
41. Bourgain, J.: Green's Function Estimates for Lattice Schrödinger Operators and Applications. Annals of Mathematics Studies, vol. 158. Princeton University Press, Princeton (2005)

42. Bourgain, J.: Anderson-Bernoulli models. Mosc. Math. J. **5**, 523–536 (2005)
43. Bourgain, J.: Anderson localization for quasi-periodic lattice Schrödinger operators on $\mathbb{Z}^d$, d arbitrary. Geom. Funct. Anal. **17**, 682–706 (2007)
44. Bourgain, J.: An approach to Wegner's estimate using subharmonicity. J. Stat. Phys. **134**, 969–978 (2009)
45. Bourgain, J., Goldstein, M.: On nonperturbative localization with quasi-periodic potential. Ann. Math. **152**, 835–879 (2000)
46. Bourgain, J., Jitomirskaya, S.: Absolutely continuous spectrum for 1D quasiperiodic operators. Invent. Math. **148**, 453–463 (2002)
47. Bourgain, J., Kenig, C.E.: On localization in the continuous Anderson-Bernoulli model in higher dimension. Invent. Math. **161**, 389–426 (2005)
48. Bourgain, J., Goldstein, M., Schlag, W.: Anderson localization for Schrödinger operators on $\mathbb{Z}^2$ with quasi-periodic potential. Acta Math. **188**, 41–86 (2002)
49. Boutet de Monvel, A., Chulaevsky, V., Suhov, Y.: Wegner-type bounds for a two-particle Anderson model in a continuous space (2008). Preprint, `arXiv:math-ph/0812.2627`
50. Boutet de Monvel, A., Chulaevsky, V., Stollmann, P., Suhov, Y.: Wegner-type bounds for a multi-particle continuous Anderson model with an alloy-type external potential. J. Stat. Phys. **138**, 553–566 (2010)
51. Bratteli, O., Robinson, D.W.: Operator Algebras and Quantum Statistical Mechanics. Springer, New York (1987)
52. Carmona, R.: One-dimensional Schrödinger operators: a survey, Acta Appl. Math. **4**, 65–91 (1985)
53. Carmona, R., Lacroix, J.: Spectral Theory of Random Schrödinger Operators. Birkhäuser, Boston (1990)
54. Carmona, R., Klein, A., Martinelli, F.: Anderson localization for Bernoulli and other singular potentials. Commun. Math. Phys. **108**, 41–66 (1987)
55. Chulaevsky, V.: A Wegner-type estimate for correlated potentials. Math. Phys. Anal. Geom. **11**, 117–129 (2008)
56. Chulaevsky, V.: A remark on charge transfer processes in multi-particle systems (2010). Preprint, `arXiv:math-ph/1005.3387`
57. Chulaevsky, V.: On resonances in disordered multi-particle systems. C. R. Acad. Sci. Paris I **350**, 81–85 (2011)
58. Chulaevsky, V.: Direct scaling analysis of localization in disordered systems. II. Multi-particle lattice systems (2011). Preprint, `arXiv:math-ph/1106.2234`
59. Chulaevsky, V.: Anderson localization for generic deterministtic potentials. J. Funct. Anal. **262**, 1230–1250 (2011)
60. Chulaevsky, V.: Direct scaling analysis of localization in single-particle quantum systems on graphs with diagonal disorder. Math. Phys. Anal. Geom. **15**, 361–399 (2012)
61. Chulaevsky, V.: From fixed-energy MSA to dynamical localization: a continuing quest for elementary proofs (2012). Preprint, `arXiv:math-ph/1205.5763`
62. Chulaevsky, V.: Fixed-energy multi-particle MSA implies dynamical localization (2012). Preprint, `arXiv:math-ph/1206.1952`
63. Chulaevsky, V.: On the regularity of the conditional distribution of the sample mean (2013). Preprint, `arXiv:math-ph/1304.6913`
64. Chulaevsky, V., Delyon, F.: Purely absolutely continuous spectrum for almost Mathieu operators. J. Stat. Phys. **55**, 1279–1284 (1989)
65. Chulaevsky, V., Dinaburg, E.: Methods of KAM theory for long-range quasi-periodic operators on $\mathbb{Z}^n$. Pure point spectrum. Commun. Math. Phys. **153**, 559–577 (1993)
66. Chulaevsky, V., Sinai, Y.: Anderson localization for the $1D$ discrete Schrödinger operator with two-frequency potential. Commun. Math. Phys. **125**, 91–112 (1989)
67. Chulaevsky, V., Sinai, Y.: Anderson localization and KAM-theory. Analysis, et cetera, Res. Pap. in Honor of J. Moser's 60th Birthd., pp. 237–249 (1990)
68. Chulaevsky, V., Spencer, T.: Positive Lyapunov exponents for a class of deterministic potentials. Commun. Math. Phys. **168**, 455–466 (1995)

69. Chulaevsky, V., Suhov, Y.: Anderson localisation for an interacting two-particle quantum system on $\mathbb{Z}$ (2007). arXiv:math-ph/0705.0657
70. Chulaevsky, V., Suhov, Y.: Eigenfunctions in a two-particle Anderson tight binding model. Commun. Math. Phys. **289**, 701–723 (2009)
71. Chulaevsky, V., Suhov, Y.: Multi-particle Anderson localisation: induction on the number of particles. Math. Phys. Anal. Geom. **12**, 117–139 (2009)
72. Chulaevsky, V., Boutet de Monvel, A., Suhov, Y.: Dynamical localization for a multi-particle model with an alloy-type external random potential. Nonlinearity **24**(5), 1451–1472 (2011)
73. Chulaevsky, V., Boutet de Monvel, A., Suhov, Y.: Multi-particle dynamical localization in a Euclidean space with a Gaussian random potential (in preparation)
74. Combes, J.-M., Thomas, L.: Asymptotic behaviour of eigenfunctions for multiparticle Schrödinger operators. Commun. Math. Phys. **34**, 251–270 (1973)
75. Combes, J.-M., Hislop, P.D., Klopp, F.: An optimal Wegner estimate and its application to the global continuity of the integrated density of states for random Schrödinger operators. Duke Math. J. **140**(3), 469–498 (2007)
76. Combes, J.-M., Germinet, F., Hislop, P.: Conductivity and the current–current correlation measure. J. Phys. A **43**, 474010 (2010)
77. Cycon, H.L., Froese, R.G., Kirsch, W., Simon, B.: Schrödinger Operators. Springer, Berlin (1987)
78. Damanik, D., Stollmann, P.: Multi-scale analysis implies strong dynamical localization. Geom. Funct. Anal. **11**(1), 11–29 (2001)
79. Del Rio, R., Jitomirskaya, L., Last, Y., Simon, B.: Operators with singular continuous spectrum, IV. Hausdorff dimensions, rank one perturbations, and localization. J. Anal. Math. **69**, 163–200 (1996)
80. Dinaburg, E.I., Sinai, Y.G.: On the spectrum of a one-dimensional Schrödinger operator with a quasi-periodic potential. Funct. Anal. Appl. **9**, 8–21 (1975)
81. Disertori, M., Kirsch, W., Klein, A., Klopp, F., Rivasseau, V: Random Schrödinger Operators. Panoramas et Synthèses, vol. 25. Société Mathématique de France, Paris (2008)
82. Ekanga, T.: On two-particle Anderson localization at low energies. C. R. Acad. Sci. Paris I **349**(3–4), 167–170 (2011)
83. Ekanga, T.: Anderson localization in the multi-particle tight-binding model at low energies or with weak interaction (2012). Preprint, arXiv:math-ph/1201.2339
84. Elgart, A., Tautenhahn, M., Veselić, I.: Anderson localization for a class of models with a sign-indefinite single-site potential via fractional moment method. Ann. Henri Poincaré **12**(8), 1571–1599 (2010)
85. Eliasson, L.H.: Floquet solutions for the 1-dimensional quasi-periodic Schrödinger equation. Commun. Math. Phys. **146**, 447–482 (1992)
86. Eliasson, L.H.: Discrete one-dimensional quasi-periodic Schrödinger operators with pure point spectrum. Acta Math. **179**, 153–196 (1997)
87. Eliasson, L.H.: On the discrete one-dimensional quasi-periodic Schrödinger equation and other smooth quasi-periodic skew products. In: Hamiltonian Systems with Three or More Degrees of Freedom (S'Agaró, 1995). NATO Advanced Science Institute Series, vol. 533, pp. 55–61. Kluwer, Dordrecht (1999)
88. Enss, V.: Asymptotic completeness for quantum-mechanical potential scattering. Short-range potentials. Commun. Math. Phys. **61**, 285–291 (1978)
89. Exner, P., Helm, M., Stollmann, P.: Localization on a quantum graph with a random potential on edges. Rev. Math. Phys. **19**, 923–939 (2007)
90. Exner, P., Keating, J.P., Kuchment, P., Sunada, T., Teplyaev, A. (eds.): Analysis on Graphs and Its Applications. Proceedings of Symposia in Pure Mathematics, vol. 77. American Mathematical Society, Providence (2008)
91. Figotin, A., Pastur, L.: An exactly solvable model of a multidimensional incommensurate structure. Commun. Math. Phys. **95**, 401–425 (1984)
92. Fischer, W., Leschke, H., Müller, P.: Spectral localization by Gaussian random potentials in multi-dimensional continuous space. J. Stat. Phys. **101**(5/6), 935–985 (2000)

93. Fishman, S., Grempel, D., Prange, R.: Localization in a d-dimensional incommensurate structure. Phys. Rev. B **194**, 4272–4276 (1984)
94. Fleishman, L., Anderson, P.W.: Interactions and the Anderson transition. Phys. Rev. B **21**, 2366–2377 (1980)
95. Fröhlich, J., Spencer, T.: Absence of diffusion in the Anderson tight-binding model for large disorder or low energy. Commun. Math. Phys. **88**, 151–184 (1983)
96. Fröhlich, J., Martinelli, F., Scoppola, E., Spencer, T.: Constructive proof of localization in the Anderson tight-binding model. Commun. Math. Phys. **101**, 21–46 (1985)
97. Fröhlich, J., Spencer, T., Wittwer, P.: Localization for a class of one-dimensional quasi-periodic Schrödinger operators. Commun. Math. Phys. **132**, 5–25 (1990)
98. Germinet, F., De Bièvre, S.: Dynamical localization for discrete and continuous random Schrödinger operators. Commun. Math. Phys. **194**, 323–341 (1998)
99. Germinet, F., Klein, A.: Bootstrap multi-scale analysis and localization in random media. Commun. Math. Phys. **222**, 415–448 (2001)
100. Germinet, F., Klein, A.: A comprehensive proof of localization for continuous Anderson models with singular random potentials. J. Eur. Math. Soc. (2011, to appear). `arxiv:math-ph/1105.0213`
101. Goldsheid, I.Y., Molchanov, S.A.: On Mott's problem. Sov. Math. Dokl. **17**, 1369–1373 (1976)
102. Goldsheid, I.Y., Molchanov, S.A., Pastur, L.A.: A pure point spectrum of the one-dimensional Schrödinger operator. Funct. Anal. Appl. **11**, 1–10 (1977)
103. Gordon, A.Y.: On the point spectrum of the one-dimensional Schrödinger operator. (Russian) Uspekhi Matem. Nauk **31**, 257–258 (1976)
104. Gordon, A.Y., Jitomirskaya, S., Last, Y., Simon, B.: Duality and singular continuous spectrum in the almost Mathieu equation. Acta Math. **178**, 169–183 (1997)
105. Gornyi, I.V., Mirlin, A.D., Polyakov, D.G.: Interacting electrons in disordered wires: Anderson localization and low-temperature transport. Phys. Rev. Lett. **95**, 206603 (2005)
106. Graf, G.M., Vaghi, A.: A remark on the estimate of a determinant by Minami. Lett. Math. Phys. **79**, 17–22 (2007)
107. Grempel, D., Fishman, S., Prange, R.: Localization in an incommensurate potential: an exactly solvable model. Phys. Rev. Lett. **49**, 833 (1982)
108. Harper, P.G.: Single band motion of conducting electrons in a uniform magnetic field. Proc. Phys. Soc. Lond. A **68**, 874–878 (1955)
109. Hofstadter, D.R.: Energy levels and wavefunctions of Bloch electrons in rational and irrational magnetic fields. Phys. Rev. B **14**, 2239–2249 (1976)
110. Hundertmark, D.: A short introduction to Anderson localization. In: Mörters, P., et al. (ed.) Analysis and Stochastics of Growth Processes an Interface Models. Oxford University Press (2008). http://dx.doi.org/10.1093/acprof:oso/9780199239252.001.0001
111. Jitomirskaya, S.Y.: Metal-insulator transition for the almost Mathieu operator. Ann. Math. **150**, 1159–1175 (1999)
112. Kato, T.: Perturbation Theory for Linear Operators. Springer, New York (1976)
113. Kirsch, W.: An invitation to random Schrödinger operators. With an appendix by F. Klopp, in Ref. [81], pp. 1–119 (2008)
114. Kirsch, W.: A Wegner estimate for multi-particle random Hamiltonians. Zh. Mat. Fiz. Anal. Geom. **4**, 121–127 (2008)
115. Kirsch, W., Stollmann, P., Stolz, G.: Anderson localization for random Schrödinger operators with long range interactions. Commun. Math. Phys. **195**, 495–507 (1998)
116. Klein, A.: Absolutely continuous spectrum in the Anderson model on the Bethe lattice. Math. Res. Lett. **1**, 399–407 (1994)
117. Klein, A.: Extended states in the Anderson model on the Bethe lattice. Adv. Math. **133**(1), 163–184 (1998)
118. Klein, A.: Multiscale analysis and localization of random operators. In: Ref. [81], pp. 121–159
119. Klein, A., Molchanov, S.: Simplicity of eigenvalues in the Anderson model. J. Stat. Phys. **122**, 95–99 (2006)

120. Klopp, F., Zenk, H.: The integrated density of states for an interacting multiparticle homogeneous model and applications to the Anderson model. Adv. Math. Phys. **2009**, 1–15 (2009). Art. ID 679827
121. Kohn, W.: Theory of the insulating state. Phys. Rev. **133**, A171–A181 (1964)
122. Kravchenko, S.V., Sarachik, M.P.: A metal–insulator transition in 2D: established facts and open questions. Preprint, `arXiv:math-ph/1003.2968`; also, In: Abrahams, E. (ed.) 50 Years of Anderson Localization, p. 473. World Scientific, Singapore (2010); reprinted in Int. J. Mod. Phys. B **24**, 1640–1663 (2010)
123. Kunz, H., Souillard, B.: Sur le spectre des opérateurs aux différences finies aléatoires. Commun. Math. Phys. **78**, 201–246 (1980)
124. Kunz, H., Souillard, B.: The localization transition on the Bethe lattice. J. Phys. Lett. **44**, 411–414 (1983)
125. Lagendiik, A., van Tiggelen, B., Wiersma, D.S.: Fifty years of Anderson localization. Phys. Today **62**, 24–29 (2009)
126. Lifshitz, I.M.: Structure of the energy spectrum of the impurity bands in disordered solids. Sov. Phys. JETP **17**, 1159–1170 (1963)
127. Lifshitz, I.M.: The energy spectrum of disordered systems. Adv. Phys. **13**, 483–536 (1964)
128. Lifshitz, I.M., Gredescul, S.A., Pastur, L.A.: Introduction to the Theory of Disordered Systems. Wiley, New York (1988)
129. Martinelli, F.: A quantum particle in a hierarchical potential: a first step towards the analysis of complex quantum systems. In: Phénomènes critiques, systèmes aléatoires, théories de jauge, Les Houches, 1984, pp. 1197–1199. North-Holland, Amsterdam (1986)
130. Martinelli, F., Holden, H.: On absence of diffusion near the bottom of the spectrum for a random Schrödinger operator on $L^2(\mathbb{R}^d)$. Commun. Math. Phys. **93**, 197–217 (1984)
131. Martinelli, F., Scoppola, E.: Absence of absolutely continuous spectrum in the Anderson model for large disorder or low energy. In: Infinite-Dimensional Analysis and Stochastic Processes, Bielefeld, 1983. Research Notes in Mathematics, vol. 124, pp. 94–97. Pitman, Boston (1983)
132. Martinelli, F., Scoppola, E.: Remark on the absence of absolutely continuous spectrum for d-dimensional Schrödinger oerators with random potential for large disorder or low energy. Commun. Math. Phys. **97**, 465–471 (1985)
133. Minami, N.: Local fluctuation of the spectrum of a multidimensional Anderson tight-binding model. Commun. Math. Phys. **177**, 709–725 (1996)
134. Molchanov, S.A.: Structure of eigenfunctions of one-dimensional unordered structures. (Russian) Math. USSR Izv. **42**, 70–100 (1978)
135. Molchanov, S.A.: The local structure of the spectrum of the one-dimensional Schrödinger operator. Commun. Math. Phys. **78**, 429–446 (1981)
136. Moser, J., Pöschel, J.: An extension of a result by Dinaburg and Sinai on quasi-periodic potentials. Comment. Math. Helv. **59**, 39–85 (1984)
137. Mott, N.F.: Metal–insulator transition. Rev. Mod. Phys. **40**, 677–683 (1968)
138. Mott, N.F., Twose, W.D.: The theory of impurity conditions. Adv. Phys. **10**, 107–163 (1961)
139. Nakano, F.: The repulsion between localization centers in the Anderson model. Commun. Math. Phys. **123**(4), 803–810 (2006)
140. Nakano, F.: Distribution of localization centers in some discrete random systems. Rev. Math. Phys. **19**, 941–965 (2007)
141. Novikov, S.P.: Periodic problem for the Korteveg–de Vries equation. Funct. Anal. Appl. **8**, 54–66 (1974)
142. Pankrashkin, K.: Quasiperiodic surface Maryland models on quantum graphs. J. Phys. A **42**, 265–304 (2009)
143. Pastur, L., Figotin, A.: An exactly solvable model of a multidimensional incommensurate structure. Commun. Math. Phys. **95**, 401–425 (1984)
144. Pastur, L., Figotin, A.: Spectra of Random and Almost-Periodic Operators. Springer, Berlin (1992)

145. Puig, J.: Cantor spectrum for the almost Mathieu operator. Commun. Math. Phys. **244**, 297–309 (2004)
146. Reed, M., Simon, B.: Methods of Modern Mathematical Physics, vol. 1. Academic, New York (1980)
147. Ruelle, D.: A remark on bound states in potential scattering theory. Nouvo Cimento **61A**, 655–662 (1969)
148. Sabri, M.: Anderson localization for a multi-particle quantum graph. Rev. Math. Phys. (2012, to appear). Preprint, `arXiv:math-ph/1201.6247`
149. Shepelyansky, D.L.: Coherent propagation of two interacting particles in a random potential. Phys. Rev. Lett. **73**, 2607–2610 (1994)
150. Shnol, I.: On the behaviour of the Schrödinger equation. (Russian) Mat. Sb. **42**, 273–286 (1957)
151. Simon, B.: Schrödinger semigroups. Bull. Am. Math. Soc. **7**, 447–526 (1983)
152. Simon, B.: Almost periodic Schrödinger operators. IV: The Maryland model. Ann. Phys. **159**, 157–183 (1985)
153. Simon, B., Wolff, T.: Singular continuous spectrum under rank-one perturbations and localization for random Hamiltonians. Commun. Pure Appl. Math. **39**, 75–90 (1986)
154. Sinai, Y.G.: Anderson localization for one-dimensional difference Schrödinger operator with quasi-periodic potential. J. Stat. Phys. **46**, 861–909 (1987)
155. Spencer, T.: The Schrödinger equation with a random potential. A mathematical review. In: Critical Phenomena, Random Systems, Gauge Theories. Proc. Summer Sch. Theor. Phys. Sess., vol. 43, pp. 895–942. Les Houches, France 1984, Pt. 2 (1986)
156. Spencer, T.: Localization for random and quasi-periodic potentials. J. Stat. Phys. **51**, 1009–1019 (1988)
157. Stollmann, P.: Wegner estimates and localization for continuum Anderson models with some singular distributions. Arch. Math. **75**, 307–311 (2000)
158. Stollmann, P.: Caught by Disorder. Birkhäuser, Boston (2001)
159. Suhov, Y., Kelbert, M.: Probability and Statistics by Example. Markov Chains: A Primer in Random Processes and Their Applications, vol. 2. Cambridge University Press, Cambridge (2007)
160. von Dreifus, H.: On effect of randomness in ferromagneic models and Schrödinger operators. PhD dissertation, New York University, New York (1987)
161. von Dreifus, H., Klein, A.: A new proof of localization in the Anderson tight-binding model. Commun. Math. Phys. **124**, 285–299 (1989)
162. von Dreifus, H., Klein, A.: Localization for random Schrödinger operators with correlated potentials. Commun. Math. Phys. **140**, 133–147 (1991)
163. Wegner, F.: Bounds on the density of states of disordered systems. Z. Phys. **B44**, 9–15 (1981)

Index

V. Chulaevsky and Y. Suhov, *Multi-scale Analysis for Random Quantum Systems with Interaction*, Progress in Mathematical Physics 65, DOI 10.1007/978-1-4614-8226-0,

Printed in the United States
By Bookmasters